# The Superfund Manual

## A PRACTITIONER'S GUIDE TO CERCLA LITIGATION

PETER L. GRAY



AMERICAN BAR ASSOCIATION
Section of Environment, Energy, and Resources

Cover design by Mary Anne Kulchawik/ABA Design



Printed in the United States of America.

20 19 18 17 16    5 4 3 2 1

**Library of Congress Cataloging-in-Publication Data**

Names: Gray, Peter L. (Peter Lawrence), 1957- author.
Title: Superfund handbook / Peter L. Gray.
Description: Chicago : American Bar Association, 2016.
Identifiers: LCCN 2016009634 | ISBN 9781634252614
Subjects: LCSH: Liability for hazardous substances pollution damages—United States. | Liability for environmental damages—United States. | Hazardous waste sites—Law and legislation—United States. | United States. Comprehensive Environmental Response, Compensation, and Liability Act of 1980.
Classification: LCC KF1299.H39 G73 2016 | DDC 344.7304/62—dc23
LC record available at http://lccn.loc.gov/2016009634

Discounts are available for books ordered in bulk. Special consideration is given to state bars, CLE programs, and other bar-related organizations. Inquire at Book Publishing, ABA Publishing, American Bar Association, 321 N. Clark Street, Chicago, Illinois 60654-7598.

www.ShopABA.org

# CONTENTS

# OVERVIEW

The Comprehensive Environmental Response, Compensation, and Liability Act of 1980 (CERCLA or Superfund) was signed into law by President Jimmy Carter on December 11, 1980, after passage in the closing days of a lame-duck session of Congress.[1] CERCLA was intended to fill a significant gap in the nation's environmental protection laws: cleanup of abandoned contaminated properties. The passage of the Resource Conservation and Recovery Act (RCRA) in 1976 established a comprehensive scheme for regulating the management and disposal of hazardous waste by active facilities. Notably, RCRA did not provide the United States Environmental Protection Agency (EPA) with the necessary authority or funding to address contamination from countless abandoned facilities, a legacy of the industrial revolution. This problem came into sharp relief following public outrage over the apparent inability of EPA to respond promptly to the Love Canal disaster.

The infamous 16-acre site was a long-forgotten burial ground for industrial wastes. From 1942 to 1953, over 20,000 tons of chemicals were dumped into the landfill, after which the property was sold to the Niagara Falls school district for one dollar. The 99th Street School was built over Love Canal a year later, and within the next decade, residential neighborhoods followed. Love Canal made national headlines in 1978. Local reporters had detected contaminants in the water supply, and after conducting a survey of local residents found alarming evidence of birth defects, abnormally high incidence of miscarriages, and leukemia; these findings

1. Pub. L. No. 96-510, 94 Stat. 2767 (1980) (codified as amended at 42 U.S.C. §§ 9601–9675).

were confirmed a year later by EPA, but the agency lacked express legal authority to take remedial action.

Local residents, frustrated with the federal government's failure to respond immediately to the situation, went so far as to take two EPA officials hostage in May 1980. Ultimately, the federal government did respond by purchasing the property in and around Love Canal, relocating approximately 950 families, and capping the site.

Love Canal became a catalyst for legislative action. In congressional hearings, it became clear that Love Canal was not an isolated incident but rather was typical of a "pervasive national problem."[2] EPA estimated that there were 30,000 to 50,000 unregulated hazardous waste sites and that 1,200 to 2,000 of these sites posed a "serious risk to public health."[3] To address this looming public health threat, several bills were introduced to provide EPA with express authority to respond to and remediate abandoned contaminated sites and to establish a massive new fund (a "Superfund") to cover the costs. Among the many sticking points was whether to impose joint and several liability for EPA's cleanup costs on responsible parties, whether to impose such liability retroactively, whether to include personal injury and property damage within the ambit of the liability scheme, and whether to exclude petroleum spills from the liability scheme. With time running out on the lame-duck session, a bipartisan group of senators hammered out a compromise bill, the now-famous Stafford-Randolph Compromise.[4]

Under the Stafford-Randolph Compromise, much of the contentious language from earlier bills—the words "joint and several," the liability provisions for personal injury or property or income loss and oil spill cleanup authority—were removed. The Senate passed this compromise

2. *Congress Clears "Superfund" Legislation*, 36 Cong. Q. Almanac 584, 585–86 (1980).

3. H.R. Rep. No. 96-1016, pt. 1, at 18 (1980), *reprinted in* 1980 U.S.C.C.A.N. 6120–6121.

4. S. 1480, 96th Cong., 2d Sess. (1980).

bill after limited debate[5] and then shuttled the legislation to the House for a "take-it-or-leave-it" vote.[6] They took it.[7]

CERCLA enables the federal government to respond to releases or threatened releases of hazardous substances from abandoned and leaking hazardous waste disposal sites. Specifically, EPA is authorized to utilize congressionally appropriated funds—the Hazardous Substances Superfund (Superfund or the Fund)—to remediate sites[8] that the agency places on the National Priorities List (NPL), a list of the nation's most hazardous sites.[9]

Congress appropriated $1.6 billion for the first five years of the Superfund program; 87.5 percent of this revenue came from a tax on chemical and petroleum feedstocks, and the remaining 12.5 percent came from Treasury appropriations. Funding rose to $8.5 billion in 1986, when Congress passed the Superfund Amendments and Reauthorization Act of 1986 (SARA or the 1986 amendments).[10] An underlying principle of Superfund, however, is to shift the costs of remediation from taxpayers to the parties whose operations caused the contamination—the "polluters pay" principle.[11] To that end, Congress authorized EPA to seek recovery of cleanup costs from any responsible party.[12] Alternatively, EPA may order a responsible party to clean up a hazardous waste site

5. 126 Cong. Rec. 5, 14,905, 15,009 (daily ed. Nov. 24, 1980).

6. On December 1, 1980, Senators Stafford and Randolph sent a letter to Congressman Florio, the sponsor of the House bill that the Senate had replaced, and warned against any tinkering, noting: "That the bill passed at all is a minor wonder. Only the frailest, moment-to-moment coalition enabled it to be brought to the senate floor and considered. Indeed, within a matter of hours that fragile coalition began to disintegrate to the point that, in our judgment, it would now be impossible to pass the bill again, even unchanged. . . ." Mark Reisch, Committee on Environment and Public Works, A Legislative History of the Comprehensive Environmental Response, Compensation and Liability Act of 1980 (Superfund), Pub. Law 96-510, at 774–75 (1983).

7. For a comprehensive summary of the legislative history of CERCLA, see Frank P. Grad, *A Legislative History of the Comprehensive Environmental Response, Compensation, and Liability ("Superfund") Act of 1980*, 8 Colum. J. Envtl. L. (1980).

8. CERCLA § 104(a).

9. CERCLA § 105(b). EPA is also authorized to remediate non-NPL sites using the fund.

10. Pub. L. No. 99-499, 100 Stat. 1613 (1986).

11. *See, e.g.*, H.R. Rep. No. 99-253 (1985), *reprinted in* 1986 U.S.C.C.A.N. 3038, 3038 ("CERCLA has two goals: (1) to provide for clean-up if a hazardous substance is released into the environment . . . and (2) to hold responsible parties liable for the costs of these clean-ups.").

12. CERCLA § 107.

where the site presents an "imminent and substantial endangerment to the public health or welfare or the environment."[13]

As noted above, Congress chose to omit from the final text of CERCLA the words "joint and several" or any other words clarifying the liability standard. Nevertheless, courts have interpreted the legislative history of CERCLA as intending liability to be strict, joint, and several among and between four classes of potentially responsible parties (PRPs).[14] PRPs face liability without regard to fault, without regard to the fact that the disposal activity now giving rise to such liability occurred decades ago, and without regard to the fact that such past practices were not only lawful but also often directed, permitted, or at least known by state officials. CERCLA liability is not predicated on some form of noncompliant behavior by PRPs. Moreover, under the principle of joint and several liability, one PRP can in theory be held liable for the entire cost of cleaning up a site, notwithstanding that many other parties share responsibility for the contamination—and the costs can be enormous.[15]

The harsh realities of CERCLA's liability provisions have met with considerable, if mostly unsuccessful, resistance from PRPs. Nevertheless, as the Superfund cleanup program has evolved and developed, and as PRPs have gained experience, albeit painful, with the intricacies and realities of Superfund litigation, insights and lessons have emerged.

One goal of this book is to identify and develop these insights and lessons. They require a full understanding of the major elements of the CERCLA program. To that end, this book describes EPA's response and enforcement authorities; EPA procedures for identifying cleanup standards and selecting remedial plans that achieve those standards; the statutory liability and judicial review provisions; PRP organization and

13. *See* CERCLA § 106.

14. We discuss these four categories of PRPs in Chapter 4, but briefly, they are as follows: (1) the current owners and operators of a contaminated site; (2) former owners and operators who were present when the contamination occurred; (3) parties that arranged for treatment or disposal of a hazardous substance at the site; and (4) transporters who selected the site for treatment or disposal of the hazardous substance. *See* CERCLA § 107(a)(1)–(4).

15. The Government Accountability Office (GAO) reported that the average cleanup cost for a site on the NPL is $26 million. *See* U.S. Gov't Accountability Office, GAO-96-125, Superfund—Barriers to Brownfield Redevelopment 3 (1996).

settlement issues; litigation, both as between PRPs and the United States and as among PRPs; and a handful of other significant issues arising out of CERCLA cleanup activities. From this discussion, two central themes should become apparent.

First, there has been, and continues to be, an endless appetite to litigate the contours of CERCLA liability. Sadly, Congress *knew* it was creating a litigation behemoth when it passed Superfund. Consider the following remark by Representative Harsha on passage of the Stafford-Randolph Compromise:

> We are establishing civil liability and criminal penalties in this legislation, and numerous questions have been raised as to what we are doing to common law with this new statute. These are not spurious issues. They are going to be litigated and the courts are going to have a field day in ridiculing the Congress on passing laws that are vague, internally inconsistent, and using tools such as superseding laws which are in conflict without any further guidance. This bill is not a superfund bill—it is a welfare and relief act for lawyers.[16]

One need look no further than the nation's highest court for confirmation of the extraordinary endurance of Superfund litigation. Since 2000—the third decade of the Superfund program—the U.S. Supreme Court has reviewed CERCLA decisions on four separate occasions.[17] In large measure, Congress is to blame for the litigation that has dogged Superfund since its inception. The statute, even as amended, is often vague and/or contradictory, leaving many important issues to be resolved by the courts.[18] The legislative history of CERCLA is scarce, as there were no committee or subcommittee hearings and no committee reports on

16. 126 Cong. Rec. 31,970 (1980).

17. The four cases are as follows: *Cooper Indus., Inc. v. Aviall Serv., Inc.*, 543 U.S. 157 (2004); *United States v. Atlantic Research Corp.*, 551 U.S. 128 (2007); *Burlington N. & Santa Fe Ry. Co. v. United States*, 556 U.S. 599 (2009); and *CTS Corp. v. Waldburger*, 134 S. Ct. 2175 (2014), *reh'g denied*, 135 S. Ct. 23 (2014).

18. Indeed, as the U.S. Supreme Court has noted, the provisions of CERCLA are not "model[s] of legislative draftsmanship," and its statutory language is "at best inartful and at worst redundant." *Exxon Corp. v. Hunt*, 475 U.S. 355, 363 (1986).

the Stafford-Randolph Compromise that became law—nothing resembling the typical legislative process. All discussions and negotiations took place behind closed doors.[19] The legislative history we have is dominated by contradictory interpretations of the bill that various senators offered during limited floor debate.[20]

The huge sums of money so routinely at risk in CERCLA cleanups, combined with the innumerable vagaries surrounding the statutory scheme, virtually ensure that the private parties who face Superfund liability will continue to look to the courts to protect their interests. For that reason, this book analyzes key issues primarily from the perspective of PRPs who become enmeshed in the web of Superfund claims and litigation and, where appropriate, describes the current and/or evolving regulations, guidance, and judicial interpretations.

The second theme is that, notwithstanding the historic and continuing high levels of CERCLA litigation, negotiation and settlement are, more often than not, the only means by which PRPs can gain any measure of success in mitigating their liabilities. Hardball litigation tactics, often seen by litigators as the road to success, simply do not play well in the Superfund litigation arena. Thus, we devote considerable attention to the provisions of CERCLA that are designed to enhance the prospect of settlement, as well as the considerations that bear on the decision of whether to settle or litigate. (See Chapter 5.)

## Disclaimers

Three disclaimers are in order. First, this book provides a general explanation of CERCLA and the most significant issues that have arisen during the past 35 years of its implementation and administration by EPA. It is not intended to represent an exhaustive analysis of every issue that has been litigated nor a comprehensive repository of every case that addresses these issues.

---

19. *See* Alfred R. Light, *Clean Up of a Legislative Disaster: Avoiding the Constitution under the Original CERCLA*, 37 Environs: Envtl. L. & Pol'y J. 197–214 (2014).

20. *See* 126 Cong. Rec. 30,873, 30,916–76 (daily ed. Nov. 24, 1980).

Second, it goes without saying that this book should not be considered as providing legal advice. It is intended solely as a treatise summarizing the state of the law under CERCLA. Although there are numerous practice tips scattered throughout the text, they do not constitute legal advice. To provide legal advice, an attorney must first understand the particular problem and facts for which advice is needed—only then can the attorney apply the law to those facts and offer legal advice. Thus, this book should not be used as a substitute for obtaining legal advice on any issue.

Third, any treatise analyzing a statutory program that is implemented by a federal agency through regulation and guidance—and that is the subject of substantial case law interpretation—is, inevitably, a reflection of the program's evolution at the moment of publication. To illustrate, as recently as 2007, the U.S. Supreme Court changed what had been considered a bedrock principle of Superfund jurisprudence stretching back to the 1980s (ruling that PRPs may pursue recovery of cleanup costs under section 107, reversing decisions of 11 federal circuits).[21] Thus, readers should bear in mind that the contents of this book are but a snapshot of the state of the law as of 2015.

* * *

It is tempting to dismiss Superfund as a closed chapter of environmental history—a statute whose funding expired long ago[22] and whose mission largely has been accomplished. Resist the temptation. To paraphrase the immortal words of Mark Twain, the reports of Superfund's death have been greatly exaggerated. At last count, there were over 1,300 sites on the NPL.[23] There is also an immense number of non-NPL sites that have yet to be remediated. Much has been accomplished, but much remains to be done.

---

21. United States v. Atlantic Research Corp., 551 U.S. 128, 131–32 (2007).

22. The Superfund trust stopped receiving funds in 1995 after the expiration of the "polluter pays" taxes that were sustaining it.

23. *See Superfund: National Priorities List (NPL)*, EPA, http://www.epa.gov/superfund/sites/npl/ (last updated Dec. 29, 2015).

# ABOUT THE AUTHOR

Peter L. Gray is a partner in the Washington, D.C., office of Dentons US LLP. He chairs the firm's Environment & Natural Resources Practice Group. Mr. Gray has focused his practice on environmental law for 30 years. Over this period, he has handled virtually every imaginable environmental issue, including legal challenges to state and federal environmental regulations; state and federal enforcement cases involving charges of improper hazardous waste management, effluent discharges, air emissions, release reporting, and pesticide sales; arbitration between companies over compensation owed for use of scientific data supporting pesticide registrations; pre-acquisition due diligence; and climate change advocacy and planning. A constant focus throughout Mr. Gray's career, however, has been representation of companies in disputes relating to cleanup of contaminated property, which can arise under CERCLA, RCRA, and/or their state counterparts.

Mr. Gray writes extensively in the environmental area. He has authored leading treatises on environmental law, including the American Bar Association's Basic Practice Series on CERCLA (2008) and EPCRA (2002). Since 2012, he has been recognized by *Chambers USA* as a Leading Environmental Attorney in the Washington, D.C., area and by *Super Lawyers* as a leader in the field of environmental law. Mr. Gray is a cofounder of the Association of Climate Change Officers and is a member of the Environmental Law Institute Leadership Council. He received his undergraduate degree in chemistry from The University of Maryland, College Park; his master's degree in environmental science from George Washington University; and his law degree from the George Washington National Law Center (Order of the Coif). He is admitted to the bar of the District of Columbia and numerous federal circuit courts.

# CHAPTER 1

# Government Response Authority under Section 104 of CERCLA

## I. INTRODUCTION

As noted in the Overview, the Comprehensive Environmental Response, Compensation, and Liability Act of 1980 (CERCLA or Superfund) was a legislative response to the U.S. Environmental Protection Agency's (EPA's) lack of authority to address significant historical environmental contamination problems in the United States, typified by Love Canal. With the passage of CERCLA in 1980, Congress gave federal and state governments broad authority to respond to environmental contamination (under CERCLA § 104) and the right to recover the costs of such action from certain categories of responsible parties (under CERCLA § 107). This broad response authority, however, is subject to a number of important limitations and prerequisites, which have been extensively litigated. This chapter explores both the reach and limitations of response authority under CERCLA.

## II. FEDERAL AGENCIES AND DEPARTMENTS THAT MAY BE INVOLVED IN CERCLA CLEANUPS

Section 104 of CERCLA empowers the president to respond to environmental contamination. Pursuant to section 115 of CERCLA, the president has issued several executive orders that delegate this response authority to a number of federal agencies, but principally to EPA. The other federal agencies that were delegated response authority included the U.S. Coast Guard, which can respond to releases of hazardous substances that contaminate coastal zones, ports, harbors, or the Great Lakes; the Department of Defense, which is authorized to respond to releases on military bases; and the Department of Energy, which is authorized to respond to releases at facilities within its jurisdiction.[1]

Congress established special procedures for cleanup of federal facilities, the vast majority of which are facilities owned by the Department of Energy and Department of Defense. Under section 120 of CERCLA, federal agencies must enter into interagency agreements with EPA that establish their respective roles in the cleanup of the agencies' federal facilities. Additionally, section 120 spells out detailed procedures that federal agencies must follow to complete cleanup of federal facilities in a timely manner.[2]

Another federal agency that has responsibilities under CERCLA is the Agency for Toxic Substances and Disease Registry (ATSDR). Congress established the ATSDR in 1980 (within the Public Health Service), among other reasons, to prepare toxicological profiles of hazardous substances found at Superfund sites and to assess potential health risks at sites being cleaned up by EPA.[3] Although the ATSDR's findings may be used by EPA in deciding how to clean up a site, the ATSDR has no authority to dictate how a site should be cleaned up.

## III. STATE AUTHORITY UNDER SUPERFUND

Response authority under CERCLA is not limited to the federal government. States also have authority to respond to releases of hazardous substances under section 104 of CERCLA.[4] Indeed, states *must*

1. *See* Exec. Order 12,580, 52 Fed. Reg. 2923 (Jan. 23, 1987), which superseded Exec. Order 12,316, 46 Fed. Reg. 42,237 (Aug. 14, 1981).

2. Cognizant of the fact that military facilities composed the majority of contaminated federal facilities, Congress in 1986 established a separate, stand-alone program for cleanup of military facilities. Specifically, section 211 of the Superfund Amendments and Reauthorization Act of 1986 established the Defense Environmental Restoration Program, which is to be carried out by the secretary of defense. *See* 10 U.S.C. §§ 2701 *et seq.*

3. *See* CERCLA § 104(i).

4. Indian tribes also possess response authority to address releases within tribal lands. In this section, however, we focus on state authority.

participate in any remedial action within their boundaries—either by cooperating with EPA on cleanup where EPA is the lead agency or by taking the lead themselves.

### A. Required State Participation

Before undertaking any remedial action under section 104, EPA is obligated to consult with the affected state or states.[5] EPA must also enter into a cooperative agreement with the state under which the state guarantees the maintenance aspects of the remedial action, as well as the availability of an authorized treatment, storage, and disposal facility for wastes generated as a result of the remedial action.[6] The state must also agree to fund 10 percent of the remedial action, unless the site being cleaned up is owned by the state or a political subdivision, in which case the state must fund 50 percent of the remedial action.[7] The state's role in remedy selection is discussed in chapter 2.

### B. State Authority to Lead Remedial Action

In addition to the foregoing mandatory participation, states may act as the lead agency in responding to releases of hazardous substances within their borders. To that end, EPA is authorized to enter into a cooperative agreement with a state under which the state is designated the lead agency in implementing a remedial action (either at a specific National Priorities List [NPL] site or all NPL sites statewide). State-led cleanups of NPL sites must be conducted in accordance with EPA's National Contingency Plan (NCP), which, as discussed below, provides procedures for conducting removal and remedial actions. States may obtain reimbursement from the Fund, subject to the above-described cost-sharing provisions of section 104(c)(3) of CERCLA.[8] To the extent a state expends

5. CERCLA § 104(c)(2).

6. CERCLA § 104(c)(3); *see also* § 104(c)(9), under which states are required to provide assurances that they have adequate capacity to destroy, treat, or secure disposition of all hazardous wastes that are reasonably expected to be generated within the state during the 20-year period following execution of the cooperative agreement.

7. CERCLA § 104(c)(3)(B) & (C).

8. CERCLA § 104(d).

funds on non-NPL sites, it cannot obtain reimbursement from the Fund, but can recover such response costs from responsible parties.[9]

## IV. THE STATUTORY PREREQUISITES TO GOVERNMENT RESPONSE

Section 104(a) of CERCLA enumerates a number of prerequisites to the federal government's exercising its authority to remediate environmental contamination. Under section 104(a), the president is authorized "to remove or arrange for removal of, and provide for remedial action relating to" a "release or substantial threat of release into the environment" of a "hazardous substance" or "any pollutant or contaminant which may present an imminent and substantial endangerment." As discussed below, each of these phrases has been subject to exhaustive judicial interpretation.[10]

### A. Release or Threat of Release

Congress defined "release" very broadly to encompass virtually any activity that causes or is likely to cause hazardous substances to enter the environment:

> "[R]elease" means any spilling, leaking, pumping, pouring, emitting, emptying, discharging, injecting, escaping, leaching, dumping, or disposing into the environment (including the abandonment or discarding of barrels, containers and other closed receptacles containing any hazardous substance or pollutant or contaminant)....[11]

Although the term "release" suggests a sudden rush of contaminants into the environment, courts have construed the term to include the

9. *See* CERCLA § 107(a)(4).

10. Although this chapter concerns government response authority and its limitations, we have included references to cases between private parties where relevant to the body of jurisprudence on the meanings of these key terms.

11. CERCLA § 101(22).

gradual seepage or leaching of hazardous substances from storage areas or spills.[12]

A release need not be the original disposal activity to be subject to CERCLA response authority. Courts have held that subsequent redisposal of released contaminants may likewise constitute a release covered by CERCLA.[13] Indeed, courts have ruled that the current property owner's grading of the site constituted "disposal" because it dispersed hazardous substances that a prior owner had released on-site.[14]

A hazardous substance need not have been actually released for EPA to have response authority under section 104(a). EPA is also authorized to respond to a substantial threat of release. Though "substantial threat of release" is not defined under CERCLA, several courts have interpreted the phrase as being akin to a menace. In *United States v. Northernaire Plating Co.*,[15] for example, the United States commenced a CERCLA action against an owner of a locked plant containing drums and tanks of chemicals. After ruling that an actual release had occurred, the court

---

12. *See, e.g.*, Emhart Indus., Inc. v. Duracell Int'l, Inc., 665 F. Supp. 549 (M.D. Tenn. 1987); United States v. Vertac Chem. Corp., 671 F. Supp. 595 (E.D. Ark. 1987), *vacated on other grounds*, 855 F.2d 856 (8th Cir. 1988); New York v. Shore Realty Corp., 759 F.2d 1032, 1045 (2d Cir. 1985). It bears mentioning that many courts have held that passive migration of previously released contaminants does not constitute a release or disposal, but the context of those rulings was different. Those courts were *not* determining whether EPA had authority to respond to a release, but rather whether a defendant is liable as a former owner-operator under CERCLA § 107(A)(2), which requires proof that the defendant owned or operated the facility at the time of disposal. *See, e.g.*, Bob's Beverage, Inc. v. Acme, Inc., 264 F.3d 692 (6th Cir. 2001) (passive migration of hazardous substances on property during period of defendant's ownership did not constitute release or disposal); Carson Harbor Vill., Ltd. v. Unocal Corp., 270 F.3d 863 (9th Cir. 2001) (affirming district court's grant of summary judgment that the gradual spread of highly viscous tar-like material did not constitute disposal). Additional cases are discussed in chapter 4.

13. *See, e.g.*, Missouri v. Indep. Petrochemical Corp., 610 F. Supp. 4 (E.D. Mo. 1985) (waste generators held liable for wastes that were transshipped for the original disposal site to a second site); United States v. Conservation Chem. Co., 619 F. Supp. 162, 234 (W.D. Mo. 1985) (same).

14. *See, e.g.*, Ashley II of Charleston, LLC v. PCS Nitrogen, Inc., 791 F. Supp. 2d 431 (D.S.C. 2011); Redwing Carriers, Inc. v. Saraland Apartments, 94 F.3d 1489, 1511–12 (11th Cir. 1996) ("[A] disposal may occur when a party disperses contaminated soil during the course of grading and filling a construction site.").

15. 670 F. Supp. 742 (W.D. Mich. 1987), *aff'd sub nom.* United States v. R.W. Meyer, Inc., 889 F.2d 1497 (6th Cir. 1989).

went on to discuss whether a "substantial threat of release" had been established as well:

> The evidence shows that there were hazardous substances at the Northernaire site and that these substances, individually and collectively, would pose a threat to the population of the Cadillac, Michigan area if they were to be released into the environment. The evidence further demonstrates that none of the defendants were willing to take responsibility for ensuring that no such release would occur. The evidence of the presence of hazardous substances at the facility, when combined with the evidence of the unwillingness of any party to assert control over the substances, amounts to a threat of a release.[16]

Note that the court does not mention the statutory adjective "substantial." Regardless of whether the court's omission of "substantial" was intentional or accidental, the court's analysis suggests that a "substantial threat of release" need not constitute an imminent threat of release to be actionable.[17]

On the other hand, the mere presence of a hazardous substance in a building or structure, standing alone, may not be sufficient to qualify as posing a threat of release.[18]

Congress expressly excluded the following activities and releases from the CERCLA definition of a release:

16. *Id.* at 747.

17. Other courts have similarly interpreted the term "substantial threat of release." *See, e.g.*, Dedham Water Co. v. Cumberland Farms Dairy, 889 F.2d 1146 (1st Cir. 1989) (threatened release of hazardous substances that caused the plaintiff "reasonably" to incur response costs satisfies causation requirement under CERCLA; actual contamination is not a prerequisite to recovery); *Emhart*, 665 F. Supp. at 559 (PCBs on concrete flooring can "leach into the concrete floor and ultimately the ground"); Amland Props. Corp. v. Aluminum Co. of Am., 711 F. Supp. 784 (D.N.J. 1989).

18. Sycamore Indus. Park Assocs. v. Ericsson, Inc. 546 F.3d 847 (7th Cir. 2008) (presence of asbestos insulation in building does not constitute a "threatened release" because any emission would be confined within the building and thus would not reach the "outside environment").

(A) [A]ny release which results in exposure to persons solely within a workplace, with respect to a claim which such persons may assert against the employer of such persons,
(B) [E]missions from engine exhaust of a motor vehicle, rolling stock, aircraft, vessel, or pipeline pumping station engine,
(C) [R]elease of source, by-product, or special nuclear material from a nuclear incident, as those terms are defined in the Atomic Energy Act of 1954 . . . and
(D) [T]he normal application of fertilizer.[19]

These exclusions from the definition of release are narrow, and courts have rarely found a release to fit within them. For example, the exclusion for a release of "source, by-product, or special nuclear material" was at issue in *Pennsylvania v. Lockheed Martin Corp.*[20] In this case, Pennsylvania sought to recover from Lockheed the costs associated with cleanup of radioactive strontium, which had been used in a state-owned facility that manufactured thermoelectric generators in the 1960s. Pennsylvania had performed the cleanup pursuant to a decommissioning plan that had been submitted to and approved by the Nuclear Regulatory Commission. Lockheed sought dismissal of Pennsylvania's CERCLA cost recovery action on the ground that strontium is excluded radioactive by-product material under CERCLA § 101(22)(c). The district court denied the motion, reasoning that it was unclear under CERCLA's statutory language and EPA's implementing regulations whether a release of radioactive by-product material such as strontium could be considered "disposal" subject to CERCLA. The court deferred to EPA's expertise on this issue, and declined to dismiss the case.[21]

Another failed attempt to characterize a release as excluded from CERCLA concerned the exclusion for releases that constitute the "normal application of fertilizer." In *City of Tulsa v. Tyson Foods, Inc.*,[22] the defendant poultry manufacturers filed a motion for summary judgment

19. *Id.*
20. 684 F. Supp. 2d 564 (M.D. Pa. 2010).
21. *Id.* at 576.
22. 258 F. Supp. 2d 1263, 1288–89 (N.D. Okla. 2003).

arguing that land application of chicken manure qualified as the "normal application of fertilizer." The court rejected the motion, concluding that the defendants had failed to provide evidence that their application of manure could be considered "normal" application of fertilizer.[23]

### B. Environment

Under section 101(22) of CERCLA, releases are subject to government response authority only if they reach the environment. Congress defined "environment" very broadly to include

> (A) the navigable waters, the waters of the contiguous zone, and the ocean waters of which the natural resources are under the exclusive management authority of the United States . . . and (B) any other surface water, ground water, drinking water supply, land surface or subsurface strata, or ambient air within the United States or under the jurisdiction of the United States.[24]

As broadly as Congress defined the environment, courts have gone even further. On the related issue of interpreting "disposal into . . . the environment," several courts have held that EPA's response authority extends to releases that remain entirely within a building.[25] Similarly, EPA

23. *See also* City of Waco v. Schouten, 385 F. Supp. 2d 595 (W.D. Tex. 2005) (rejecting defendant's argument that cow manure disposal constituted "normal application of fertilizer," citing releases from improper manure storage as a cause of phosphorus releases that contaminated Lake Waco).

24. CERCLA § 101(8).

25. *See, e.g.*, BCW Assocs. Ltd. v. Occidental Chem. Corp., Civ. No. 86-5947, 1988 WL 102641 (E.D. Pa. Sept. 29, 1988) (placement of hazardous wastes inside an enclosed manufacturing facility may constitute disposal of such wastes into or on any land so as to satisfy the CERCLA definition); *Emhart*, 665 F. Supp. 549; *cf.* New York v. Gen. Elec. Co., 592 F. Supp. 291, 296 (N.D.N.Y. 1984) ("Indeed, it appears that Congress sought to deal [in CERCLA] with every conceivable area where hazardous substances come to be located . . . ."). *But see Sycamore Indus. Park*, 546 F.3d at 851–53 (presence of asbestos insulation inside a building does not constitute "disposal" because there is no exposure to the environment); The Fertilizer Inst. v. EPA, 935 F.2d 1303 (D.C. Cir. 1991) (a release into an impervious, unenclosed containment structure does not constitute a release into the environment for purposes of release reporting under section 103(a) of CERCLA).

takes the position that hazardous substances have been released into the environment even if they remain entirely on plant or facility grounds.[26]

### C. Hazardous Substance, Pollutant, or Contaminant

Section 104(a) references two categories of substances as to which the government is authorized to perform a response action: (1) hazardous substances; and (2) pollutants or contaminants. As discussed below, there are significant differences between these two categories of substances.

#### 1. *Hazardous Substance*

Section 101(14) of CERCLA defines "hazardous substance" by reference to lists of hazardous substances developed under a number of other federal environmental statutes, specifically, sections 311 and 307 of the Clean Water Act of 1977, and section 112 of the Clean Air Act. EPA has created a master list of these hazardous substances, which is codified at 40 C.F.R. § 302.4. In addition to these specifically designated hazardous substances, section 101(14)(C) of CERCLA defines hazardous substances to include any waste that EPA has specifically listed as a hazardous waste,[27] and any waste that exhibits one or more of the characteristics of hazardous wastes (ignitability, corrosivity, reactivity, and toxicity) established by EPA under the Resource Conservation and Recovery Act (RCRA).[28]

#### 2. *Pollutant or Contaminant*

In addition to responding to a release or substantial threat of release of a hazardous substance, the government may respond to the release or substantial threat of release of "any pollutant or contaminant" that

26. 50 Fed. Reg. 13,462 (Apr. 4, 1985).

27. EPA has promulgated several lists of hazardous wastes. The so-called "F-List" identifies wastes from common manufacturing and industrial processes, such as solvents that have been used in cleaning or degreasing operations. The F-List of hazardous wastes is set forth at 40 C.F.R. § 261.31. The K-List of hazardous wastes is a source-specific list that identifies the waste by reference to specific industries, such as petroleum refining or pesticide manufacturing. The K-List is set forth at 40 C.F.R. § 261.31. The P-List and U-List are lists of commercial chemical products that are deemed hazardous waste when discarded. *See* 40 C.F.R. § 261.33.

28. The four hazardous waste characteristics are set forth at 40 C.F.R. §§ 261.21–.24.

may present an "imminent and substantial danger to the public health or welfare."[29] The phrase "pollutant or contaminant" is defined as any material "which after release into the environment . . . may reasonably be anticipated to cause death, disease, behavioral abnormalities, cancer, genetic mutation, physiological malfunctions . . . or physiological deformations."[30]

Note that there is a heightened standard applicable to the government's decision to respond to a release of a pollutant or contaminant that is not a hazardous substance. It may only do so if it determines that the release may present an "imminent and substantial danger to public health [or] welfare." There is no such requirement for the government to respond to a release of a hazardous substance.[31] Another important difference between a hazardous substance and a pollutant/contaminant concerns liability. Section 107 does not impose liability on responsible parties for a release of a pollutant or contaminant that is not a hazardous substance.

### D. Petroleum Exclusion

The response authorities of CERCLA do not extend to releases of petroleum. The CERCLA definition of "hazardous substance" excludes "petroleum, including crude oil or any fraction thereof which is not otherwise specifically listed or designated as a hazardous substance under subparagraphs (A) through (F) of this paragraph."[32] The definition of "pollutant or contaminant" contains parallel petroleum exclusion language.[33] (Both definitions also exclude "natural gas, liquefied natural gas, or synthetic gas of pipeline quality or mixtures of natural gas and such synthetic gas.")

There have been countless cases examining the scope of the petroleum exclusion, mostly in the context of asserting the exclusion as a defense to liability (rather than as challenges to EPA response authority).

29. CERCLA § 104(a)(1)(B).
30. CERCLA § 101(33).
31. *See* United States v. Marisol, Inc., 725 F. Supp. 833, 837 (M.D. Pa. 1989).
32. *See* CERCLA § 101(14).
33. *See* CERCLA § 101(33).

In most cases, courts have read the exclusion narrowly, and imposed liability on the defendants.[34] One of the first successful applications of the petroleum exclusion was *Wilshire Westwood Associates v. Atlantic Richfield Corp.*[35] In this case, two development companies sought to recover cleanup costs associated with a spill of leaded gasoline. The developers asked the Ninth Circuit to interpret the petroleum exclusion narrowly, so that the exclusion did not apply to petroleum containing CERCLA hazardous substances such as lead. The court rejected this approach, noting that the developers' interpretation "renders the petroleum exclusion a nullity," and concluded that "the petroleum exclusion in CERCLA [applies] to unrefined and refined gasoline even though certain of its indigenous components and certain additives during the refining process have themselves been designated as hazardous substances within the meaning of CERCLA."[36] A number of courts have followed the reasoning of *Wilshire*.[37]

34. *See, e.g.*, Franklin Cnty. Convention Facilities Auth. v. Am. Premier Underwriters, Inc., 240 F.3d 534 (6th Cir. 2001) (affirming lower court's ruling that there was insufficient evidence to show that a mixture of benzene and creosote was in fact petroleum excluded from the CERCLA definition of hazardous substances); Cose v. Getty Oil Co., 4 F.3d 700 (9th Cir. 1993) (crude oil tank bottoms are not petroleum or petroleum "fractions" because they are not a product of the refining process, but are simply discarded as wastes); United States v. Gurley, 43 F.3d 1188, 1199 (8th Cir. 1994) (exclusion does not apply to waste oil containing PCBs and sulfuric acid); Marrero v. Esso Standard Oil Co., 597 F. Supp. 2d 272, 289–90 (D.P.R. 2009) (oil company's motion for summary judgment denied because oil company could not show that the release of a refined petroleum product from a corroded tank did not also contain hazardous substances from the corrosion); Acme Printing Ink Co. v. Menard, Inc., 881 F. Supp. 1237 (E.D. Wis. 1995) (exclusion does not apply to waste oil contaminated with substances that are not constituents of petroleum); Licciardi v. Murphy Oil USA, Inc., Civ. A. No. 93-0490, 1994 WL 285051 (E.D. La. June 20, 1994) (waste materials such as refinery sludge do not fall within the petroleum exclusion); United States v. W. Processing, 761 F. Supp. 713, 721 (W.D. Wash. 1991) ("tank bottom sludge is a contaminated waste product, and not a petroleum fraction, as that term is used in [CERCLA]").

35. 881 F.2d 801, 804 (9th Cir. 1989).

36. *Id.* at 810.

37. *See, e.g.*, Carriddi v. Consol. Aluminum Corp., 478 F. Supp. 2d 150 (D. Mass. 2007) (the "petroleum exclusion" includes mineral spirits distilled from petroleum, as well as a mixture of heavy oils and mineral spirits); Foster v. United States, 922 F. Supp. 642 (D.D.C. 1996) (plaintiff failed to establish that PAHs, TPHs, and kerosene fell outside the petroleum exclusion, and therefore no liability imposed for related contamination); United States v. Atlas Minerals & Chems., Inc., Civ. A. No. 91-5118, 1995 WL 510304 (E.D. Pa. Aug 22, 1995) (asphalt condensate deemed to be covered by the petroleum exclusion); Textron Inc. v. Barber-Colman Co., 903 F. Supp. 1558, 1567 (W.D.N.C. 1995) (adding nonhazardous additives to substances

There is a split of authority on the question of whether used oil is covered by the petroleum exclusion when there is no evidence that the oil was contaminated with hazardous substances. In *Marmon Group, Inc. v. Rexnord, Inc.*,[38] the district court held that, because cutting oil is a form of oil that is covered by the petroleum exclusion, so too is "waste cutting oil." The decision was reversed on other grounds by the Seventh Circuit, but in its opinion the Seventh Circuit recites the fact that the "district court had dismissed the CERCLA count on the ground that the cutting oil was not a substance regulated by CERCLA."[39] A number of other courts have similarly deemed used oil to be covered by the petroleum exclusion.[40] These courts appear to presume that used oil does not contain nonindigenous hazardous substances.

Several courts, however, appear to presume precisely the opposite—namely, that used oil is contaminated with nonindigenous substances as a result of use and thus is not covered by the petroleum exclusion absent contrary evidence.[41]

---

that are covered by the petroleum exclusion, such as kerosene, does not remove the product from the scope of the exclusion). *See also* Niecko v. Emro Mktg. Co., 769 F. Supp. 973, 982 (E.D. Mich. 1991), *aff'd*, 973 F.2d 1296 (6th Cir. 1992) (petroleum exclusion applies to gasoline leaking from gas station underground storage tank); Bunger v. Hartman, 797 F. Supp. 968, 972–73 (S.D. Fla. 1992) (same).

38. No. 85-C-7838, 1986 WL 7070 (N.D. Ill. June 16, 1986), *rev'd on other grounds*, 822 F.2d 31 (7th Cir. 1987).

39. 822 F.2d at 33.

40. *See, e.g.*, S. Pac. Trans. Co. v. Cal. Dep't of Transp., 790 F. Supp. 983 (C.D. Cal. 1991) (holding that used petroleum products mixed with soil are within the petroleum exclusion, noting that "EPA has consistently maintained that used petroleum products are covered by the petroleum exclusion," but also noting EPA's position that hazardous substances added to petroleum products that become more concentrated during use are not covered by the petroleum exclusion); *Niecko*, 769 F. Supp. at 982 (same).

41. *See, e.g.*, Esso Standard Oil Co. v. Rodriguez Perez, No. 01-2012, 2004 WL 2238894 (D.P.R. Oct. 1, 2004) (petroleum exclusion not applicable where defendants had disposed of waste oil containing hazardous substances at levels above what would be considered indigenous to a petroleum product); Ekotek Site PRP Comm. v. Self, 881 F. Supp. 1516 (D. Utah 1995) (petroleum exclusion does not cover used oil containing nonindigenous hazardous substances or indigenous substances, which are present at elevated concentrations); *W. Processing Co.*, 761 F. Supp. 713 (absent evidence to the contrary, there is a presumption that oil sludge from petroleum storage tanks contain hazardous substances from walls of tanks); *see also* City of New York v. Exxon Corp., 766 F. Supp. 177, 186 (S.D.N.Y. 1991) ("By its plain language, this exclusion for petroleum does not include waste oil," but also finding that levels of lead, cadmium, and chromium had increased in the oil emulsion because of industrial processes).

### E. Other Limitations on Response Authority

Under section 104(a)(3) of CERCLA, the government is prohibited from responding to a release or threat of release:

(A) of a naturally occurring substance in its unaltered form, or altered solely through naturally occurring processes or phenomena, from a location where it is naturally found;
(B) from products which are part of the structure of, and result in exposure within, residential buildings or business or community structures; or
(C) into public or private drinking water supplies due to deterioration of the system through ordinary use.[42]

Notably, however, if EPA determines that a public health or environmental emergency exists, and no one else will respond to the emergency in a timely manner, EPA is authorized to respond to these otherwise excluded releases.[43]

## V. GOVERNMENT RESPONSE AUTHORITIES

Where the statutory prerequisites to response action are satisfied—that is, where EPA has evidence of a release or substantial threat of release into the environment of a hazardous substance, pollutant, or contaminant—EPA is authorized to act, consistent with the NCP,[44] to take short-term "removal" actions that enable it to respond quickly, as well as to take longer-term "remedial" actions that are intended to be a permanent solution to the contamination.[45]

Before EPA may take the longer-term remedial action at a contaminated site, it must place the site on its master list of sites eligible for funding by the Hazardous Substances Superfund. This list is known as the NPL.[46]

42. CERCLA § 104(a)(3)(A)–(C); 40 C.F.R. § 300.400(b).
43. CERCLA § 104(a)(4); 40 C.F.R. § 300.400(b).
44. 40 C.F.R. pt. 300.
45. CERCLA § 104.
46. CERCLA § 105(a)(8)(B).

Each of the procedural and substantive elements of EPA's response authority is discussed below.

## A. National Contingency Plan (NCP)

The NCP was originally established under the Clean Water Act to address and respond to oil spills affecting waters of the United States. Section 105(b) of CERCLA directs the president to revise and republish the NCP to reflect CERCLA's new removal and remedial authorities in a new section of the NCP referred to as the National Hazardous Substance Response Plan.

The NCP has been referred to as the "road map" for the government's hazardous waste cleanup program. The NCP provides guidelines and procedures for discovering hazardous waste sites, for identifying the most hazardous among those sites and creating a priority list for action (the NPL), and for selecting and implementing removal and remedial actions. Set forth below is a summary of key aspects of the NCP.

### *1. Site Identification*

Obviously, before EPA can respond to a release or threatened release of a hazardous substance, it must identify or learn of the release. The release reporting provisions of CERCLA are the primary source of this information. Section 103 requires responsible parties to report the existence of hazardous waste sites and releases from hazardous waste facilities to EPA, as discussed in chapter 11. Based on this information, EPA determines which sites to include on the Comprehensive Environmental Response, Compensation, and Liability Information System (CERCLIS) computer list—the national inventory of potential NPL sites. CERCLA also authorizes private persons who are, or may be, affected by a release or threatened release of a hazardous substance, pollutant, or contaminant to petition the president to perform a risk assessment on any release or threatened release.[47]

47. CERCLA § 105(d); 40 C.F.R. § 300.405(a)(6).

### 2. *National Priorities List*

Once a potential site has been identified, it is subjected to a series of investigations and reviews to determine if the site should be placed on the NPL. The first of these is the preliminary assessment (PA), which is a review of available information to assess the site and the hazardous substances present, and to determine if any potentially responsible parties (PRPs) can be identified.[48] The second step is a site inspection (SI), which consists of a visual inspection of the site and routinely includes collection of samples.[49]

The results of the PA/SI are used by EPA to rank the site under the agency's Hazard Ranking System (HRS). Pursuant to section 105(a)(8) of CERCLA, EPA developed the HRS (which is described in Appendix A to the NCP) to assess the risk or danger to public health or the environment posed by a site. Using criteria identified in the statute, EPA uses the HRS to assign a numerical value to a given site, which represents the risks posed by the site. By regulation, EPA has established that sites with an HRS score of at least 28.50 are eligible for inclusion on the NPL.[50] Based on the HRS score assigned to a site, EPA proposes the site for inclusion on the NPL. Following receipt of comments, the site is listed on the NPL or dropped from further consideration for remedial action.[51]

There are two other mechanisms for placing a site on the NPL. First, each state has a one-time opportunity to include on the NPL a site that presents "the greatest danger to public health or welfare or the environment among the known facilities in such State."[52] Second, EPA may list a site on the NPL if (1) the ATSDR issues a health advisory that recommends removing individuals from the vicinity of the release; (2) EPA determines that the release poses a significant threat to public health; and

48. *See* 40 C.F.R. § 300.420(b).

49. *See* 40 C.F.R. § 300.420(c).

50. *See* 40 C.F.R. § 300, App. A.

51. EPA promulgated an original NPL list of 406 sites. 48 Fed. Reg. 40,658 (Sept. 8, 1983). The NPL must be—and has been—updated at least once annually by EPA. A complete listing of the NPL sites can be found at EPA, Superfund, NPL Site Status Information, http://www2.epa.gov/superfund/npl-site-status-information.

52. CERCLA § 105(a)(8)(B); *see also* 40 C.F.R. § 300.425(c)(2).

(3) EPA further determines that it will be more cost-effective to use its remedial authority than its removal authority to respond to the release.[53]

Once a site has been included on the NPL, it has proven very difficult to successfully challenge its inclusion.[54] That is not to say that NPL listings are immune from challenge. For example, in *Mead Corporation v. Browner*,[55] the D.C. Circuit vacated the NPL listing of the "Tennessee Products" site, which included three noncontiguous areas, one of which was a mile away from the others. The court stated that the inclusion of each area must be individually justified, and that the agency's aggregating a low-risk site with high-risk sites under the agency's so-called "Aggregation Policy"[56] was unlawful. (Chapter 6 provides a comprehensive discussion of judicial challenges to NPL listings.)

Only sites listed on the NPL qualify for Superfund-financed remedial action. Inclusion on the NPL, however, is not a precondition either for Superfund-financed removal action or for agency action under section 106, as discussed in chapter 3.

### 3. *Remedial Investigation/Feasibility Study*

Following addition of a site to the NPL, the government must perform a remedial investigation/feasibility study (RI/FS).[57] The purpose of a remedial investigation (RI) is to verify the existence of a release

53. 40 C.F.R. § 300.425(c).

54. *See, e.g.*, U.S. Magnesium, LLC v. EPA, 630 F.3d 188 (D.C. Cir. 2011) (challenge to NPL listing based on argument that HRS scoring artificially inflated the site's score rejected because argument was in essence a challenge to underlying regulation, which could no longer be asserted); RSR Corp. v. EPA, 102 F.3d 1266 (D.C. Cir. 1997) (challenge to NPL listing ruled untimely because it was raised three years after listing); Eagle-Picher Indus., Inc. v. EPA, 822 F.2d 132 (D.C. Cir. 1987) (upholding NPL listing of sites at which mining wastes and power plant ash were disposed, as well as uranium mills, notwithstanding that such wastes are expressly excluded from the CERCLA definition of hazardous substance, because EPA may respond to releases of "pollutants and contaminants").

55. 100 F.3d 152 (D.C. Cir. 1996).

56. 48 Fed. Reg. 40,658, 40,663 (Sept. 8, 1983).

57. A detailed discussion of the various steps involved in the RI/FS process is beyond the scope of this book. The regulations governing the performance of an RI/FS may be found in the NCP at 40 C.F.R. § 300.430. In addition, EPA elaborates on the procedural requirements in its interim final Guidance for Conducting Remedial Investigations and Feasibility Studies under CERCLA, Directive No. 9355.3-01, Office of Emergency and Remedial Response, EPA (Oct. 1988).

and to document its environmental and health effects. During the RI, a full characterization of the site is performed. Field investigations are conducted to evaluate the physical characteristics of the site, including soils, geology, and hydrogeology; contamination of air, surface water, and groundwater; nature of the wastes disposed on-site; and the potential exposure pathways. A baseline risk assessment is conducted that includes an assessment of the toxicity of chemicals at the site and the potential for exposure. Based on the foregoing information, the government identifies the standards that the selected remedy must be designed to achieve. (Remedy selection standards are discussed in chapter 2.)

The RI is also designed to provide information necessary to perform the feasibility study (FS). Through the FS, the government develops and screens potential remedial alternatives, ranging from a treatment alternative that, to the degree possible, would eliminate the need for long-term management at the site, to a containment option that involves little or no treatment, to a "no action" alternative. The most promising of these remedial alternatives are analyzed and compared against one another based on their relative strengths and weaknesses as measured against specific remedy selection criteria, also discussed in chapter 2. Thus, the RI/FS is the fundamental document on which EPA bases its decision to select a specific remedial action for a site.

### B. Removal Actions versus Remedial Actions

As already noted, response actions implemented at a contaminated site can take two forms: removal action or remedial action. As discussed in more detail below, each term has a specific statutory definition. Conceptually, a removal action should be thought of as a short-term, interim response action taken to prevent imminent harm or, as some courts have held, an action "taken in response to an immediate threat."[58] By contrast, a remedial action is the long-term, permanent solution or remedy for the contamination.

58. Versatile Metals, Inc. v. Union Corp., 693 F. Supp. 1563, 1577 (E.D. Pa. 1988); *see also* Gen. Elec. Co. v. Litton Indus. Automation Sys., Inc., 920 F.2d 1415, 1419 (8th Cir. 1990), *cert. denied*, 490 U.S. 937 (1991); Con-Tech Sales Defined Benefit Trust v. Cockerham, Civ. A. No. 87-5137, 1991 WL 209791, at *7 (E.D. Pa. Oct. 9, 1991).

Examination of the definitions of removal and remedial action, however, reveal that there is no bright-line distinction in terms of timing or even permanence. Case law has shown that removal actions can last decades and embody the permanent remedy for a site. The blurred distinction between removal and remedial action has engendered litigation. Different statutes of limitation apply to actions for recovery of costs of removal and remedial action. Moreover, different procedural requirements must be satisfied for costs of removal and remedial action to be recoverable from liable parties. Thus, improper characterization of an action as removal or remedial action can have significant consequences for a party seeking recovery of the associated costs. We begin our discussion by examining the underlying statutory definitions of removal and remedial action.

*1. Definitional Differences between Removal and Remedial Actions*

The text of the definitions of removal action and remedial action follow a similar pattern. Each begins with a general statement of principle, and follows that with examples. The principles of each definition are clearly different, but the examples provided after the principle statement appear to overlap. In this regard, a "removal action" is defined as

> the cleanup or removal of released hazardous substances from the environment, such actions as may be necessary taken [sic] in the event of the threat of release of hazardous substances into the environment, such actions as may be necessary to monitor, assess, and evaluate the release or threat of release of hazardous substances, the disposal of removed material, or the taking of such other actions as may be necessary to prevent, minimize, or mitigate damage to the public health or welfare or to the environment, which may otherwise result from a release or threat of release. The term includes, in addition, without being limited to, security fencing or other measures to limit access, provision of alternate water supplies, temporary evacuation and housing of threatened individuals not otherwise provided for, action taken under [section 104(b)] of this

> title, and any emergency assistance which may be provided under the Disaster Relief and Emergency Assistance Act.[59]

Thus, the definition of "removal" includes not only actions needed to address the immediate threat posed by a release, such as providing "temporary evacuation and housing" and "alternate water supplies," but also actions that may prove to be the long-term remedy, such as removing and disposing of contaminated material.

By contrast, CERCLA defines the terms "remedy" or "remedial action" as meaning "those actions consistent with permanent remedy taken instead of or in addition to removal actions in the event of a release or threatened release of a hazardous substance into the environment, to prevent or minimize the release of hazardous substances so that they do not migrate to cause substantial danger to present or future public health or welfare or the environment."[60] The reference to "permanent remedies" embodies the broad principle, but the examples provided thereafter in the definition include some of the very same actions found in the definition of removal:

> The term includes, but is not limited to, such actions at the location of the release as storage, confinement, perimeter protection using dikes, trenches, or ditches, clay cover, neutralization, cleanup of released hazardous substances and associated contaminated materials, recycling or reuse, diversion, destruction, segregation of reactive wastes, dredging or excavations, repair or replacement of leaking containers, collection of leachate and runoff, onsite treatment or incineration, provision of alternative water supplies, and any monitoring reasonably required to assure that such actions protect the public health and welfare and the environment. The term includes the costs of permanent relocation of residents and businesses and community facilities where the President determines

59. CERCLA § 101(23).
60. CERCLA § 101(24).

> that, alone or in combination with other measures, such relocation is more cost-effective than and environmentally preferable to the transportation, storage, treatment, destruction, or secure disposition offsite of hazardous substances, or may otherwise be necessary to protect the public health or welfare; the term includes offsite transport and offsite storage, treatment, destruction, or secure disposition of hazardous substances and associated contaminated materials.[61]

As is apparent, the definitions of removal action and remedial action both refer to the "cleanup of released hazardous substances" and encompass off-site disposal or destruction of the hazardous substances. There are other areas of overlap (e.g., both definitions refer to "providing alternative water supplies"), but the most significant overlap is that both definitions refer to "cleanup" actions. These overlapping definitions create the possibility of a response action being characterized as either a removal or a remedial action. For example, the excavation of buried drums may be deemed a removal because it "may be necessary to prevent, minimize or mitigate" the spread of contamination that would occur if the drums remained buried. On the other hand, excavating buried drums may be the only action required to clean up a site, and thus would be the permanent remedy, the hallmark of a remedial action. Similarly, depending on the circumstances, installing a cap over contaminated soil could qualify as either a removal action where "necessary to prevent, minimize, or mitigate" the spread of contamination, or a remedial action where it constitutes the permanent remedy selected for a site.[62]

As discussed below, there are significant substantive and procedural differences between the requirements applicable to removal actions and remedial actions. Thus, it is commonplace for parties to disagree over

61. CERCLA § 101(24).

62. *See, e.g.*, *Versatile Metals*, 693 F. Supp. at 1576 (cap could be either removal or remedial depending on whether its purpose was abatement or permanent cover).

the characterization of an action as a removal or remedial action, which often can be dispositive of a cost recovery claim.[63]

## 2. *Regulatory Differences under the NCP*

As already noted, the regulatory implications of the distinction between removal and remedial actions are not trivial. The NCP establishes when EPA may initiate a removal or a remedial action. Removal actions are authorized whenever EPA determines that there is a threat to public health or welfare or the environment, based upon a review of the following factors:

> (i) [A]ctual or potential exposure to hazardous substances or pollutants or contaminants by nearby populations, animals, or food chain;
> (ii) [A]ctual or potential contamination of drinking water supplies or sensitive ecosystems;
> (iii) [H]azardous substances or pollutants or contaminants in drums . . . or other bulk storage containers, that may pose a threat of release; [and] . . .
> (iv) [T]hreat of fire or explosion. . . .[64]

On the other hand, Superfund-financed remedial actions may only be undertaken after a site has been listed on the NPL.[65] Accordingly, the universe of sites at which EPA itself may initiate remedial action is vastly smaller than the universe of sites at which it may initiate removal action. However, EPA can require PRPs to initiate remedial actions at non-NPL sites¸ including through the issuance of unilateral orders under section 106 of CERCLA.

The procedural requirements applicable to a remedial action are far more demanding than those applicable to a removal action. The NCP

63. *See, e.g.*, *Gen. Elec. Co.*, 920 F.2d at 1419 ("Any distinction between 'excavation' of contaminated soils [listed as a remedial action] and 'removal' of contaminated soils is one that eludes us.").

64. *See* 40 C.F.R. § 300.415(b)(2).

65. 40 C.F.R. § 300.425(b)(1).

establishes a detailed procedural framework for selecting a remedial action for sites on the NPL. As noted earlier, before any remedy may be selected, an RI/FS must be conducted in which the nature and extent of contamination is determined, the health risk posed by the contamination is quantified, and the remedial alternatives for mitigating those health risks are judged comparatively, using remedy selection criteria.[66] Remedial actions must be cost-effective, while removals need not be.[67] Moreover, the public must be provided with an opportunity to comment on the proposed remedy selection for NPL sites.[68] By contrast, removal actions need only comply with the relatively simple NCP requirements, specifically (1) undertake a removal site evaluation;[69] (2) select the appropriate removal action based on the site evaluation;[70] and (3) determine whether responsible parties "can and will perform the necessary removal action promptly and properly."[71] Of particular significance to PRPs, removal actions are to be terminated after $2 million has been obligated for the action or 12 months have elapsed from the date of initial response, unless certain requirements are met.[72] (This limitation is not applicable to investigations and monitoring done pursuant to section 104(b) of CERCLA.) EPA often uses those circumstances to justify longer-term and more expensive removal actions.

### 3. *Different Statutes of Limitation*

Another difference between removal and remedial actions is that each has a different statute of limitation for recovery of response costs from liable parties. For a remedial action, the statute of limitations for recovery of

66. *See* 40 C.F.R. 300.430(a)(2).
67. *See* 40 C.F.R. 300.430(f)(1)(ii)(D); *see also* United States v. Am. Cyanamid Co., 786 F. Supp. 152, 162 (D.R.I. 1992).
68. 40 C.F.R. § 300.430(f)(2).
69. 40 C.F.R. § 300.410.
70. 40 C.F.R. § 300.415(a)(1).
71. 40 C.F.R. § 300.415(a)(2).
72. CERCLA § 104(c). The foregoing limits on EPA's removal authority may be exceeded, however, if the immediate and imminent harm continues, or if the continued response is otherwise appropriate and consistent with the remedial action to be taken. *Id.* EPA has stated that the latter basis will apply only at a site listed on the NPL. *See* EPA, OSWER Directive No. 9360.0-12A, Final Guidance on Implementation of the "Consistency" Exemption to the Statutory Limits on Removal Actions (June 12, 1989).

response costs is "6 years after initiation of physical on-site construction of the remedial action."[73] By contrast, for a removal action the statute of limitations for recovery of response costs is "3 years after completion of the removal action," except that removal costs may be recovered as part of a cost recovery action for remedial costs if the remedial action is commenced within three years of completion of the removal action.[74]

The differing limitation periods for removal actions and remedial actions are at the heart of a broad, deep vein of CERCLA case law in which courts adjudicating cost recovery actions have had to decide whether the response was a removal action subject to a three-year limitations period or remedial action subject to a six-year limitations period.

One group of these cases have focused on the question of when "physical on-site construction" commences, which is the event that triggers the six-year limitations period for a remedial action.[75] In a second, larger group of cases, the plaintiffs attempt to link a series of cleanup actions that may span decades as one long removal action, because the three-year limitations period for a removal action commences upon "the completion of a removal action." The question in these cases is whether the limitations period applies to each response action individually or to all actions collectively. For example, in *Kelly ex rel. Michigan v. E.I. du Pont de Nemours & Co.*,[76] the initial removal action—namely, excavation and disposal of drums and contaminated soil—was completed in 1986. The contractor was paid, and state officials announced that the removal action was finished. More than a year later, four more drums were uncovered during the RI/FS, and were excavated. Although this

73. CERCLA § 113(g)(2)(B).

74. CERCLA § 113(g)(2)(A); *see also* Agere Sys. Inc. v. Advanced Envtl. Tech. Corp., 602 F.3d 204, 229–33 (3d Cir. 2010).

75. *See, e.g.*, United States v. Akzo Nobel Coatings, Inc., 990 F. Supp. 897, 905 (E.D. Mich. 1998) (six-year limitations period for filing cost recovery action was not triggered by construction of storage pads, drum excavations, and delivery of equipment since none of those actions were "an initiation of physical on-site construction"); United States v. Atl. Richfield, 147 F. Supp. 2d 214 (S.D. Tex. 2001) (six-year limitations period was not triggered by installing a fence, clearing the site, and widening the access road, or by installing temporary trailers, electrical generators, and air monitoring stations; each activity was part of the removal action, not a remedial action).

76. 17 F.3d 836 (6th Cir. 1994).

later drum removal was performed by a different contractor, the court held that it was part of the original removal. Thus, the removal was not complete and the statute of limitations did not begin to run until the last drum was removed. This is just one of a number of cases involving the "never-ending removal."[77]

In yet another group of removal versus remedial action cases, the court had to decide whether a response action was properly characterized as a removal where there is a question as to whether the release in question posed an emergency.[78]

In sum, despite decades of jurisprudence, courts continue to struggle with the distinction between a removal action and a remedial action.[79]

## C. Ancillary Response Authorities

As noted, EPA's removal and remedial authority authorize EPA to permanently relocate residents, businesses, and community facilities in those cases in which it determines that relocation is (1) more cost-effective than, and environmentally preferable to, off-site disposal of the hazardous substances, or (2) necessary to protect the public health or welfare.[80]

A related response authority was added to section 104 of CERCLA by the Superfund Amendments and Reauthorization Act of 1986. Under section 104(j), EPA may use Superfund money to acquire interests in real

77. *See, e.g.*, State of New York v. Next Millenium Realty, LLC, 732 F.3d 117 (2d Cir. 2013) (removal began in 1989 with discovery of groundwater contamination, continued in 1990 with installation of activated carbon absorption system to remove contaminants and with installation of air stripper in 1997; record of decision was issued in 2003); United States v. Rohm & Haas, 790 F. Supp. 1255 (E.D. Pa. 1992) (removal not complete even after excavation of all contaminated soil because EPA continued to monitor and assess the release), *rev'd on other grounds*, 2 F.3d 1265 (3d Cir. 1993).

78. *See, e.g.*, United States v. Horne, No. 05-0497-CV-W-NKL, 2006 WL 2506447 (W.D. Mo. Aug. 29, 2006) (three-year limitation period for removal applied because EPA sought an "emergency" waiver of removal requirements); OBG Tech. Servs., Inc. v. Northrop Grumman Space & Mission Sys. Corp., 503 F. Supp. 2d 490 (D. Conn. 2007) (plaintiff's action characterized as remedial action because it did not involve an emergency situation, there was no threat of release off-site, no imminent danger to the public, and the action taken was the permanent, long-term remedy; consequently, plaintiff's claim was dismissed as untimely).

79. Additional cases adjudicating the applicable limitation period on the basis of an analysis of a response action as removal or remedial are summarized in chapter 9, note 37.

80. CERCLA § 101(24); 40 C.F.R. § 300.6. EPA used this authority in the cleanup of the Times Beach, Missouri, Superfund site, where waste oil mixed with dioxins had been sprayed on rural roads as a dust suppressant.

estate where a response action would constitute a taking of property. Land may be acquired where it is needed to carry out remedial actions, but only on condition that the state takes title to the property upon completion of the remedial action. Notably, section 104(j) expressly excludes any private right of action to force the government to acquire property.

CERCLA also authorizes EPA to place a lien on property to recover amounts spent on its cleanup.[81] This lien is established by operation of law, and becomes effective upon EPA's filing a notice of lien. Notably, however, in *Reardon v. United States*,[82] the First Circuit held this lien provision unconstitutional on the ground that it deprived property without due process. (We discuss this case in more detail in chapter 6.)

## VI. INFORMATION-GATHERING AUTHORITY

### A. Information Requests

Before EPA can exercise its response authority, the agency must have evidence of a release or threatened release sufficient to satisfy the statutory prerequisites. For this reason, Congress gave EPA extensive authority to obtain information and records from any person it believes may have a relationship to a site.[83] EPA is empowered to require detailed information that is relevant to the following:

> (A) The identification, nature, and quantity of materials which have been or are generated, treated, stored, or disposed of at a vessel or facility . . .
> (B) The nature or extent of a release or threatened release of a hazardous substance . . . [and]
> (C) Information relating to the ability of a person to pay for or to perform a cleanup.[84]

81. CERCLA § 107(l).
82. 947 F.2d 1509 (1st Cir. 1991).
83. CERCLA § 104(e).
84. CERCLA § 104(e)(2).

EPA generally exercises this authority through information requests. Typically, such requests are extremely broad. The recipient must identify the nature and amount of wastes sent to the site, the transporters utilized to convey such wastes to the site, and generally must provide EPA with "all documents in your possession relating to the site." Failure to comply with an information request can result in substantial penalties—up to $37,500 for each day of noncompliance.[85] An extreme example of the potential financial exposure for resisting an EPA information request is *United States v. Gurley*,[86] in which the court upheld EPA's imposition of a $1.9 million civil penalty against the defendant for a seven-year delay in responding to a 1992 information request.[87]

EPA is also authorized to collect information by entering a PRP's facility for purposes of inspecting and copying documents or records relating the three above-referenced types of information bearing on the existence, nature, and quantity of hazardous substance releases.[88]

A company receiving an information request from EPA must carefully consider its response, with regard both to timing and content. Regarding the timing of response, EPA information requests generally allow 30 days for compiling and submitting the requested information. Given that information requests often contain scores of "interrogatories," the first order of business upon receiving an information request is for the PRP to assess the level of effort necessary to comply with the request, including the search for old, often difficult to locate records. In many instances, a PRP may need—and should request—an extension of time to respond.

As to information a PRP provides to EPA, the PRP must understand that the information may be made public. Accordingly, the PRP should

85. CERCLA § 104(e)(5)(B). The above-quoted penalty figure reflects an inflation adjustment of the statutory $25,000 amount, which is required by section 4 of the Federal Civil Penalties Inflation Adjustment Act of 1990, 28 U.S.C. 2461. *See also* 78 Fed. Reg. 66,643 (Nov. 6, 2013).

86. 384 F.3d 316 (6th Cir. 2004).

87. *See also* United States v. 718 W. Wilson Ave., Glendale, Cal., 778 F. Supp. 2d 1067, 1073 (C.D. Cal. 2011) (civil penalties awarded against landowner who provided a tardy, inadequate response to an EPA information request).

88. *See* CERCLA § 104(e)(2) & (3).

determine whether any of the information is protectable as a trade secret. A PRP may assert a claim of confidentiality over data if it can establish that (1) it has not released the information to anyone other than specified government officials under a confidentiality agreement; (2) the information is not "required to be disclosed . . . to the public under any other Federal or State Law"; (3) if the information sought is a specific chemical identity, it is "not readily discoverable through reverse engineering"; and (4) disclosure of the information is likely to cause "substantial harm" to the claimant's "competitive position."[89]

## B. Entry and Access

In addition to EPA's authority to obtain information from PRPs through information requests, EPA has broad power to enter private property to a site to determine if the hazardous substances on property pose a threat to public health. Specifically, EPA may enter private property on which a hazardous substance may be or has been generated, stored, treated, disposed of, or transported from; any place a hazardous substance has been or may have been released; any place that is or may be threatened by the release of a hazardous substance; or any place where entry is needed to determine the need for a response action, including property adjacent to the actual site.[90] Pursuant to this authority, EPA may enter private property for the purpose of taking samples at "any location of any suspected hazardous substance or pollutant or contaminant."[91] If EPA takes samples, it must provide a receipt, offer to split the samples, and provide a copy of the analytical results to the property owner.[92]

In addition to entering private property to determine if hazardous substances pose a threat, courts have confirmed that the United States may access private property for the purpose of beginning response activities.[93] In *United States v. Charles George Trucking Co.*,[94] the court held that

89. CERCLA § 104(e)(7)(E)(ii), (iv).
90. CERCLA § 104(e)(3)(A)–(D).
91. CERCLA § 104(e)(4)(A).
92. CERCLA § 104(e)(4)(B).
93. *See, e.g.*, United States v. W.R. Grace & Co., 134 F. Supp. 2d 1182 (D. Mont. 2001).
94. 682 F. Supp. 1260 (D. Mass. 1988).

where the owner of a hazardous waste site refuses to grant permission for EPA to enter the site and adjacent property for the purpose of beginning a remedial action, section 104(e)(5) of CERCLA allows the United States to come directly to the district courts to enjoin interference with EPA's entry on the property.[95] Alternatively, EPA may issue an administrative order requiring the property owner to allow entry and then seek court enforcement of that order.[96] The court also ruled that the right of entry under section 104(e) of CERCLA extends equally to adjacent property and is not limited to the boundaries of the hazardous waste site, so long as the entry is necessary to effectuate the response action.[97]

Before EPA may enter private property, EPA must have a "reasonable basis to believe there may be a release or threat of release."[98] Moreover, courts may require EPA to spell out its intended actions relating to entry with a reasonable degree of specificity.[99] Once a release has been identified and EPA has specified the response actions it intends to take, the agency enjoys a broad right of access to implement the response actions.

EPA's broad right of access to implement response actions, however, is not without limits. EPA exceeded those limits in *United States v. Omega Chemical Corp.*,[100] where the agency had requested written consent for unconditional access from the owner of contaminated property to implement the cleanup. When the owner refused, EPA sought civil penalties. The Ninth Circuit held that the owner was not subject to civil penalties because the owner had consistently allowed EPA to enter its facility and conduct investigative activities during the period for which penalties were assessed. Moreover, the court held that neither the statute nor its implementing regulations contain language requiring unconditional written consent. Accordingly, the owner's refusal to sign a formal written consent giving EPA unconditional access to its facility did not constitute a violation of the statute.

95. *Id.* at 1267.
96. *Id.*
97. *Id.* at 1273.
98. CERCLA § 104(e)(5).
99. *See* United States v. Charles George Trucking Co., 682 F. Supp. 1260 (D. Mass. 1988).
100. 156 F.3d 994 (9th Cir. 1998).

By the same token, property owners cannot impose restrictive conditions before permitting EPA to access property being remediated. In *United States v. Iron Mountain Mines, Inc.*,[101] the court held that defendant property owners could not require representatives of EPA to sign waivers of liability or a code of conduct as a condition of gaining access to the property.

An open and important issue is whether EPA's access authority can be used by PRPs that have agreed to implement a remedial action. The language of section 104 does not expressly allow EPA to use its access authority to secure entry for third parties. In *B.F. Goodrich Co. v. Murtha*,[102] EPA sought access to a site on behalf of a group of generator PRPs who were obligated to conduct EPA's selected remedy pursuant to the terms of a consent decree that the PRPs and EPA had entered. The court avoided the section 104(e) issue by premising its ruling—requiring the landowner to permit access—on EPA's injunctive authority under section 106.[103]

## C. Responding to EPA Notice Letters

If EPA determines that there has been a release of a hazardous substance from a facility, and identifies PRPs associated with that facility, the agency will often contact the PRP through a notice letter. There are two varieties of notice letters: "general notice" letters and "special notice" letters. The procedural distinctions between general and special notice are addressed in chapter 5. Both kinds of notice, however, will state:

1. The company has been identified by EPA as potentially responsible for the release or threatened release of hazardous substances at a particular site;
2. the government is prepared to take remedial action;

101. Civ. No. 5-87-1189 MLS, 1987 WL 46792 (E.D. Cal. Sept. 18, 1987).
102. 697 F. Supp. 89 (D. Conn. 1988).
103. *Id.* at 93–94.

3. the company should indicate within a specified period of time whether it is willing to undertake part or all of the response actions; and
4. the recipient may wish to contact other PRPs who have been identified by the agency to coordinate a cleanup effort.

A notice letter often includes an information request, discussed above, to elicit further information regarding the nature and amount of hazardous substances sent to the site by the PRP, as well as the identity of other PRPs.

Upon receipt of a notice letter, a PRP must decide how to respond. In light of both the broad array of the government's response authorities, as discussed above, and of the statute's liability provisions, discussed in chapter 4, it is incumbent upon the PRP to develop a broad response strategy. The PRP should first inform itself as to the facts relating to liability by gathering and organizing company data relating to the generation and shipment of wastes to the site. In this regard, the company should consider:

1. appointing counsel to supervise document retrieval and internal investigations to perfect the attorney-client and work-product protection over such documents, to the extent they may apply;
2. identifying and interviewing personnel (both current and former) responsible for relevant facility waste management;
3. collecting and analyzing documents pertaining to shipments of waste to the site;
4. determining the volume of material sent to site, the nature of material, and the contractual provisions relating to off-site transport of materials; and
5. identifying and reviewing insurance policies and contacting carriers.

In addition, the PRP should devise a strategy for communicating with other companies identified by EPA as PRPs at the site. In this regard, a PRP must consider whether to participate in a PRP group. The numerous considerations relating to this decision are discussed in chapter 7.

Occasionally, albeit infrequently, a PRP will discover exculpatory information demonstrating at some level of proof that no hazardous substances were sent to the site. Upon discovery of such information, the PRP may wish to consider a proactive strategy designed to convince EPA that the company should be deleted from the site list of PRPs. To that end, upon discovering exculpatory information, the PRP should consider an informal meeting with EPA regional personnel to bring this information to the agency's attention.

# CHAPTER 2

# The Substance and Procedure of Remedy Selection

## I. INTRODUCTION

This chapter describes the administrative process that the government must follow to decide on a course of action for responding to a release of hazardous substances.

At the heart of EPA's selection of a plan for cleaning up a Superfund site is the pivotal question of the appropriate cleanup standard: How clean is clean? Much of the litigation that has mired the Superfund program has been driven by this question, because the more stringent the cleanup standard, the more costly the cleanup. Potentially responsible parties (PRPs), confronted with tens of millions of dollars in potential cleanup costs, have an incentive to litigate when financial exposure is so large and uncertain.

Justice Stephen Breyer captured the conundrum of "how clean is clean" in his 1993 book *Breaking the Vicious Circle*, in which he describes the *Ottati & Goss* litigation:

> Let me provide some examples. The first comes in a case in my own court, *United States v. Ottati & Goss*, arising out of a ten-year effort to force cleanup of a toxic waste dump in southern New Hampshire. The site was mostly cleaned up. All but one of the private parties had settled. The remaining private party

> litigated the cost of cleaning up the last little bit, a cost of about $9.3 million to remove a small amount of highly diluted PCBs and "volatile organic compounds" (benzene and gasoline components) by incinerating the dirt. How much extra safety did this $9.3 million buy? The forty-thousand page record of this ten-year effort indicated (and all the parties seemed to agree) that, without the extra expenditure, the waste dump was clean enough for children playing on the site to eat small amounts of dirt daily for 70 days each year without significant harm. Burning the soil would have made it clean enough for the children to eat small amounts daily for 245 days per year without significant harm. But there were no dirt-eating children likely to appear there, for future building seemed unlikely. The parties also agreed that at least half of the volatile organic chemicals would likely evaporate by the year 2000. To spend $9.3 million to protect non-existent dirt-eating children is what I mean by the problem of "the last 10 percent."[1]

This chapter describes the statutory and regulatory framework for identifying cleanup standards for Superfund sites, as well as for selecting a remedy that achieves the applicable cleanup standard.[2]

## II. REMEDY SELECTION STANDARDS

Of all the changes to CERCLA that Congress made when it amended the statute in 1986, arguably none was more significant than the addition of section 121, which greatly expanded upon the rather skeletal cleanup guidance originally provided for by section 104(c)(4) of CERCLA. Responding to intense pressure from environmental and citizen groups, Congress incorporated into the law sweeping and stringent minimum cleanup standards, and "preferences" concerning types of cleanups, applicable to all CERCLA remedial actions. By all estimates, the cost impact of these cleanup standards, coupled with other provisions

1. STEPHEN BREYER, BREAKING THE VICIOUS CIRCLE: TOWARD EFFECTIVE RISK REGULATION 12 (Harvard Univ. Press 1993).

2. Judicial review of EPA's remedy selection decision is discussed in chapter 6.

of the Superfund Amendments and Reauthorization Act (SARA), such as an increased involvement of state governments and citizen groups in the selection of remedial actions, substantially increased the cost of the average cleanup over the cost of pre-SARA cleanups.

Section 121 of CERCLA establishes general standards for determining the necessary degree of cleanup of a Superfund site and for selecting a remedial action. Section 121(b)(1) expresses Congress's preference for remedial actions implementing treatment "which permanently and significantly reduces the volume, toxicity or mobility of the hazardous substances, pollutants, and contaminants" at the site.[3] Not surprisingly, Congress expressed a disdain for remedial actions involving the "offsite transport and disposal of untreated hazardous substances or contaminated materials."[4] Overall, EPA must

> select a remedial action that is protective of human health and the environment, that is cost effective, and that utilizes permanent solutions and alternative treatment technologies or resource recovery technologies to the maximum extent practicable.[5]

Deceptively straightforward at first glance, these three elements—protectiveness of human health and the environment, cost-effectiveness, and permanency—are actually filled with a variety of complex and sometimes interrelated issues.

## A. Protection of Human Health and the Environment

A remedial action must be determined to be "protective of human health and the environment" before it can be selected and implemented. To ensure that this objective is met, section 121(d) of CERCLA requires EPA to identify "applicable or relevant and appropriate requirements" (ARARs) for the hazardous substances, pollutants, or contaminants that

3. CERCLA § 121(b)(1).
4. *Id.*
5. *Id.*

will remain at the site. Unless a particular ARAR has been waived,[6] the selected remedial action must then achieve compliance with each ARAR.[7] In the absence of an identifiable ARAR for a substance that will remain at the site, information generated during the risk assessment process will also be used to set chemical-specific standards that are "protective of human health and the environment." Like ARARs, these standards must be met by the selected remedial action.

### 1. *ARARs*

#### a. General Considerations for Identification

How to determine when a particular standard derived from federal or state laws is an ARAR under the circumstances of a particular site cleanup is, perhaps, the most nettlesome of all the section 121 cleanup issues. The actual process of identifying an ARAR is drawn out over the various stages involved in the development of the remedial investigation/feasibility study (RI/FS). Basic factual information about the site (e.g., types of substances at the site, location of the site, potential remedies) and considerations like those discussed below will help to define an initial universe of *potential* ARARs. By the end of the RI/FS process, each remedial alternative still under consideration will be assigned a final set of ARARs culled from the original universe of potential ARARs. The record of decision (ROD) itself will explain *how* the selected remedial alternative will meet its identified ARARs.

In addition to the fact-specific circumstances of each site, a number of general considerations will always be pertinent to the process of defining the universe of potential ARARs and winnowing this universe

6. So long as the selected remedy assures adequate "protection of human health and the environment," the identified ARARs need not be satisfied if EPA makes one of the following findings: (1) the cleanup action is only part of a total cleanup that will attain the ARAR; (2) compliance "will result in greater risk to human health and the environment"; (3) compliance is "technically impracticable from an engineering perspective"; (4) an equivalent standard will be reached through another method; (5) "with respect to a State standard . . . the State has not consistently applied . . . the standard . . . in similar circumstances at other remedial actions within the State"; or (6) compliance will not provide a balance between the need to protect health, welfare, and the environment at the site and the availability of fund money to clean up other sites. CERCLA § 121(d)(4)(A)–(F).

7. *See id.* § 121(d)(2)(A).

to a final set of ARARs for each remedial alternative. Among the more important of these considerations are the interplay between federal and state law, the distinction between standards that are "applicable" requirements and standards that are "relevant and appropriate," the federal and state standards that are not legally binding, and the distinction between the substantive and procedural requirements of state and federal environmental laws.

For those hazardous substances, pollutants, or contaminants that will remain on-site, section 121(d)(2)(A) of CERCLA requires that the selected remedy attain "legally applicable or relevant and appropriate" requirements of any federal environmental law including, but not limited to, the Toxic Substances Control Act (TSCA), the Safe Drinking Water Act (SDWA), the Clean Air Act (CAA), the Clean Water Act (CWA), the Marine Protection, Research, and Sanctuaries Act (MPRSA), or the Resource Conservation and Recovery Act (RCRA). If a state environmental or facility siting law provides a standard *more stringent* than any federal standard, the selected remedy must meet the requirements of the state standard.[8] Special limitations on state standards pertaining to land disposal may exist in some circumstances.[9]

Beginning with the promulgation of the 1985 National Contingency Plan (NCP), EPA has drawn a distinction between applicable requirements and relevant and appropriate requirements. In that regard, EPA guidance published subsequent to the enactment of section 121 notes that a standard must be either applicable or relevant and appropriate, but need not be both.[10] An "applicable requirement" is defined as "those cleanup standards, standards of control, and other substantive requirements, criteria, or limitations promulgated under federal environmental

8. CERCLA § 121(d)(2)(A)(ii).

9. *Id.* § 121(d)(2)(C).

10. EPA, Office of Emergency & Remedial Response, OSWER Directive No. 9234.1-01, CERCLA Compliance with Other Laws Manual—Part I (Aug. 1988), at xiii. In addition to providing general principles for determining ARARs, CERCLA Compliance with Other Laws Manual Part I also contains specific discussions of RCRA, SDWA, and the CWA. In 1989, EPA published Part II of this ARARs Manual, which covers the CAA, the Toxic Substances Control Act (TSCA), the Federal Insecticide Fungicide and Rodenticide Act (FIFRA), and a number of resource protection statutes such as the Endangered Species Act.

or state environmental or facility siting law that specifically address a hazardous substance, pollutant, contaminant, remedial action, or other circumstance at a CERCLA site."[11] By contrast, "relevant and appropriate requirements" are defined as "cleanup standards, standards of control, and the substantive environmental protection requirements, criteria, or limitations under federal or state law" that, while not "applicable" to a particular hazardous substance or remedial action or location, address situations sufficiently similar to the particular CERCLA site that their use is "well suited to the particular site."[12] Determination of whether a standard is a "relevant and appropriate requirement" is a two-step process; the standard must first be determined to be relevant and, if so, appropriate under the site-specific circumstances.[13]

Section 121(e)(1) of CERCLA provides that "[n]o Federal, State or local permit shall be required for the portion of" a remedial action "conducted entirely onsite."[14] Recognizing the more general distinction between the substantive and procedural requirements of environmental laws, EPA has interpreted this provision as exempting *all* administrative requirements, whether or not they are actually styled as "permits," from consideration as ARARs.[15] Notably, this interpretation does not excuse the need to comply with permitting and other administrative requirements for *off-site* actions.

Finally, ARARs may not be based on state and federal criteria, guidance, and proposed standards that have not been promulgated and, thus, are not legally binding. Nevertheless, these nonbinding criteria or guidelines are "to be considered" (TBCs); they are to be evaluated along with ARARs as part of the risk assessment conducted for each CERCLA site and may be used to determine the necessary level of cleanup in the absence of an identifiable ARAR.[16]

11. 40 C.F.R. § 300.5.
12. *Id.*
13. *See* 40 C.F.R. § 300.400(g)(2).
14. *See also* 40 C.F.R. § 300.400(e)(1), which implements this statutory mandate.
15. *See* OSWER Directive No. 9234.1-01, *supra* note 10, at 1-11.
16. 40 C.F.R. § 300.400(g)(3); *see also* OSWER Directive No. 9234.1-01, *supra* note 10, at 1-76.

**b. Types**

During the process of defining the universe of potential ARARs and arriving at a final set of ARARs for each remedial alternative under consideration for the site, three basic types of ARARs are likely to be identified: chemical-specific requirements that are usually expressed as health- or risk-based numerical requirements (e.g., SDWA maximum contaminant levels); location-specific requirements that place restrictions on the conduct of activities because of their location (e.g., floodplains, wetlands, historic places); and action-specific requirements that are usually technology-based (e.g., RCRA management requirements). Closer examination of a few of the basic issues associated with chemical-specific and action-specific ARARs in particular reveals how such standards, when they are available, can drive the remedy.

At many sites, potential action-specific ARARs are likely to be supplied by Subtitle C of RCRA, which regulates the treatment, storage, and disposal of hazardous waste.[17] PRPs should always check to see whether such standards are applicable to the particular on-site remedial alternative in question by determining whether a RCRA hazardous waste is present at the site.[18] A more specific issue of concern within the RCRA context will be the applicability of the RCRA section 3004 land disposal prohibitions, which could greatly impede the use of a proposed remedy calling for excavation, on-site treatment, and return of treated waste to the excavated area.[19] The issue is so complex that from 1989 to 1991 EPA published nearly a dozen guidance documents devoted exclusively to determining when and how the RCRA land disposal restrictions (LDRs) apply to CERCLA response actions.[20]

17. *See, e.g.*, 40 C.F.R. pt. 264.

18. Even if the particular Subtitle C management standard is not "applicable," it is still possible that it may be "relevant and appropriate."

19. *See, e.g.*, 54 Fed. Reg. 41,566 (Nov. 17, 1986).

20. *See* Superfund LDR Guide #1, Overview of RCRA Land Disposal Restrictions (July 1989); Superfund LDR Guide #2, Complying with the California List Restrictions under Land Disposal Restrictions (July 1989); Superfund LDR Guide #3, Treatment Standards and Minimum Technology Requirements under Land Disposal Restrictions (July 1989); Superfund LDR Guide #4, Complying with the Hammer Restrictions under Land Disposal Restrictions (July 1989) (no longer in effect); Superfund LDR Guide #5, Determining When Land Disposal Restrictions (LDRs) Are *Applicable* to CERCLA Response Actions (July 1989); Superfund LDR

As for chemical-specific ARARs, a frequently encountered groundwater issue pertaining to levels of certain chemical substances revolves around SDWA maximum contaminant levels (MCLs) and SDWA maximum contaminant level goals (MCLGs) that have been promulgated for such substances.[21] MCLGs are developed under the SDWA as chemical-specific health goals used in setting MCLs for public water supply systems. Significantly, MCLGs are based entirely on health considerations and do not take cost or feasibility into account. MCLs and MCLGs that have been promulgated for noncarcinogens are usually identical. But for a carcinogen, the MCL will be set at a higher acceptable concentration than an MCLG, which will be set at zero, a virtually unattainable goal at most sites. Although section 121(d)(2)(A)(i) of CERCLA requires a level or standard of control that at least attains MCLGs, EPA has allowed MCLs to suffice for carcinogens, reasoning that such levels fall within the acceptable risk range of $10^{-4}$ to $10^{-7}$ (one cancer case in 10,000 to one cancer case in 10,000,000), the same range of risk used by EPA to determine acceptable levels of residual carcinogens following site cleanup.[22]

### c. When and Where Compliance Must Be Achieved

Stringency will not be the only aspect of an ARAR that will drive the remedy selection process for a particular site. When and where compliance with a particular ARAR must be achieved will also influence remedy selection. Section 121(d)(2)(A) of CERCLA states that ARARs must be achieved "at the completion of the remedial action"; the statute

Guide #6A (2d ed.), Obtaining a Soil and Debris Treatability Variance for Remedial Actions (Sept. 1990); Superfund LDR Guide #6B, Obtaining a Soil and Debris Treatability Variance for Removal Actions (Sept. 1990); Superfund LDR Guide #7, Determining When Land Disposal Restrictions (LDRs) Are *Relevant and Appropriate* to CERCLA Response Actions (Dec. 1989); Superfund LDR Guide #8, Compliance with Third Third Requirements under the LDRs (Oct. 1990); ARARs Q's & A's: The Fund-Balancing Waiver (Jan. 1991).

21. MCLs and MCLGs have been promulgated for a relatively small portion of the universe of chemicals that are likely to be found at Superfund sites. For chemicals that do not have MCLs or MCLGs, the risk assessment process discussed in section II.A.2 of this chapter, *infra*, will supply groundwater cleanup standards when groundwater ARARs for such chemicals cannot be otherwise identified.

22. *See* 52 Fed. Reg. 32,496, 32,499 (Aug. 27, 1987).

does not require ARAR compliance during the performance of those activities that precede completion of the remedial action (e.g., installation and construction of the remedy). Extending the requirement that ARARs must be met during the stages of the remedial action that lead to completion makes sense only for certain action-specific ARARs, such as RCRA treatment, storage, and disposal standards; air emission standards under the CAA; and effluent discharge limitations under the CWA.

CERCLA itself does not specify the geographic point of ARAR compliance. Where compliance must be achieved is an issue that is typically most important for chemical-specific ARARs. As a matter of policy, EPA states that chemical-specific ARARs must be attained "at all points of potential exposure."[23] For instance, EPA has stated that MCLs must be met in the groundwater itself, even though such standards technically apply "at the tap." Also with respect to groundwater ARARs, EPA has taken the position that "remediation levels should generally be attained throughout the contaminated plume, or at and beyond the edge of the waste management area when waste is left in place."[24] Ordinarily, compliance with a chemical-specific groundwater ARAR will be far easier to achieve at or beyond the edge of the waste management area. In this respect, the way in which the boundaries of the waste management area are defined during the RI/FS process could be of critical importance.

### 2. *Baseline Risk Assessment*

A key element of EPA's process for selecting a remedial action for a site is the "baseline risk assessment." A baseline risk assessment, which is always performed as part of the RI/FS, evaluates the potential threat to human health and the environment that a contaminated site poses in the absence of any remedial action.[25] Specifically, the NCP states that the baseline risk assessment should "characterize the current and potential threats to human health and the environment that may be posed by

23. OSWER Directive No. 9234.01, *supra* note 10, at 1-8.

24. *Id.*

25. *See* 55 Fed. Reg. 8634 (Mar. 8, 1990) (promulgating 40 C.F.R. § 300.430(d)(4)); *see also* EPA, Office of Emergency & Remedial Response, EPA 540-R-97-013, Rules of Thumb for Superfund Remedy Selection (Aug. 1997).

contaminants migrating to ground water or surface water, releasing to air, leaching through soil, remaining in the soil, and bio-accumulating in the food chain."[26] EPA has published voluminous guidance on the protocol for performing a baseline risk assessment, both with respect to human and environmental risks.[27] A comprehensive discussion of risk assessment process is beyond the scope of this book; we focus only on the basic elements of EPA's approach to assessing human health risks.

To perform a baseline risk assessment, EPA first develops a "conceptual site model," which essentially is a three-dimensional picture of site conditions. The conceptual site model identifies the contaminant sources present at the site, exposure pathways, migration routes, and potential human and ecological receptors. Based on the conceptual site model, EPA conducts an "exposure assessment" in which the agency estimates the magnitude of actual and/or potential human exposures. To quantify exposure to each contaminant, EPA will evaluate "reasonable maximum exposure assumptions" for each potential exposure pathway (oral, dermal, and inhalation).[28] EPA must also perform a toxicity assessment in which it identifies the types of adverse health effects associated with exposure to each identified contaminant, the relationship between the magnitude of exposure and adverse effects, and related uncertainties. This analysis consists of two elements: the "hazard assessment," which identifies the adverse effects associated with exposure to a contaminant, and the dose-response assessment, which characterizes the relationship between the dose administered and the incidence of adverse effects in an exposed population.

26. 40 C.F.R. § 300.430(d)(4).

27. In this regard, EPA's Risk Assessment Guidance for Superfund: Volume I, Human Health Evaluation Manual—Part A, EPA/540/1-89/002 (Dec. 1989), provides guidance on how to conduct the human health portion of the baseline risk assessment. Volume II of the Risk Assessment Guidance for Superfund, the Environmental Evaluation Manual (EPA/540/1-89/001) (Mar. 1989), and the companion manual, Ecological Assessment of Hazardous Waste Sites: A Field and Laboratory Reference, EPA/600/3-89/013 (1989), provide guidance on conducting the environmental portion of the baseline risk assessment. Other pertinent guidance includes, EPA/540/G-89/004, Guidance for Conducting Remedial Investigations and Feasibility Studies under CERCLA (Oct. 1988), which describes how the baseline risk assessment fits into the overall RI/FS process.

28. The "reasonable maximum exposure" is the highest exposure that is reasonably expected to occur at a site; it is intended to be a conservative estimate of exposure.

The final piece of the baseline risk assessment is the risk characterization in which EPA combines the output of the exposure assessment and toxicity assessment to determine if the potential levels of exposure to contaminants present on-site pose a significant risk of adverse human health effects. Generally, remedial action is not warranted where the baseline risk assessment indicates that the cumulative site risk, using reasonable maximum exposure assumptions, poses less than a $1 \times 10^{-4}$ lifetime excess cancer risk (that is, one additional incident of cancer in a population of 10,000 due to lifetime exposure to a contaminant).[29]

The vast majority of sites will contain multiple contaminants, some of which will not have chemical-specific ARARs. For a suspected or known carcinogenic substance that does not have a chemical-specific ARAR, the remedies under consideration will have to reduce potential exposure to the contaminant to select a level associated with a carcinogenic risk range of $10^{-4}$ to $10^{-6}$—with $10^{-6}$ being regarded as the "point of departure" for deciding the appropriate level of risk.[30] For a noncarcinogenic substance that does not have a chemical-specific ARAR, concentrations will have to be reduced to acceptable exposure levels, that is, concentration levels to which a human population, including sensitive subgroups, may be exposed without adverse effects during a lifetime or part of a lifetime, incorporating an adequate margin of safety.[31] Generally, concentrations of carcinogens will have to be reduced to lower levels than noncarcinogens. Thus, at the typical site containing a potpourri of carcinogens and noncarcinogens for which chemical-specific ARARs do not exist, the carcinogens will usually drive the degree of cleanup required of the remedial alternatives under consideration.

## B. Cost-effectiveness and Permanence

In addition to always being protective of human health and the environment, the selected remedial action must also be cost-effective and, to the

29. *See* EPA, OSWER Directive No. 9355.0-30, Role of the Baseline Risk Assessment in Superfund Remedy Selection Decisions (Apr. 22, 1991).
30. 40 C.F.R. § 300.430(e)(2)(i)(A)(2).
31. *Id.* § 300.430(e)(2)(A)(1).

maximum extent practicable, must utilize permanent solutions.[32] This presents a significant problem in implementing the Superfund program. There is a natural tension between the desire to achieve permanent solutions and the typically higher costs of implementing such permanent solutions. In some instances, a permanent solution may be so high as to render virtually meaningless the cost-effectiveness criterion.[33]

Yet beyond stating a "preference" for permanent remedies, neither the language of CERCLA nor the conference report accompanying SARA indicate the relative weight and emphasis that these two elements are to be afforded in the ultimate selection of the remedy.[34] In this regard, the statute does not address the issue of whether, or to what extent, an available permanent solution may be so expensive—particularly in relation to other nonpermanent solutions—that EPA may reject that permanent solution because it is not cost-effective. To make matters worse, legislative debate on SARA after adoption of the conference report produced conflicting characterizations of the relative importance of cost-effectiveness and permanence considerations in making site-specific remedy selections.[35]

The conflict reflected in the legislative history has continued in a site-specific context, with public interest groups often advocating for adoption of permanent remedies at any cost, while PRPs argue that alternative remedial actions providing greater cost-effectiveness and practicability should be selected if adequately protective of human health and the

32. CERCLA § 121(b)(1).

33. For example, in April 2014, EPA proposed a bank-to-bank dredging of the Passaic River that would remove more than 4 million cubic yards of contaminated sediment from the river bottom at a cost of $1.7 billion—an amount that rivals the gross national product of some countries.

34. The conference report accompanying SARA indicates that EPA is to consider cost-effectiveness after determining ARARs for the hazardous substances, pollutants, or contaminants that may remain at the site at the completion of cleanup. H.R. Conf. Rep. No. 962, 99th Cong., 2d Sess. 245 (1986), *reprinted in* 1986 U.S.C.C.A.N. 3276, 3338.

35. Generally, members of the Senate who commented on the issue sought to subordinate the role of cost-effectiveness to the preference for permanent remedies, while members of the House stated that where remedies were not cost-effective, they could not be selected. *See, e.g.*, 132 Cong. Rec. S14,925 (daily ed. Oct. 3, 1986) (cmts. of Sen. Chaffee); 132 Cong. Rec. H9566 (daily ed. Oct. 8, 1986) (cmts. of Rep. Lent).

environment. At many of these sites, EPA clearly has included consideration of cost-effectiveness in the final remedy selection.

EPA has attempted to establish a middle ground in defining a policy for selecting remedies based on these conflicting criteria. For guidance, EPA stated:

> The selected remedy should represent the best balance across all the effectiveness, implementability, and cost factors examined in the detailed analysis. In making this selection, the decision-maker must consider the statutory preference for treatment which permanently and significantly reduces the toxicity, mobility or volume of the waste.[36]

The directive to balance the relevant factors, however, coupled with special emphasis on consideration of the preference for permanence, does not clarify the process to be followed in selection of a remedy.

Although permanence is defined in absolute terms, EPA's implementation of the remedy selection process will continue to reflect the fact that the preference for permanent remedies is a relative concept that will be defined on a case-by-case basis. Permanence is relative because, inter alia, many mixed wastes will not lend themselves to complete and total destruction. For example, incinerating a waste laden with organics and metals will destroy most organics, but will do little to the metals. Similarly, permanence is relative because permanent remedies either do not exist for specific wastes, or where they do exist, there is insufficient capacity.

In sum, the remedy selection criteria under section 121 do not compel EPA to select a remedy that achieves absolute permanence in every instance.[37] Rather, for some sites the remedy that is selected following consideration of all factors may be one that achieves the necessary levels

36. EPA, OSWER Directive No. 9355.0-19, Interim Guidance on Superfund Selection Remedy (Dec. 24, 1986), at 7.

37. While this position will generally inure to the benefit of PRPs in terms of immediately identifiable cost considerations, it must also be recognized that a remedy that leaves hazardous substances at the site will have to be reviewed once every five years. CERCLA § 121(c).

of protection of human health and the environment while balancing cost-effectiveness and the preference for permanent remedies.

### C. Innovative Technologies

As part of EPA's mandate under section 121(b)(1) to select a remedy that will permanently treat the conditions at a site, EPA also is required to assess "alternative treatment technologies or resource recovery technologies." This requirement will necessitate that EPA and PRPs be creative in selecting alternative technologies that are not only cost-effective and lead to a permanent treatment, but also are innovative.

EPA is directed to establish an alternative or innovative treatment technology research and demonstration program.[38] In accordance with this directive, EPA formed the Superfund Innovative Technology Evaluation (SITE) Strategy and Program Plan. The purpose of the SITE program is to enhance the development and demonstration of alternatives to land disposal, and thereby establish their commercial availability. Under the SITE program, technologies are to be chosen according to their potential for treating "high priority" wastes, which includes (1) wastes for which there are few or only costly treatment options and (2) frequently occurring large-volume wastes. Demonstration projects may be conducted at National Priorities List (NPL) sites, state or private sites, or EPA test facilities. During the RI/FS process, EPA will consider any combination of technologies included in the site program to remedy the conditions of a particular site.

## III. REMEDY SELECTION PROCEDURES

### A. Proposed Remedial Action Plan

In chapter 1, we explained that the first step in selecting a remedy for addressing a release of hazardous substances is to perform a remedial investigation/feasibility study (RI/FS). As noted, the RI characterizes the nature and extent of contamination. The FS identifies and compares

38. *See id.* § 311(b)(1).

various alternative potential remedies in terms of cost-effectiveness and permanence.[39]

Upon completion of the RI/FS process, EPA formulates a recommended remedial action plan to present to the community for comment. To inform the public of progress concerning the remedy selection procedure, section 117(a) of CERCLA requires EPA, or the state, to publish notice and an analysis of the proposed remedial action plan and provide an opportunity for comment prior to adoption of the plan. EPA generally satisfies this statutory requirement through the issuance of a proposed remedial action plan (PRAP). This document describes the remedial alternatives considered for a site and identifies EPA's (and, where appropriate, the state's) preferred remedial alternative with a rationale for the agency's preference. The PRAP is distributed to solicit public comments on all of the remedial alternatives evaluated and the preferred alternative.[40]

At this point, PRPs and other interested parties have an opportunity to comment on the remedial alternatives being considered and to advocate the remedy they support.[41] This comment period represents the last opportunity for PRPs to comment on the remedial alternatives being considered prior to EPA's final selection of a remedy for the site. Once the final remedy is selected public notice of the final remedy must be provided to the public, along with an explanation of any "significant" differences from the proposed remedy, and a response to each "significant" public comments on the proposed remedy.[42]

## B. Record of Decision

Following receipt of comments on the PRAP, EPA formalizes its final remedy selection through publication of the record of decision (ROD).[43]

39. *See* 40 C.F.R. § 300.430(d) & (e).

40. *See* 40 C.F.R. § 300.430(f)(2).

41. In extreme cases, PRPs have conducted a second RI/FS, even while EPA did its own, and have submitted the second RI/FS as comments on the PRAP. While this approach may not change EPA's decision, it will create a record for later use, if necessary.

42. CERCLA § 117(b).

43. *See* 40 C.F.R. § 300.430(f)(4).

As explained by EPA in its interim guidance document on the RI/FS process, the ROD is

> the centerpiece of the administrative record against which the Agency's decision making may be judged by the courts. In addition to containing an accurate and complete summary of the site, the threat it poses, and the selected remedy, the ROD must describe the relative strengths and weaknesses of each alternative considered and offer a clear justification for the final decision that is made.[44]

In principle, the ROD describes the site, identifies the threat that the site poses to human health and the environment, and explains how the remedy that has been selected will minimize such threats. To that end, the ROD must describe (1) how the selected remedy is protective of human health and the environment, explaining how it eliminates, reduces, or controls exposure to hazardous substances, pollutants or contaminants; (2) the federal and state ARARs that the remedy attains; (3) the federal and state ARARs that the remedy will not meet, as well as the basis for waiving such ARARs; (4) how the remedy is cost-effective; (5) how the remedy utilizes permanent solutions and alternative treatment technologies or resource recovery technologies to the maximum extent practicable; and (6) whether the reference for treatment that permanently and significantly reduces the toxicity, mobility or volume of hazardous substances was satisfied, and if not, why not.[45] For Fund-financed remedial actions, the ROD should include a written concurrence from the state.[46] Section 117(b) of CERCLA requires publication of the ROD before commencement of the remedial action.

### C. Administrative Record

Once EPA has issued its ROD, section 113(h) of CERCLA generally precludes preenforcement judicial review of the remedial action selected.

44. OSWER Directive No. 9355.0-19, *supra* note 36, at 8.
45. 40 C.F.R. § 300.430(f)(5)(ii).
46. OSWER Directive No. 9355.0-19, *supra* note 36, at 8.

PRP challenges to remedy selections are limited to section 106 enforcement or section 107 cost recovery actions. Pursuant to section 113(j), judicial review is limited to the administrative record developed in the remedy selection process. Thus, PRPs must take an early, active role in the remedy selection process to include all favorable material and to preserve their challenges in the administrative record. PRPs that merely sit back and await EPA's decision must depend entirely on convincing the court that review of the response selection should not be confined to the agency record. (See more detailed discussion in chapter 6.)

### D. State Participation in Remedy Selection

As CERCLA was originally enacted, states were given a somewhat limited role in selecting remedies at Superfund sites. Section 104(c)(2) of CERCLA required EPA to consult with a state before selecting a particular remedial action for a site in that state. Section 104(c)(3) of CERCLA also required states to pay 10 percent of the costs of fund-financed remedial actions and to enter into a contract or cooperative agreement assuring "all future maintenance" of removal actions and the availability of a permitted hazardous waste disposal facility. In addition, section 104(d)(1) allowed states to take the lead agency role upon completion of a cooperative agreement with EPA. But for the most part, the EPA regions were reluctant to share their Superfund authority with the states.

This state of affairs changed greatly when Congress amended CERCLA in 1986, and gave states a more prominent role in decision making at Superfund sites. The 1986 amendments significantly enlarged the state "consultation" requirement and gave states a substantial role in evaluating and approving remedial actions.[47] At an early stage, states also must receive notice of the negotiations with PRPs and have an opportunity to participate in those negotiations.[48] Most importantly, states now have the opportunity to review and comment on the RI/FS and other technical matters previously beyond the scope of a state's involvement.[49]

47. CERCLA § 121(f).
48. *Id.* § 121(f)(1)(F).
49. *Id.* § 121(f)(1)(E).

As noted *supra*, remedial actions must not only meet federal ARARs, but also state ARARs that are more stringent than federal ARARs.[50] If EPA decides to select a remedy that does not comply with state ARARs, the state may intervene in any judicial proceeding to approve a consent decree. For remedial actions secured through a consent decree to be entered under section 106 of CERCLA, a state that does not concur with the waiver of an ARAR under section 121(d)(4) and seeks to have the remedial action conform to the waived ARAR may intervene, as a matter of right, in the underlying section 106(a) action.[51] If the state can establish, on the basis of the administrative record, that the finding of EPA was not supported by "substantial evidence," the remedial action will be required to conform with such standard.[52] Although section 121(f)(2)(B) has been successfully used by states to intervene in an underlying section 106(a) action, state challenges to EPA's decisions regarding ARAR waivers have met with little success.[53] Nevertheless, PRPs involved in a proposed consent decree under section 106(a) must be mindful of the possibility that a state may attempt to impose more stringent state ARARs that EPA waived.

### E. Five-Year Review

Where EPA selects a remedial action that results in any hazardous substance, pollutant, or contaminant remaining on-site, the government must review the remedial action at least once every five years.[54] The purpose of the five-year review is to "assure that human health and the environment are being protected by the remedial action being implemented."[55]

EPA has narrowed this obligation slightly, stating that remedial actions that result in any hazardous substances, pollutants, or contaminants remaining at the site above levels that allow for "unlimited use and

50. *See id.* § 121(d)(2)(A)(ii), and discussion *supra*.
51. *Id.* § 121(f)(2)(B).
52. *Id.*
53. *See, e.g.*, United States v. Akzo Coatings of Am., Inc., 719 F. Supp. 571 (D. Mich. 1989), *aff'd*, 949 F.2d 1409 (6th Cir. 1991).
54. *See* CERCLA § 121(c).
55. *Id.*

unrestricted exposure" be reviewed every five years to ensure protection of human health and the environment.[56] "Unlimited and unrestricted exposure" means that the selected remedy will place no restrictions on the potential use of land and/or other natural resources.[57] According to EPA policy, five-year reviews "may no longer be needed when no hazardous substances, pollutants or contaminants remain on-site above levels that allow for unlimited use and unlimited exposure."[58]

56. *See* 40 C.F.R. § 300.430(f)(4)(ii).

57. *See* EPA, Office of Emergency & Remedial Response, OSWER Directive No. 9355.7-03B-P, Comprehensive Five-Year Review Guidance (June 2001), at 1–2.

58. *Id.* at 1-4.

# CHAPTER 3

# Abatement Authority under Section 106

## I. INTRODUCTION

Section 106(a) of CERCLA authorizes EPA to compel responsible parties to respond to hazardous substance releases that may endanger public health and welfare or the environment. EPA may exercise this "abatement" authority in two ways. It may institute a civil action in federal district court in which it seeks to have the court enjoin a responsible party to perform a particular response activity. Alternatively, EPA may issue an administrative order requiring the responsible party to perform the response activity in accordance with EPA's preselected methodology.

Although seldom used in the early years of Superfund enforcement, section 106(a)—in particular, the section 106(a) administrative order—has become one of EPA's most important enforcement tools.[1] Instead of having to deplete the Fund to finance cleanups and rely on cost recovery actions against potentially responsible parties (PRPs) for reimbursement, section 106(a) orders enable EPA to force PRPs to perform cleanups in the first instance. The agency's "big stick" is that a PRP who fails to comply with a section 106(a) order faces daily penalties of $25,000[2]

1. *See Guidance on CERCLA Section 106(a) Unilateral Administrative Orders for Remedial Designs and Remedial Actions*, 20 Envtl. L. Rep. 35,253 (1990).

2. CERCLA § 106(b)(1).

*and* potential treble damages[3] where such noncompliance is determined to have occurred without "sufficient cause."

This chapter reviews the triggering elements of EPA's section 106(a) authority and discusses issues common to section 106(a) civil actions and administrative orders. This chapter also reviews issues unique to these two types of actions and the strategic options available to PRPs.

## II. OVERVIEW OF SECTION 106(a)

Section 106(a) of CERCLA authorizes EPA, acting under a delegation of authority from the president,[4] to bring a civil action for injunctive relief to abate an "imminent and substantial endangerment" to the public health or welfare or the environment caused by the release or threatened release of a hazardous substance from a facility. In pertinent part, section 106(a) provides:

> In addition to any other action taken by a State or local government, when the President determines that there may be an imminent and substantial endangerment to the public health or welfare or the environment because of an actual or threatened release of a hazardous substance from a facility, he may require the Attorney General of the United States to secure such relief as may be necessary to abate such danger or threat. . . .[5]

Section 106(a) also authorizes EPA to unilaterally issue an administrative order to protect public health and welfare and the environment:

> The President may also, after notice to the affected State, take other action under this section including, but not limited to, *issuing such orders as may be necessary to protect public health and welfare and the environment.*[6]

3. CERCLA § 107(c)(3).
4. *See* Exec. Order No. 12,316, 46 Fed. Reg. 42,237 (Aug. 14, 1981).
5. CERCLA § 106(a).
6. *Id.* (emphasis added).

As with EPA's authority to seek injunctive relief in federal district court, EPA's administrative order authority is subject to the proviso that the release or threatened release that is the subject of the order present an "imminent and substantial endangerment."[7]

Failure to comply with a section 106 administrative order can have draconian consequences. First, under subsection (b)(1), any person who, "without sufficient cause, willfully violates, or fails or refuses to comply with" a section 106(a) administrative order may be fined up to $37,500 for each day of noncompliance.[8] Second, under section 107, the general liability section of CERCLA, if the recipient of an administrative order fails to comply "without sufficient cause," that person may be liable for "punitive damages in an amount at least equal to, and not more than three times, the amount of any costs incurred by the Fund as a result of such failure to take proper action."[9]

The use and application of the penalty provision differs from that of the treble damages provision. EPA may seek penalties in district court in an action to enforce an administrative order. By contrast, EPA may not recover treble damages unless and until it incurs response costs. In other words, EPA must implement a Superfund-led cleanup of the site pursuant to section 104 as a prerequisite to recovery of section 107 treble damages.

### A. Prerequisites for Assertion of Authority under Section 106(a)

Whether through a civil action or an administrative order, EPA may assert its abatement authority under section 106(a) only when the "triggering" elements identified in the statute have been satisfied.[10] EPA must determine that (1) an actual or threatened release (2) into the environment (3) of a hazardous substance (4) from a facility (5) may present an imminent and substantial endangerment to the public health or welfare

7. *Id.*

8. The above-quoted penalty figure reflects an inflation adjustment of the statutory $25,000 amount, which is required by section 4 of the Federal Civil Penalties Inflation Adjustment Act of 1990, 28 U.S.C. 2461. *See also* 78 Fed. Reg. 66,643 (Nov. 6, 2013).

9. CERCLA § 107(c)(3).

10. *See, e.g.*, United States v. Outboard Marine Corp., 556 F. Supp. 54, 57 (N.D. Ill. 1982).

or the environment. The first four elements are essentially the same elements that trigger the government's response authority under section 104 (see chapter 1). The only difference is that EPA's section 106(a) authority extends only to releases of "hazardous substances." It does *not* extend to releases of "pollutants and contaminants." Thus, section 106 is applicable to a smaller universe of releases than section 104.

The fifth element, "imminent and substantial endangerment," is peculiar to section 106(a). Preliminarily, note that the United States need not prove that an imminent and substantial endangerment actually exists. The statute authorizes the United States to obtain relief when there *may* be an imminent and substantial endangerment.[11] Note also that human endangerment is not a prerequisite to section 106(a) authority; relief is authorized where there may be an endangerment to "the public health or welfare *or the environment*" (emphasis added).[12]

"Imminent and substantial endangerment" is not defined in CERCLA and its legislative history is less than illuminating. Consequently, courts have attempted to interpret the phrase by considering its meaning in other "imminent hazard" statutes.

### *1. "Imminent"*

A number of courts have examined the legislative history of section 7003 of the Resource Conservation and Recovery Act (RCRA)—an "imminent hazard" provision that is analogous to section 106(a)—to interpret the term "imminent."[13] These courts have focused on the following congressional passage:

> "Imminence" in this section applies to the nature of the threat rather than identification of the time when the endangerment

11. *See* United States v. Conservation Chem. Co., 619 F. Supp. 162, 192 (W.D. Mo. 1985).

12. Taken to its logical extreme, section 106(a) injunctive authority extends to releases or threatened releases that cause purely aesthetic injury. The court in *Conservation Chemical* noted that "[t]he term 'public welfare' is exceptionally broad, and encompasses 'health and safety, recreational, *aesthetic*, environmental and economic interests.'" 619 F. Supp. at 192 (emphasis added) (citation omitted).

13. *See, e.g.*, United States v. Price (*Price I*), 688 F.2d 204, 213 (3d Cir. 1982); United States v. Waste Indus., Inc., 734 F.2d 159, 166 (4th Cir. 1984).

initially arose. The section, therefore, may be used for events which took place at some time in the past but which continue to present a threat to the public health or the environment.[14]

*B.F. Goodrich v. Murtha*[15] illustrates how courts have interpreted the term "imminent" in the context of section 106(a) of CERCLA. The court held that hazardous substances released from a landfill posed an imminent threat to public health because there was a significant risk of those hazardous substances migrating through groundwater to nearby residential wells and brooks where they might be ingested by humans and animals, respectively.

### 2. *"Substantial"*

The second term, "substantial," relates to the degree of risk or harm imposed by the release. The court in *Conservation Chemical* stated:

> [A]n endangerment is substantial if there is reasonable cause for concern that someone or something may be exposed to a risk of harm by a release or a threatened release of a hazardous substance if remedial action is not taken. . . ."[16]

Based on this interpretation, "substantial" is not synonymous with significant. Instead, it merely requires that the possibility of a release causing harm to human health or the environment is realistic.

### 3. *"Endangerment"*

Finally, as the preceding passage suggests, the word "endangerment" has been interpreted as including not only actual harm, but also a "threatened or potential harm."[17]

---

14. Subcomm. on Oversight and Investigation, House Comm. on Interstate and Foreign Commerce, 96th Cong., 1st Sess., Report on Hazardous Waste Disposal 32 (Comm. Print No. 31, 1979).

15. 697 F. Supp. 89, 95 (D. Conn. 1988).

16. 619 F. Supp. at 194.

17. *Id.* at 192; *B.F. Goodrich Co.*, 697 F. Supp. at 96 (D. Conn. 1988) ("'endangerment' is not actual harm, but a threatened or potential harm" (quoting *Conservation Chem.*,

In sum, the showing required to establish that a release or threat of a release poses an "imminent and substantial endangerment" is a relatively modest one.

*4. Is Section 106(a) an "Emergency Power"?*

Although most courts have broadly construed the "imminent and substantial endangerment" requirement, they are divided on whether EPA may apply its section 106(a) powers to nonemergencies. Typical of the judicial reasoning to limit the use of section 106(a) to emergency situations is the following dicta from the court's opinion in *United States v. Wade*:

> A straightforward reading of this language requires that I conclude that Congress intended Section 106(a) to be used in emergency situations where hazardous waste was currently being discharged or threatened to be discharged "from a facility" and where such discharge could be stopped by an injunction.[18]

On the other hand, several courts have rejected the premise that section 106(a) and other imminent hazard provisions are limited to emergencies. In response to arguments that EPA improperly invoked section 106(a) in a suit to obtain access to a site to implement a response action,' the court in *B.F. Goodrich Co. v. Murtha* ruled that the imminent and substantial endangerment provision "is not limited to emergency situations."[19] Similarly, the Fourth Circuit in *United States v. Waste Industries, Inc.*[20] held that the analogous imminent hazard language of section 7003 of RCRA, which authorizes injunctive relief when waste

619 F. Supp. at 192)); United States v. Ottati & Goss, Inc. 630 F. Supp. 1361, 1394 (D.N.H. 1985) ("Endangerment means a threatened or potential harm and does not require proof of actual harm.").

18. 546 F. Supp. 785, 794 (E.D. Pa. 1982); *see also* Outboard Marine Corp. v. Thomas, 773 F.2d 883, 890 (7th Cir. 1985), *vacated on other grounds*, 479 U.S. 1002 (1986) (section 106 "applies only in emergency situations"); United States v. NEPACCO, 579 F. Supp. 823, 839 n.17 (W.D. Mo. 1984) ("Section 106(a) is an emergency provision, not a general cleanup provision"), *aff'd*, 810 F.2d 726 (8th Cir. 1986).

19. 697 F. Supp. at 96.

20. 734 F.2d at 165.

handling practices may present an imminent and substantial endangerment, was not limited to emergency situations.[21]

### B. Parties Subject to Section 106(a)

Like many other provisions of CERCLA, section 106(a) is not a model of clarity. Here, Congress failed to specify who may be subject to EPA's authority under section 106(a). Because of Congress's silence on this issue, a majority of courts have looked to section 107—CERCLA's only list of liable parties—as the appropriate reference for determining potentially liable parties under section 106(a). As discussed in chapter 4, section 107(a) identifies four categories of liable parties: (1) the current owner or operator of a facility from which there is a release or threatened release of a hazardous substance; (2) the owner or operator of the facility at the time of disposal of the hazardous substance; (3) any person who "arranged for disposal or treatment" of hazardous substances at the site (often referred to as "generators"); and (4) any person who accepts hazardous substances for transport to a facility selected by such person.[22]

Courts supporting the majority view that section 106(a) should be read as applying to parties designated as liable under section 107 have done so on several grounds: (1) section 107 is the main liability provision of CERCLA and was meant to apply to all CERCLA enforcement actions;[23] (2) section 107 is titled "Liability," which implies an intention to define liability for the whole act;[24] (3) the lack of any limitation on classes of persons subject to section 106(a) makes it dependent on section 107 liability provisions;[25] and (4) not reading sections 104, 106, and 107 together would "emasculate the purpose of CERCLA and the intent of Congress."[26]

Some courts have questioned the view that persons defined as liable parties under section 107 are per se liable parties under section 106. For

21. *See also Price I*, 688 F.2d at 213.
22. CERCLA § 107(a)(1)–(4).
23. United States v. Outboard Marine Corp., 556 F. Supp. 54, 57 (N.D. Ill. 1982).
24. United States v. Price (*Price II*), 577 F. Supp. 1103, 1113 (D.N.J. 1983).
25. United States v. A & F Materials Co., 578 F. Supp. 1249, 1257 (S.D. Ill. 1984).
26. *NEPACCO*, 579 F. Supp. at 839.

instance, the court in *United States v. Stringfellow* found no relationship between section 106(a) and section 107 regarding the scope of liability, reasoning that section 107 refers to *past* hazardous waste handling practices, whereas section 106 considers *current* releases and threats of harm.[27] However, the *Stringfellow* court stopped short of indicating who should be a liable party under section 106(a).[28]

With the foregoing split of authority in mind, some specific conclusions regarding the applicability of section 106(a) authority to section 107 liable parties may be drawn.

### 1. *Owners and Operators*

Courts have consistently held that the current owner or operator of a site is subject to section 106(a) orders for the abatement of hazardous substance releases. Courts have based this both on common law nuisance grounds, holding that the site represents a nuisance condition that must be abated by those in control of the property,[29] and on statutory grounds, holding that the liability scheme of section 107(a) is to be read into section 106(a).[30]

Although no court has so ruled, prior site owners and operators may also be subject to section 106(a) orders for abatement of a hazardous substance release if such a release occurred when they occupied or controlled the site. Again, courts would rely on common law nuisance principles and on the incorporation of section 107 liability principles into section 106(a). (See "majority view" discussed under "Generators," below.) However, to the extent that no causation exists between the prior owner/operator's activities at the site and, the hazardous substance release, it is arguable that section 106(a) authority should not attach based on either theory of liability.

27. No. CV-83-2501-MML, 1984 WL 3206 (C.D. Cal. Apr. 5, 1984).

28. *See also Wade*, 546 F. Supp. at 794 (Congress intended section 106(a) "to be used in emergency situations where hazardous waste was currently being discharged or threatened to be discharged . . . and where such discharge could be stopped by an injunction.").

29. *See, e.g.*, United States v. Ottati & Goss, Inc. 630 F. Supp. 1361, 1406–07, 1410–12 (D.N.H. 1985).

30. *See, e.g.*, *Price II*, 577 F. Supp. at 1113–14.

### 2. *Generators*

A frequently litigated issue is whether off-site "generators" of hazardous substances are subject to section 106(a). Courts following the majority view that section 107 liability principles apply to section 106(a) have found generators liable under section 106(a),[31] while courts following the minority view construe the reach of section 106(a) more narrowly than that of section 107. For example, in *United States v. Wade*,[32] the court rejected the government's argument that because Congress intended to hold off-site generators liable under sections 104 and 107, it also intended to hold them liable under section 106(a). The court concluded that section 106(a) was not intended to be used to enjoin off-site generators because (1) there was no evidence of legislative intent to extend section 107 liability to section 106(a), and (2) sections 104 and 107 provide another carefully delineated route to reach the same cleanup result.[33]

### 3. *Transporters*

There are no court decisions regarding whether transporters may be held liable under section 106(a). Thus, it is difficult to predict with any degree of certainty the potential liability of transporters. It is likely, however, that transporter liability under section 106(a) could turn on whether the reviewing court follows the majority or minority view. Courts following the majority view would probably hold transporters liable, since transporters are identified in section 107(a)(4) as a class of liable parties.[34] Courts following the minority view, which predicates liability under section 106(a) on nuisance theory, might not impose liability on a transporter unless it is demonstrated that the transporter either exercised control over the hazardous substance or was able to exercise control and either knew or should have known about a hazardous substance release.[35]

31. *See id.* at 1111–13 (section 106 applies to past, off-site generators); *see also* United States v. Conservation Chem. Co., 589 F. Supp. 59, 62 (W.D. Mo. 1984).
32. 546 F. Supp. 785.
33. *Id.* at 794.
34. *See, e.g., NEPACCO*, 579 F. Supp. at 839, 844.
35. *See, e.g., Ottati & Goss, Inc.*, 630 F. Supp. at 1406–07, 1410–12.

## C. Standard of Liability

The standard of liability under section 106(a) is another area that has precipitated differing judicial interpretations, at least during the early years of Superfund. A threshold issue is whether section 106(a) is a liability-creating provision (i.e., substantive) or merely enables courts to assert jurisdiction in a section 106(a) action (i.e., procedural). Most courts to rule on this issue have held that the liability standard of section 107—uniformly interpreted as strict, joint, and several—applies to section 106(a).[36] The majority reasons that the statutory language of section 106(a) does not support any interpretation of liability beyond authorizing injunctive relief and administrative orders.[37] As a result, these courts have viewed section 106(a) as jurisdictional only and have looked elsewhere in CERCLA (i.e., to section 107) for defining the standards of liability.

A small minority support application of a federal common law nuisance standard; they view the equitable principles contained in the injunctive relief portion of section 106(a) as evidence of congressional intent for courts to apply the equitably oriented nuisance standard.[38] The minority view, however, appears to be an anachronism of early case law, and may no longer be viable.

## D. Statute of Limitations

Once EPA determines that a release or threatened release presents an imminent and substantial endangerment, must an injunctive action to abate the release or threatened release be filed within a specified period of time (a "limitations" period) or can EPA exercise its authority so long as the endangerment exists? The statute does not answer this question,

36. *See, e.g.*, United States v. Outboard Marine Corp., 556 F. Supp. 54 (N.D. Ill. 1982); *Price II*, 577 F. Supp. 1103.

37. The court in *United States v. Outboard Marine Corp.* noted that a plain reading of section 106(a) "does not appear to create liability in any party. It authorizes lawsuits and injunctions, but it does not indicate who may be sued or enjoined; also, it does not specify what one must do to be subject to suit or injunction." 556 F. Supp. at 55.

38. *See* United States v. Reilly Tar & Chem. Corp., 546 F. Supp. 1100 (D. Minn. 1982) (noting that "equitable principles of federal common law nuisance actions may apply to determine what remedies are appropriate"); *Stringfellow*, 1984 WL 3206; *Wade*, 546 F. Supp. 785 (rejecting imposition of strict liability on past off-site generator under section 106(a)).

as neither section 106 nor section 107 contains specific statute of limitations provisions.

Prior to the 1986 amendments to CERCLA, section 112(d) was the only CERCLA provision with a limitations period. That section referred to actions "for damages," and courts interpreted the phrase as applying to claims against the Fund under section 111 and to judicial actions for damages to natural resources under section 107.[39]

The 1986 amendments to CERCLA added section 113, which specifies the limitations period for actions for natural resource damages, cost recovery, contribution, and indemnification. However, these amendments do not specify a limitations period for a section 106(a) action.

Several limitations periods could apply to a section 106(a) action, depending on how it is viewed. If section 106(a) is viewed as a codification of the federal common law of nuisance, then the federal tort statute of limitations of three years would apply.[40] Alternatively, if a section 106 action can be regarded as quasi-contract action, as EPA has argued,[41] then the federal contract statute of limitations of six years should apply.[42]

Another approach would be for a court to look to state statutes of limitations or state common law of nuisance and apply the limitations period of the state where the release occurred. Because states have differing limitations periods, however, EPA could be precluded from using its section 106 authority merely because the release occurred in a state with a short limitations period. Such disparate results make it unlikely that a federal court would look to state or common law to determine whether the action can be maintained.

## III. SECTION 106 ADMINISTRATIVE ORDERS

As stated at the outset, EPA may compel responsible parties to undertake response actions to abate an "imminent and substantial endangerment"

39. *See, e.g.*, United States v. Conservation Chem. Co., 619 F. Supp. 162, 213 (W.D. Mo. 1985).

40. 28 U.S.C. § 2415(b).

41. *See* EPA, Guidance Memorandum on Pursuing Cost Recovery Actions under CERCLA (Aug. 26, 1983).

42. 28 U.S.C. § 2415(a).

in one of two ways: (1) EPA may unilaterally issue an administrative order directing the PRPs to undertake the response action; or (2) EPA may file a civil action in federal district court seeking injunctive relief. With respect to the first option, EPA has aptly described the section 106(a) administrative order as "one of the most potent administrative remedies available to the Agency under any existing environmental statute."[43] The recipient of an administrative order is truly between the proverbial rock and a hard place.

### A. Constitutionality of Section 106(a) Orders

Generally, the recipient of a section 106(a) administrative order (referred to as a unilateral administrator order, or UAO) will be directed to undertake a specific response action or actions at a site—routinely costing hundreds of thousands, if not millions, of dollars—and will be given a short time to commit to complying with the UAO. The UAO will usually provide the recipient with an "opportunity to confer" about the order prior to its effective date, but this opportunity to confer is far from a due process hearing. Rather, its purpose is to "help EPA and [PRPs] reach a common understanding of how the Order should be implemented or modified."[44]

In the short term, the PRP has only two choices: (1) comply with the terms of the UAO (following the pro forma conference with EPA), or (2) defy the UAO. Seeking judicial review of the UAO is usually *not* a choice, since section 113(h) of CERCLA, added to the statute in 1986, precludes federal jurisdiction to review these administrative orders under most circumstances.[45] Except for PRPs with ironclad defenses to liability, defiance of a section 106 UAO is a perilous proposition. Refusal to

43. *See* EPA, Memorandum on Use and Issuance of Administrative Orders under § 106(a) of CERCLA (Sept. 8, 1983).

44. *Id.* at 21.

45. Judicial review of administrative orders issued pursuant to section 106(a) is precluded except where EPA has initiated one of several civil actions: a cost recovery action under section 107, a contribution action under section 113, or an action to compel remedial work under section 106. Administrative orders may also be reviewed in an action by a PRP for reimbursement of response costs from the fund under section 106(b)(2). For a complete discussion of judicial review, see chapter 6.

comply without "sufficient cause" exposes the PRP to daily penalties of $25,000 and, if EPA chooses to perform the cleanup itself rather than seeking to enforce the UAO, treble damages *on top* of the cost of the cleanup.

In the early years of Superfund, many recipients of UAOs challenged the constitutionality of section 106(a) by arguing that the lack of pre-enforcement judicial or administrative review, combined with the potential imposition of daily penalties under section 106(b)(1) and treble damages under section 107(c)(3), has a chilling effect on a PRP's ability to challenge a UAO, thereby constituting a deprivation of property without due process of law.[46] The courts that have reviewed these constitutional challenges have followed *Ex parte Young*[47] in determining whether the treble damages provision in section 107(c)(3) violates due process requirements. The Supreme Court in *Ex parte Young* held that where a law imposes penalties for its disobedience by fines so enormous and imprisonment so severe as to intimidate a party subject to the law from resorting to the courts to test its validity, it is unconstitutional on its face.[48]

Applying the *Ex parte Young* test, the *Aminoil I* court analyzed whether it should preliminarily enjoin EPA from imposing the daily penalties and treble damages provisions. Its analysis turned on whether the phrase in section 107(c)(3), "without sufficient cause," provided adequate opportunity for the recipient of a section 106(a) order to contest the validity of the order. To answer this question, the court reviewed the legislative history of section 107(c)(3). The meaning of "sufficient cause" was addressed in the 1980 Senate debate on CERCLA. Senator Stafford, in a colloquy with Senator Simpson, stated:

46. *See* Aminoil, Inc. v. United States (*Aminoil I*), 599 F. Supp. 69 (C.D. Cal. 1984); *see also* United States v. Reilly Tar & Chem. Corp., 606 F. Supp. 412 (D. Minn. 1985); Wagner Elec. Corp. v. Thomas, 612 F. Supp. 736 (D. Kan. 1985); Aminoil, Inc. v. EPA (*Aminoil II*), 646 F. Supp. 294 (C.D. Cal. 1986).

47. 209 U.S. 123 (1908).

48. *Id.* at 147–48.

> We intend that the phrase "sufficient cause" would encompass defenses such as the defense that the person who was the subject of the [EPA] order was not the party responsible under the act for the release of the hazardous substance. It would certainly be unfair to assess punitive damages against a party who for good reason believed himself not to be the responsible party. For example, if there were, at the time of the order, substantial facts in question, or if the party subject to the order was not a substantial contributor to the release or threatened release, . . . punitive damages should either not be assessed or should be reduced in the interest of equity. There could also be "sufficient cause" for not complying with an order if the party . . . did not at the time have the financial or technical resources to comply or if no technological means for complying was available.[49]

Based on Senator Stafford's remarks, the court in *Aminoil I* concluded that, although the penalty provision of section 106(b) and the treble damages provision of section 107(c)(3) "would not apply to a party who could demonstrate that 'sufficient cause' existed for noncompliance with a [section 106(a)] administrative order," the situations suggested by Senator Stafford in which such a defense could be asserted appeared to be "extremely limited."[50] The court thus concluded that insufficient cause "does not appear to apply to situations in which alleged responsible parties in good faith assert a reasonable defense that is ultimately rejected by the Court."[51] Accordingly, the court held that the penalty provisions might violate due process.[52]

The PRP's victory in *Aminoil I* was short-lived, however. Although the preliminary injunction was granted upon the court's finding of the plaintiff's likelihood of success on the merits, on subsequent cross-motions for summary judgment the court ultimately construed section 107(c)

49. 126 CONG. REC. S14,967, 15,008 (daily ed. Nov. 24, 1980) (statement of Stafford).
50. 599 F. Supp. at 73 (citation omitted).
51. *Id.*
52. *Id.* at 73–75.

to provide a good-faith defense and, thereby, found that constitutional due process was satisfied.[53]

The court in *United States v. Reilly Tar & Chemical Corp.* reached the same conclusion on the ground that the defendant is able to "challenge the validity of the EPA order in the enforcement action that will take place before this court without being forced into paying exorbitant penalties if its challenge is rejected."[54] The court concluded that since the penalty provisions fell outside the types of statutes proscribed by *Ex parte Young*, they satisfied due process and, thus, were constitutional. Consequently, the court declined to enjoin the application of the penalty provisions to the defendant.

In *Wagner Electric*, the court followed the same reasoning as *Reilly Tar*, stating that "By including the phrase 'without sufficient cause,' then, Congress apparently intended to bring this statute within the scope of the 'good faith' exception to Young. . . ." The court added that even without having the legislative insight provided regarding the meaning of "sufficient cause," it would construe the phrase as referring to a "good faith" defense.[55] Hence, the court concluded that the "good faith" defense provided by the "sufficient cause" phrase in section 107(c)(3) prohibits "a reviewing court from assessing punitive damages against a party who fails to comply with an EPA order in the reasonable belief that it has a valid defense to that order."[56]

The likelihood of succeeding on a constitutionality challenge was further diminished in 1986 when Congress amended section 106 by adding section 106(b)(2), a reimbursement provision. Section 106(b)(2) provides that a PRP that complies with an administrative order may seek reimbursement for reasonable response costs incurred. It must show either that it is not liable for response costs under section 107(a) or, if liable, that the response costs incurred are the result of a response action that was arbitrary and capricious or otherwise not in accordance with law.[57]

53. *Aminoil II*, 646 F. Supp. at 299.
54. 606 F. Supp. at 422.
55. 612 F. Supp. at 745.
56. *Id.*
57. CERCLA § 106(b)(2).

The availability of reimbursement to a party that complies under protest with a section 106 order has not ended litigation over the constitutionality of UAOs issued under section 106(a).[58] In *General Electric Co. v. Jackson*,[59] GE took a slightly different constitutionality argument by challenging EPA's pattern and practice of using its UAO authority. GE argued that the threat of massive penalties for noncompliance, combined with EPA's refusal to provide a hearing before issuing a UAO, denied parties their Fifth Amendment due process rights. The result, however, was the same. The court rejected the due process challenge because the statute provides parties with the defense of sufficient cause to refuse to comply with a UAO, and the courts provide an opportunity for an evidentiary hearing before any penalty can be imposed.[60]

The EPA's pattern and practice of issuing UAOs was also challenged as unconstitutional in *City of Rialto v. West Coast Landing Corp.*[61] EPA had issued a UAO directing BF Goodrich to conduct a remedial investigation/feasibility study (RI/FS) at a contaminated site. Although BF Goodrich complied with the UAO, it alleged that EPA's pattern and practice of issuing UAOs effectively precluded judicial review of the work being required, and therefore violated Fifth Amendment due process rights of recipients. The court held that pre-enforcement review of remedial decisions is barred under section 113(h) of CERCLA, and that in any event BF Goodrich could bring a reimbursement action under section 106(b)(2) in which it may challenge the validity of the UAO.[62]

---

58. *See, e.g.*, Emp'rs Ins. of Wausau Mut. Co. v. Browner, 848 F. Supp. 1369 (N.D. Ill. 1994), *aff'd*, 52 F.3d 656 (7th Cir. 1995) (UAO is constitutional in light of availability of judicial review when EPA seeks enforcement).

59. 595 F. Supp. 2d 8 (D.D.C. 2009), *aff'd*, 610 F.3d 110 (D.C. Cir. 2010).

60. Note that the above description of the holding in *General Electric Co. v. Jackson* does not cover the appeal to the D.C. Circuit, the remand back to the district court, and yet another appeal to the D.C. Circuit. The entire judicial odyssey is described in chapter 6.

61. 581 F.3d 865 (9th Cir. 2009).

62. *Id.* at 867.

## B. Sufficient Cause

### *1. Grounds*

Sufficient cause defenses are listed below and are based on Senator Stafford's remarks during the 1980 Senate debates on CERCLA:

1. The PRP is not within one of the classes of PRPs listed in section 107(a).
2. The PRP is able to assert a section 107(b) defense (i.e., act of God, act of war, or act or omission of a third party not in contractual privity with the PRP).
3. The PRP is not a "substantial contributor" to the release or threatened release (i.e., a de minimis contributor).
4. The PRP is unable to perform the work due to the lack of technical or financial resources to comply with the order.
5. The PRP is unable to perform the work because there are no technological means available for complying.
6. The order is invalid. (Grounds of invalidity include issuance of the order by an unauthorized EPA official; failure to follow prescribed procedures; and the remedy specified in the order is not cost-effective or otherwise is inconsistent with the National Contingency Plan (NCP) or CERCLA section 121 cleanup standards.)

Unfortunately, PRPs have no legislative guidance beyond Senator Stafford's remarks to use in evaluating whether a sufficient cause defense exists. Moreover, judicial precedent is scarce. Several courts have rejected a defendant's lack of financial resources as constituting sufficient cause for noncompliance with a section 106(a) order. In *United States v. Parsons*,[63] defendants did not comply with a section 106(a) order, and EPA, therefore, undertook the cleanup itself. In the subsequent cost recovery action, EPA moved for summary judgment on its claim for treble damages, as well for response costs. All of the defendants argued that treble damages should not be awarded because EPA had commenced

63. 723 F. Supp. 757 (N.D. Ga. 1989).

cleanup of the site before the period to comply with the section 106(a) order had expired. In addition, two of the defendants argued that the government should not recover treble damages because the defendants were financially unable to clean up the site.[64]

The court rejected the inability-to-pay defense, observing that the defendants had failed to present evidence of their financial status and, moreover, that

> the Court would be reluctant to find financial inability to be a sufficient cause because, from a policy standpoint, one should consider one's financial risks before becoming involved in transporting potentially hazardous materials.[65]

There is no indication whether the court considered the legislative history of sufficient cause before rendering its opinion that inability to pay does not constitute sufficient cause. Given that the legislative history directly contradicts the court's reasoning, one might regard the court's conclusion as an aberration. Notably, however, another court reached the same conclusion.

In *United States v. LeCarreaux*,[66] a defendant facing an action by the United States for assessment of daily penalties and punitive damages argued that his lack of financial resources gave him sufficient cause not to comply with an EPA administrative order. The court rejected the defense and followed the public policy reasoning advanced by the *Parsons* court.[67] In granting summary judgment for the United States on the issue of punitive damages under section 107(c)(3) and daily penalties under section 106(b)(1), the court also rejected the defendant's other attempts to establish sufficient cause. Although held liable as a PRP in an earlier proceeding, the defendant asserted that at the time he failed to comply with EPA's order, he was relying in good faith on statutory defenses under which he believed that he was not liable for removal

---

64. *Id.* at 763–64.
65. *Id.*
66. No. Civ. 90-1672, 1991 WL 341191, at *18 (D.N.J. July 30, 1991).
67. *Id.* at *26.

costs.[68] Specifically, he believed he was not liable because a fire, as an act of God, was the sole cause of the release of hazardous substances. The court concluded that since the defendant had been found liable under CERCLA, his good-faith defense was not reasonable.[69]

### 2. *Subjective "Good Faith" versus Objective "Reasonableness"*

There is limited case law on the question of whether sufficient cause for noncompliance with a section 106(a) order may be based on a subjective or objectively reasonable belief that one has such a defense to the order. This distinction is at least theoretically an important one in that a PRP's burden of proof is greater when it must demonstrate with objective evidence the basis for its claim of sufficient cause.

The legislative history of the 1986 amendments to CERCLA—which, among other things, incorporated the phrase "sufficient cause" into the section 106(b) penalty provision—sheds some light on whether a "subjective good faith" or "objective reasonableness" standard was intended by Congress. Citing the *Wagner Electric* and *Reilly Tar* decisions as governing the interpretation of the "sufficient cause" language of section 107(c)(3), the House report states that courts should accept a sufficient cause defense "only where a party can demonstrate by objective evidence the reasonableness and good faith of a challenge to an EPA order."[70]

The House report was quoted at length in *Solid State Circuits v. EPA*.[71] The court concluded that an "objective reasonableness" test was the appropriate standard of review:

> "[S]ufficient cause" as used in CERCLA § 107(c)(3) may be constitutionally interpreted to mean that treble damages may not be assessed if the party opposing such damages had an objectively

68. *Id.* at *25.
69. *Id.* at *26.
70. H.R. Rep. No. 99-253, pt. 1, at 82, *reprinted in* 1986 U.S.C.C.A.N. 2835, 2864.
71. 812 F.2d 383 (8th Cir. 1987).

reasonable basis for believing that the EPA's order was either invalid or inapplicable to it.[72]

The court suggested guidelines for applying its objective reasonableness test, noting that unless EPA is shown to have acted arbitrarily and capriciously, the court should presume that EPA has acted correctly. The court stated that a PRP's objective reasonableness showing must be based on "the applicable provisions of CERCLA, EPA regulations and policy statements, and any formal or informal hearings or guidance the EPA may provide." In the absence of such guidance, however, equity requires the burden of proof to shift to EPA to show that the PRP lacked an objectively reasonable belief in the validity or applicability of a cleanup order.[73]

The court noted that it was interpreting the statute as necessary to determine its constitutionality; therefore, it determined only the most stringent standard that could be constitutionally applied. The court went on to state that "[a]s a matter of statutory construction, the courts or the EPA remain free to adopt a more liberal 'subjective good faith construction.'"[74]

The few courts that have addressed the sufficient cause defense since the decision in *Solid State Circuits*, however, have declined the invitation to adopt a subjective good-faith standard. In *United States v. DWC Trust Holding Co.*,[75] a defendant had refused to comply with a UAO because his attorney told him he had a defense based on his having inherited the property at issue and that other parties had dumped waste on the property before he inherited it. The court held that reliance solely on an attorney's opinion that a good-faith defense existed was not sufficient. However, the court also concluded that there was a sufficient cause defense because the facts supporting the party's belief

72. *Id.* at 391.
73. *Id.* at 392.
74. *Id.* at 391 n.11.
75. Civ. A. No. 93-2859, 1996 WL 250011 (D. Md. Jan. 22, 1996).

about the inheritance and that third parties had caused the contamination were objective.[76]

In *United States v. LeCarreaux*, *supra*, the court found that the defendants' defenses were not objectively reasonable. The defendant in *LeCarreaux* was the president of a salvage company that accepted hazardous substances for treatment and storage at a site that the defendant personally owned. The defendant claimed he had sufficient cause for refusing to comply with a UAO because either a fire or the activities of third parties had caused the releases of hazardous substances. The court rejected that defense as providing sufficient cause for noncompliance, finding that the releases had occurred several years after the fire, that there was no evidence that indicated a link between the fire and releases, and that there was no evidence that the defendant took precautions to prevent those releases. Accordingly, the court found that no objective evidence supported the defense.[77]

### C. Strategic Options When Faced with a Section 106(a) Administrative Order

A PRP's evaluation of whether or not to undertake an EPA-mandated response action at a site will usually *not* begin with the issuance of a UAO. EPA's practice is to engage in settlement negotiations with the PRPs to explore the possibility of voluntary performance of the response action by the PRPs before issuing a UAO to them. Specifically, EPA will (1) send PRPs a notice letter regarding their liability for cleanup of the site, (2) negotiate with the PRPs to voluntarily undertake the necessary response actions, and (3) obtain a judicial consent decree or administrative order embodying the terms of any agreement reached.[78] It is only when EPA concludes that no agreement can be reached that EPA

76. *Id.* at *7.

77. Other decisions holding that sufficient cause must be objectively reasonable include *United States v. Capital Tax Corp.*, 545 F.3d 525, 536 (7th Cir. 2008); *United States v. Barkman*, Civ. A. 96-6395, 1998 WL 962018, at *17 (E.D. Pa. Dec. 17, 1998). *See also* Gen. Elec. Co. v. Jackson, 610 F.3d 110, 118 (D.C. Cir. 2010) (sufficient cause defense can be met for any good-faith challenge where "reasonable grounds to contest" exist).

78. EPA, Memorandum on Use and Issuance of Administrative Orders under § 106(a) of CERCLA (Sept. 8, 1983), at 2–3.

will consider issuing a UAO.[79] Thus, PRPs must begin to evaluate their liability and decide whether to perform a response action long before EPA issues a UAO.

There are some advantages to voluntarily undertaking the response action pursuant to a judicial consent decree or administrative order on consent. For example, the government often grants contribution protection to settling PRPs, which extinguishes the rights of other PRPs to pursue contribution against them for matters covered by the settlement. In addition, the government may covenant not to sue the settlors for any other response actions later deemed necessary or response costs subsequently incurred. (Other advantages to settlement are detailed in chapter 5.) By contrast, PRPs that undertake response actions pursuant to an administrative order or court order have no similar assurance that by complying with the order their liability at the site will be capped.

Ultimately, if settlement cannot be achieved, the PRPs may face the decision of whether to comply with a UAO. The options are limited.

### 1. *Perform the Work Required in the Order*

Complying with a UAO has several advantages. First, the potential penalties and treble damages that may be imposed under sections 106(b)(1) and 107(c)(3), respectively, for noncompliance are avoided. Second, PRPs are generally able to control the costs of site investigation and remediation better than EPA. By complying with the order and performing the mandated response actions, rather than giving EPA the opportunity to perform the remedy itself and seek reimbursement, cost-conscious PRPs may save money in the long run. Third, in most cases, cooperation by the PRP will help foster good relations with the community and the regulatory agency.

On the other hand, following a compliance strategy also has distinct disadvantages. The most obvious is that by agreeing to perform EPA's selected remedy, the PRP is likely committing to a long-term, capital-intensive program with speculative chances for reimbursement or contribution. Response actions selected by EPA often call for "source control"

79. *Id.*

measures, that is, elimination of the source of hazardous substances, as well as "operation and maintenance," that is, monitoring the affected environmental media (such as groundwater) for a period of time, typically 30 years, to ensure that the source control measures were effective. Obviously, depending on the nature of the hazardous substances involved, the environmental media affected, and the remedy selected by EPA, the cost of undertaking the response action can be enormous.

Avenues for recouping response costs include an action for cost recovery or contribution from other PRPs and an action for reimbursement from the Fund (see chapter 9 for a detailed discussion of contribution actions). As to reimbursement, pursuant to section 106(b)(2) of CERCLA, the complying PRP may petition for reimbursement from the Fund for the reasonable costs of such action. If the petitioner shows by a preponderance of the evidence that it is not liable for response costs under section 107 and that its costs incurred were reasonable, then it may obtain reimbursement.[80] In the event the petitioner is liable under section 107 for response costs, it may still recover some of those costs incurred to the extent that it can demonstrate, based on the administrative record, that EPA's decision in selecting the response action ordered was arbitrary and capricious or otherwise not in accordance with law.[81]

The disadvantage with each of these reimbursement mechanisms, of course, is that to recover, the PRP must satisfy various procedural and substantive requirements and, even then, must hope that the vagaries of judicial decision making work to the PRP's advantage. Given the generally long-term nature of remediation, the PRP must be prepared to endure a protracted period in which large capital and operation and maintenance expenditures may be incurred before reimbursement, if any, is realized.

The disadvantage of complying with a UAO and then seeking reimbursement from the Fund under section 106(b)(2) is vividly illustrated by the saga of Wagner Seed Company. Wagner Seed owned a warehouse containing various agricultural chemicals, which burned to the ground

80. CERCLA § 106(b)(2).
81. *Id.*

following a lightning strike. Runoff from firefighting efforts transported chemicals to surrounding properties. Wagner immediately commenced a cleanup under the supervision of the New York State Department of Environmental Conservation. EPA, however, disputed the effectiveness of the cleanup and issued an administrative order under section 106 of CERCLA, which directed Wagner to perform additional remedial work. Wagner filed suit in district court to enjoin EPA from enforcing its order, in part because the contamination was caused by an act of God and thus Wagner was not liable. The district court, however, concluded that it lacked jurisdiction to conduct pre-enforcement judicial review. That decision was upheld by the Second Circuit.[82]

Wagner thereafter elected to comply with the order, performing the remedial work, and then petitioned EPA for reimbursement under section 106(b)(2), which had been added to the statute in 1986. EPA denied the petition, interpreting the statute as precluding reimbursement of cleanup activity that had begun prior to passage of the 1986 amendments. Wagner filed suit in the U.S. District Court for the District of Columbia, seeking reimbursement under section 106(b)(2). The district court, however, sided with EPA, stating that the agency's interpretation of the statute was reasonable.[83] The D.C. Circuit affirmed this district court's decision, reasoning that EPA was the administering agency of CERCLA and its interpretation of the statute was entitled to deference if it was a "permissible interpretation."[84] Thus, Wagner Seed ended up paying all of the cleanup costs, notwithstanding that EPA had never disputed Wagner's act of God defense, which would have relieved it of all liability.

82. Wagner Seed Co v. Daggett, 800 F.2d 310, 314 (2d Cir. 1986) (Second Circuit affirmed the district court holding that there was no unconstitutional taking, since Wagner could assert a good-faith defense in a proceeding to enforce the order).

83. Wagner Seed Co. v. Bush, 709 F. Supp. 249, 250 (D.D.C. 1989), *aff'd*, 946 F.2d 918 (D.C. Cir. 1991).

84. 946 F.2d 918, 920 (D.C. Cir. 1991).

## 2. *Do Not Perform the Work Required in the Order*

A PRP may also choose not to comply with the order. Noncompliance carries with it several disadvantages, the most obvious of which is risking the imposition of daily penalties under section 106(b)(1) and treble damages under section 107(c)(3). In addition, however, the noncomplying PRP must be prepared to incur litigation expenses in defending against subsequent EPA enforcement actions that would otherwise be avoided.

The decision in *United States v. Occidental Chemical Corp.*[85] illustrates the downside of refusing to comply with a UAO. In this case, two parties (Occidental and Ruetgers) received a "special notice" letter from EPA requesting them to voluntarily perform the remedial action that EPA had selected for the site in question and to reimburse EPA for past costs incurred by the agency. Ruetgers entered a consent decree under which it agreed to perform the work, and also to pay EPA $293,895 in past costs. Occidental, on the other hand, declined to enter a settlement with EPA, prompting the agency to issue a UAO under section 106(a) of CERCLA, which directed Occidental to jointly implement the remedial action. After Occidental did not comply with the UAO, EPA filed suit to enforce its order. Occidental moved to dismiss the complaint on the ground that EPA had already obtained "complete relief" under the Ruetgers consent decree and, therefore, is precluded from pursuing Occidental for the same relief, which it characterized as "double recovery."[86] The district court agreed, and dismissed the complaint.

The Third Circuit, however, reversed the district court decision. In so holding, the Third Circuit reasoned that the Ruetgers consent decree did not provide complete relief because Ruetgers may not be able to perform the cleanup. It agreed with EPA's view that "unless and until the endangerment providing the basis for EPA's § 106(a) authority has been abated," the agency has not obtained complete relief. Moreover, the Third Circuit noted that "by issuing administrative orders to non-settling PRPs under § 106, EPA fulfills CERCLA's objectives of promoting fairness at multi-party sites, increasing the likelihood of settlements,

85. 200 F.3d 143 (3d Cir. 1999).
86. *Id.* at 147.

and accelerating the statute's ultimate goal—site cleanup."[87] Based on the foregoing, the Third Circuit reversed the district court's dismissal of EPA's complaint, and directed that the district court proceed in a manner consistent with the ruling.[88]

With these downside risks in mind, the PRP should carefully evaluate and begin to articulate legal and technical defenses to the section 106(a) order. (See chapter 4 for a detailed discussion of defenses under CERCLA.) Obviously, as discussed above, strong defenses are a prerequisite to proceeding with a strategy of noncompliance. Presenting these defenses and objections to EPA, either formally in a conference or informally, may persuade EPA to modify the order. At a minimum, even if EPA is not dissuaded from pursuing a section 106(a) order against the PRP, the written submittal of the PRP's defenses and objections will become part of an administrative record and thereby help to perfect any sufficient cause defense in a later cost recovery action.

Under certain circumstances, PRPs may wish to offer to perform part of the ordered remedy, and refuse to perform the rest. This approach may be appropriate where the action ordered by EPA is defective or objectionable only in part. In this instance, the PRP could offer to perform the acceptable portion of the order and limit its refusal to the objectionable portion. Alternatively, the PRP could offer alternatives for part of the remedy, or propose alternatives for all parts. By demonstrating cooperation with EPA, the PRP will create a record that may help to reduce the potential damages or penalties EPA may later seek. Also, the PRP could argue in a subsequent damage action that its potential damage liability should extend only to the portion of the remedy it refused to perform.

Clearly, noncompliance strategies are risky and should only be pursued, if at all, after weighing many factors, including the preorder efforts taken by the PRP to reduce or eliminate its share of liability in the proposed order; the basis for the order (e.g., site-specific characteristics of the alleged endangerment); business factors (e.g., solvency of the PRP's

87. *Id.* at 150.

88. The issue of whether Occidental had sufficient cause not to comply with the UAO was not presented to or ruled on by the Third Circuit.

business, community relations, impact of compliance on operations); the scope of the order (e.g., size of the site investigation/cleanup and number of PRPs); and the location of the action (e.g., highly populated area, EPA regional office assigned, and judicial district affected). This diverse and complex set of factors underscores the need for the PRP to consider all avenues of action carefully before embarking on a noncompliance path. Both before an order is issued and at each step of the section 106(a) process, deliberation by the PRP will help to maximize its chances of successfully reducing or eliminating its share of liability and potential penalties and damages for noncompliance.

## IV. SECTION 106 CIVIL ACTIONS

As an alternative to its authority under section 106(a) to issue an administrative order to compel responsible parties to abate an "imminent and substantial endangerment," EPA may also commence a civil action in federal district court to secure such relief. Specifically, section 106(a) authorizes EPA to "require the Attorney General of the United States to secure such relief as may be necessary to abate" an imminent and substantial endangerment caused by a release or threatened release. Further, section 106(a) gives the federal district court "in which the threat occurs . . . jurisdiction to grant such relief as the public interest and the equities of the case may require."[89]

Over the years, EPA has developed a preference for administrative orders over civil actions. The threat of daily penalties and, potentially, treble damages (if EPA performs the response action itself rather than file suit to enforce the order) for noncompliance with a section 106(a) order gives EPA tremendous leverage. Without an airtight defense to liability, EPA recognizes that most PRPs will opt to comply with the order, rather than take their chances in court, thus saving EPA the burdensome, resource-intensive, and time-consuming task of preparing and trying a civil action.

Nevertheless, section 106(a) civil actions have been filed and, no doubt, will continue to be filed. Most section 106(a) civil actions will

89. *Id.* at 151 n.9.

be tried sequentially, with liability issues preceding any issues relating to the actual form of injunctive relief being sought.

## A. Prima Facie Case

To establish a prima facie case in a section 106(a) abatement action, the United States must establish, by a preponderance of the evidence, the "triggering" elements discussed earlier, that is, that a release or threatened release into the environment of a hazardous substance from a facility may present an imminent and substantial endangerment to public health or welfare or the environment. The United States must also establish by a preponderance of evidence that the defendant is a liable party.[90]

## B. Defenses

Defendants in abatement actions may assert the same defenses that would be available to them in cost recovery actions, such as section 107(b) defenses (i.e., act of God, act of war, or act or omission of third party), that it is not a liable party under section 107(a), and so on.[91] In addition, however, defendants in abatement actions may argue equitable defenses since, unlike section 107 cost recovery actions, a grant of relief under section 106 must take into consideration "the public interest *and the equities of the case*."[92] Accordingly, laches, unclean hands, estoppel, and waiver may be asserted as a defense to a section 106(a) civil action.[93]

Defendants in civil abatement actions have also argued that the availability of an adequate "remedy at law" should foreclose the right to injunctive relief. In *United States v. Conservation Chemical Corp.*,[94] the defendants argued that injunctive relief should be denied because the plaintiff had a remedy at law under both CERCLA and RCRA. The court rejected the defendant's argument, stating that "the mere existence of another available remedy should not preclude injunctive relief" under

90. *See* United States v. Bliss, 667 F. Supp. 1298 (E.D. Mo. 1987).
91. See discussion in chapter 4.
92. CERCLA § 106(a) (emphasis added).
93. *See* United States v. Conservation Chem. Co., 619 F. Supp. 162, 204 (W.D. Mo. 1985) ("On its face, Section 106 incorporates traditional equitable defenses.").
94. 619 F. Supp. 162.

section 106(a). The court equivocated, however, noting that in considering the equities of the case, the availability of another remedy *may* preclude injunctive relief.

### C. Standard of Review Applicable to Remedy Selected

An open and important issue relates to the standard of review applicable to the remedy EPA is seeking to compel the PRP to perform. There is some precedent, that in an action for injunctive relief, the remedy selected by EPA will be tried de novo like any other aspect of the case.[95] Another line of cases, however, limits review of EPA's selected remedy to the administrative record, based on section 113(j) of CERCLA,[96] and articulates the standard of review as arbitrary and capricious.[97] The topic of judicial review of a selected remedy is more fully discussed in chapter 6.

95. *See* United States v. Conservation Chem. Co., 661 F. Supp. 1416, 1424–31 (W.D. Mo. 1987); United States v. Hardage, No. Civ-86-1401-W, 1986 WL 647, at *2 (W.D. Okla. Dec. 11, 1986).

96. CERCLA § 113(j)(1) provides in pertinent part:

In any judicial action under this chapter, judicial review of any issues concerning the adequacy of any response action taken or ordered by the President shall be limited to the administrative record.

97. *See* United States v. Bell Petrol. Servs., Inc., 718 F. Supp. 588 (W.D. Tex. 1989); *see also* United States v. Seymour Recycling Corp., 679 F. Supp. 859 (S.D. Ind. 1987). See chapter 6 for a complete discussion of this issue.

# CHAPTER 4

# Liability

## I. INTRODUCTION

Chapters 1 through 3 of this book describe the broad remedial powers that Congress created under CERCLA to enable the government to respond to releases of hazardous substances. In this chapter, we discuss the broad liability scheme that Congress established to impose on responsible parties the costs of responding to such releases. The key elements of the liability scheme are set forth in section 107(a), which may be the most heavily litigated passage of the U.S. Code. Over the past 35 years, Congress has amended CERCLA to establish new defenses to liability, but it has never amended section 107(a). Its organization is confusing and its language is inscrutable, so much so that it deserves to be reprinted in its entirely before addressing the various elements. Set forth below is the text of section 107(a), annotated with the phraseology (in bold, bracketed print) that has become standard shorthand and is used throughout this book:

> Notwithstanding any other provision or rule of law, and subject only to the defenses set forth in subsection (b) of this section—
>
> (1) the owner and operator of a vessel or a facility [**referred to as "current owners and operators"**],
> (2) any person who at the time of disposal of any hazardous substance owned or operated any facility at which such hazardous substances were disposed of [**referred to as "former owners and operators"**],

(3) any person who by contract, agreement, or otherwise arranged for disposal or treatment, or arranged with a transporter for transport for disposal or treatment, of hazardous substances owned or possessed by such person, by any other party or entity, at any facility or incineration vessel owned or operated by another party or entity and containing such hazardous substances [**referred to as "arrangers" or "generators"**], and

(4) any person who accepts or accepted any hazardous substances for transport to disposal or treatment facilities, incineration vessels or sites selected by such person, from which there is a release, or a threatened release which causes the incurrence of response costs, of a hazardous substance [**referred to as "transporters"**] shall be liable for—

(A) all costs of removal or remedial action incurred by the United States Government or a State or an Indian tribe not inconsistent with the national contingency plan;

(B) any other necessary costs of response incurred by any other person consistent with the national contingency plan;

(C) damages for injury to, destruction of, or loss of natural resources, including the reasonable costs of assessing such injury, destruction, or loss resulting from such a release; and

(D) the costs of any health assessment or health effects study carried out under [section 104(i)] of this title.

This is a powerful liability scheme. On its face, the statute imposes strict liability—that is, liability without fault—for response costs and natural resource damages, on four broad categories of responsible parties, namely (1) the current owner or operator of a facility from which there has been a release, or a threat of release, of a hazardous substance; (2) former owners and operators whose ownership/operation coincided with the time of disposal of hazardous substances; (3) parties who "arranged for" the treatment or disposal of hazardous substances at the facility; and (4) parties who transported hazardous substances to the facility, having selected it for treatment or disposal of such hazardous substance. Moreover, courts have interpreted liability under section 107(a) as joint and several, meaning that one party can in theory be held liable for the entire cost of cleaning up a Superfund site, notwithstanding that other parties may share responsibility. Finally, such liability is retroactive, meaning that it can encompass actions that occurred before enactment of CERCLA in 1980.

The absence of clear definitions in the statute, coupled with enormous potential liability, has led to a tremendous amount of litigation concerning the scope and application of CERCLA's liability provisions. Contested issues have included: the classification of a defendant within one of the four potentially responsible party (PRP) categories established by section 107; the liability of corporate parents, successors, shareholders, and officers; lender liability; and the availability of statutory and common law defenses to liability. Each of these issues, and the way they have been addressed by the courts to date, is discussed in this chapter.

## II. NATURE OF LIABILITY—STRICT, JOINT, AND SEVERAL

As discussed below, courts have interpreted liability under CERCLA as retroactive, and strict, joint, and several. The concepts are not expressly stated in the text of CERCLA. Rather, they have emerged through judicial interpretations, which are described below.

### A. Strict Liability

Although the term "strict liability" is not used in the statute, courts have uniformly interpreted section 107 as establishing a strict liability scheme.[1] Strict liability is implied by the fact that Congress did not

1. *See, e.g.*, Pakootas v. Teck Cominco Metals, Ltd., 452 F.3d 1066, 1078 n.18 (9th Cir. 2006) ("CERCLA is a strict liability statute, and liability can attach even when the generator has no idea how its waste came to be located at the facility from which there was a release."); NutraSweet Co. v. X-L Eng'g Co., 227 F.3d 776, 784 (7th Cir. 2000); *In re* Bell Petrol. Servs., Inc., 3 F.3d 889, 897 (5th Cir. 1993) (CERCLA is a strict liability statute); United States v. R.W. Meyer, Inc., 889 F.2d 1497, 1507 (6th Cir. 1989) (imposing strict liability under CERCLA to owner of facility), *cert. denied*, 494 U.S. 1057 (1990); United States v. Aceto Agric. Chems. Corp., 872 F.2d 1373, 1377 (8th Cir. 1989); Tanglewood E. Homeowners v. Charles-Thomas,

include exercise of due care as a defense to liability. Thus, no proof of fault or negligence of a responsible party is required to impose liability, which attaches even if all due care was taken to avoid releases of hazardous substances.

## B. Retroactive Liability

As noted in the Overview, one of the primary reasons Congress passed CERCLA was to establish the government's authority to clean up preexisting inactive and abandoned disposal sites such as Love Canal. Moreover, Congress made clear that it intended the costs of such cleanup to be borne by those who were responsible for creating the contamination. Thus, it comes as no surprise that courts have uniformly held that CERCLA liability applies *retroactively*, imposing liability for disposal activity and related releases that occurred, or response costs that EPA incurred, even prior to passage of the legislation.[2]

CERCLA, however, does not expressly state that liability is retroactive, and thus PRPs have frequently challenged the constitutionality of imposing retroactive liability, arguing that it violates the due process clause and the ex post facto provision of the Constitution. Despite repeated rejection of such challenges, they found new life after the Supreme

Inc., 849 F.2d 1568 (5th Cir. 1988); United States v. Monsanto Co., 858 F.2d 160, 167 (4th Cir. 1988) ("overwhelming body of precedent . . . has interpreted § 107(a) as establishing a strict liability scheme"), *cert. denied*, 490 U.S. 1106 (1989); New York v. Shore Realty Corp., 759 F.2d 1032, 1042 (2d Cir. 1985) ("Congress intended that responsible parties be held strictly liable. . . .").

2. *See, e.g.*, United States v. Olin Corp., 107 F.3d 1506 (11th Cir. 1997); United States v. Gurley, 43 F.3d 1188 (8th Cir. 1994); *Monsanto*, 858 F.2d at 174–75; United States v. Ne. Pharm. & Chem. Co. (NEPACCO), 810 F.2d 726, 734 (8th Cir. 1986), *cert. denied*, 484 U.S. 848 (1987); Allied Signal, Inc. v. Amcast Int'l Corp., 177 F. Supp. 2d 713, 729–30 (S.D. Ohio 2001); United States v. Vertac Chem. Corp., 33 F. Supp. 2d 769 (E.D. Ark. 1998); Cont'l Tire Co. v. Peoples Gas Light & Coke Co., 959 F. Supp. 893 (N.D. Ill. 1997); Cooper Indus., Inc. v. Agway, Inc., No. 92-CV-0748, 1996 WL 550128 (N.D.N.Y. Sept. 23, 1996); Gould Inc. v. A&M Battery & Tire Serv., 933 F. Supp. 431 (M.D. Pa. 1996); Nevada *ex rel.* Dep't of Transp. v. United States, 925 F. Supp. 691 (D. Nev. 1996); United States v. Rohm & Haas Co., 939 F. Supp. 1142 (D.N.J. 1996); United States v. Alcan Aluminum Corp., Nos. 87-CV-920, 91-CV-1132, 1996 WL 637559 (N.D.N.Y. Oct. 28, 1996), *aff'd*, 315 F.3d 179 (2d Cir. 2000); Farmland Indus., Inc. v. Colo. & E. R.R. Co., 944 F. Supp. 1492 (D. Colo. 1996); Ninth Ave. Remedial Grp. v. Fiberbond Corp., 946 F. Supp. 651 (N.D. Ind. 1996); Nova Chems., Inc. v. GAF Corp., 945 F. Supp. 1098 (E.D. Tenn. 1996); Chatham Steel Corp. v. Brown, 858 F. Supp. 1130 (N.D. Fla. 1994).

Court's 1994 decision in *Landgraf v. USI Film Products*.[3] In that case, the Supreme Court articulated the following test for deciding whether a retroactive statutory obligation is constitutional. First, a reviewing court must determine whether Congress expressly stated that the statute applies retroactively. If not, a court must next determine whether the statutory text and its legislative history have "expressly prescribed" Congress's intent to apply a provision retroactively. Absent such evidence, a court is to apply the traditional presumption against retroactivity.[4] The applicability of the *Landgraf* decision to CERCLA was the focus of *United States v. Olin*.[5] The district court concluded that because there was no express statement that liability under CERCLA was retroactive, nor any clear legislative history, CERCLA's liability should not apply retroactively. The Eleventh Circuit reversed, finding that the legislative history "confirms that Congress intended to impose retroactive liability for cleanup."[6]

Although the Supreme Court has never directly addressed whether retroactive liability under CERCLA is constitutionally valid, Justice Breyer, in a dissenting opinion regarding the constitutionality of the Coal Industry Retiree Health Benefit Act of 1992, identified CERCLA as an example of a constitutionally valid retroactive statute.[7]

### C. Joint and Several Liability

It is well settled that liability under CERCLA is joint and several. Before explaining the basis for this statement, a brief primer on the terminology and philosophy of joint and several liability is warranted.

Under the common law principle of joint and severable liability, where the injury or harm to a plaintiff is indivisible, the plaintiff may recover all of the damages from any of the defendants regardless of their individual share of liability. Courts may thereafter allocate the damages among and between defendants deemed jointly and severally liable to the

3. 511 U.S. 244 (1994).
4. *Id.* at 280.
5. 107 F.3d 1506 (11th Cir. 1997).
6. *Id.* at 1514.
7. E. Enters. v. Apfel, 524 U.S. 498 (1998).

plaintiff for an "indivisible harm." Where a defendant can demonstrate that harm is divisible, however, the defendant may qualify for apportionment. The difference between apportionment and allocation is not insignificant. Allocation is the court's calculations of each defendant's share of the plaintiff's damages after determining that each defendant is jointly and severally liable. By contrast, apportionment is the remedy where a defendant is liable for only a portion of the harm. As one court notes: "To apportion is to request separate checks, with each party paying only for its own meal. To allocate is to take an un-itemized bill and ask everyone to pay what is fair."[8]

As noted in the Overview, during the 1980 deliberations over the text of CERCLA, there was strong disagreement over imposing joint and several liability on PRPs. Ultimately, in the eleventh hour Stafford-Randolph compromise, the words "joint and several" were dropped from the act, but floor statements by authors of the Stafford-Randolph compromise indicate an intent to impose joint and several liability.[9] Despite this congressional compromise, whether Superfund liability was meant to be joint and several quickly became one of the most hotly contested issues in early CERCLA litigation. The first case to address the scope of liability under CERCLA was *United States v. Chem-Dyne Corp.*[10] The *Chem-Dyne* court concluded that Congress deleted the words "joint and several" to avoid inequitable results, but that joint and several liability is imposed under CERCLA in cases in which it would be considered appropriate under traditional common law tort principles, that is, when the injury resulting from multiple causes is indivisible.[11] Since *Chem-Dyne*, courts have uniformly held that CERCLA liability is joint and several, except where harm is divisible.[12]

8. Yankee Gas Servs. Co. v. UGI Utils., Inc., 852 F. Supp. 2d 229, 242 (D. Conn. 2012).

9. For example, Sen. Stafford stated, "The changes do not reflect a rejection of the standards in the earlier bill." He added, "It is intended that issues of liability not resolved by this act, if any, shall be governed by traditional and evolving principles of common law. . . . [T]he liability of joint tortfeasors will be determined under common or previous statutory law." 126 CONG. REC. 30,932 (1980).

10. 572 F. Supp. 802, 810–11 (S.D. Ohio 1983).

11. *Id.*

12. *See* United States v. Alcan Aluminum Corp., 315 F.3d 179 (2d Cir. 2003); Metro. Water Reclamation Dist. of Greater Chi. v. N. Am. Galvanizing & Coatings, Inc., 473 F.3d 824,

As a practical matter, at multiparty sites it is often difficult to demonstrate divisibility of harm. During the early years of Superfund, courts often rejected out of hand divisibility arguments based on types of wastes or numbers of drums, for example, citing the effects of commingling, synergistic impacts, and migration of hazardous substances.[13] A series of decisions, however, slowly eroded what had been a very high threshold for overcoming joint and several liability.[14] In these cases, the courts cited section 433A, *Restatement (Second) of Torts*, which provides that a defendant can demonstrate divisibility of harm by establishing a "reasonable basis" for apportionment of liability:

> [W]hen two or more persons acting independently caused a distinct or single harm for which there was a reasonable basis for division according to the contribution of each, each was subject to liability only for the portion of the total harm that he himself caused.[15]

In 2009, the Supreme Court handed down its decision in *Burlington Northern & Santa Fe Railway Co. v. United States*,[16] upholding the apportionment of liability in this section 107 cost recovery case

827 (7th Cir. 2007); *Monsanto Co.*, 858 F.2d at 171–73; O'Neil v. Picillo, 883 F.2d 176, 178 (1st Cir. 1989) ("It is now well settled that Congress intended that the federal courts develop a uniform approach governing the use of joint and several liability in CERCLA actions."); United States v. Northernaire Plating Co., 670 F. Supp. 742, 748 (W.D. Mich. 1987), *aff'd sub nom.* United States v. R.W. Meyer, Inc., 889 F.2d 1497 (6th Cir. 1989).

13. *See, e.g.*, O'Neil v. Picillo, 682 F. Supp. 706, 724–25 (D.R.I. 1988), *aff'd*, 883 F.2d 176.

14. United States v. Twp. of Brighton, 153 F.3d 307 (6th Cir. 1998) (defendant can avoid joint and several liability if it can prove divisibility; citing RESTATEMENT (SECOND) OF TORTS § 433A, court notes that to show divisibility, one must establish a reasonable basis for determining the contribution of each cause to a single harm); United States v. Hercules, Inc., 247 F.3d 706 (8th Cir. 2001) (same); *In re* Bell Petrol. Servs., Inc., 3 F.3d 889 (defendant seeking to avoid joint and several liability must prove that its volumetric percentage of waste at the disposal site is a reasonable basis for apportioning liability); United States v. Broderick Inv. Co., 862 F. Supp. 272 (D. Colo. 1994) (court refused to impose joint and several liability for that portion of environmental harm it found to be geographically divisible); Pneumo Abex Corp. v. Bessemer & Lake Erie R.R. Co., 936 F. Supp. 1250 (E.D. Va. 1996) (in section 107 action, defendant may avoid joint and several liability if it can demonstrate either that harm is divisible or that there exists a reasonable basis for apportionment).

15. RESTATEMENT (SECOND) OF TORTS § 433A, cmt.

16. 556 U.S. 599 (2009).

based on application of section 433A, *Restatement (Second) of Torts*, and providing guidance on factors that justify apportionment. The case concerned cleanup of two adjoining parcels of land, one of which was owned by two railroad companies. A third company, Brown & Bryant (B&B), had operated an agricultural chemical distribution business on the adjoining parcels, and had spilled hazardous substances during the course of its operations. A fourth company, Shell Oil, sold fumigant to B&B, which had been spilled, both during and after delivery. B&B ultimately declared bankruptcy. The United States sought to hold Shell Oil and the two railroad companies jointly and severally liable under section 107 of CERCLA. The district court, however, concluded that although the contamination constituted a single harm, liability was reasonably capable of apportionment and therefore divisible. Based on consideration of geographic surface area, period of ownership, and volume of wastes, the district court apportioned 9 percent of the United States' cleanup costs to the railroads. Shell Oil was apportioned the remaining 91 percent of the cleanup costs.

The Ninth Circuit reversed the lower court's apportionment of liability, finding that liability should have been joint and several. The Supreme Court reversed the Ninth Circuit, and upheld the district court's apportionment as reasonable.[17]

The Court endorsed the *Restatement* approach to joint and several liability, which allows for the apportionment of costs when a "reasonable basis" for such apportionment exists:

> Under the Restatement, "when two or more persons acting independently caus[e] a distinct or single harm for which there is a reasonable basis for division according to the contribution of each, each is subject to liability only for the portion of the total harm that he has himself caused. . . . But where two or more persons cause a single and indivisible harm, each is subject to liability for the entire harm." In other words, apportionment is proper when

17. *Id.* at 617–18.

"there is a reasonable basis for determining the contribution of each cause to a single harm."[18]

The Court noted that defendants have the burden of proving that a reasonable basis exists for the divisibility of harm. Additionally, the Court explained that equitable factors do not play a role in apportionment analysis; rather, they play a role only in contribution analysis for CERCLA actions brought under section 113.[19] This passage has been endlessly debated by commentators as to whether the Court relaxed the standard for apportionment of liability or merely restated long-standing principles.[20] Some courts appear to have interpreted the "reasonable basis for apportionment" language in *Burlington Northern* as having effectively, if not expressly, reduced the burden required to avoid joint and several liability under CERCLA.[21] Other courts, however, have held that *Burlington* merely reaffirmed the 1983 holding in *Chem-Dyne*.[22] Indicative of this split of authority are *Pakootas v. Teck Cominco Metals* and *Reichold v. United States Metals*.

18. *Id.* at 614 (citations omitted).

19. *Id.* at 615 n.9.

20. For an excellent summary of competing perspectives on Burlington Northern, see Ryan Brady, *CERCLA Apportionment Following Burlington Northern: How Joint and Several Liability Still Thrives—To the Surprise of Many*, 4 WASH. J. ENV'TL L. & POL'Y 141 (2014).

21. *See, e.g.*, Ashley II of Charleston, LLC v. PCS Nitrogen, Inc., 791 F. Supp. 2d 431, 483 (D.S.C. 2011) ("Some legal commentary and case law have interpreted *Burlington Northern* to lessen the burden on defendants seeking to avoid joint and several liability by demonstrating a reasonable basis for apportionment."); Reichold, Inc. v. U.S. Metals Ref. Co., 655 F. Supp. 2d 400, 448 (D.N.J. 2009) (court apportioned liability despite concluding that defendants caused a single indivisible harm); United States v. P.H. Glatfelter Co. & NCR Corp., 768 F.3d 662 (7th Cir. 2014) (revisiting a prior ruling rejecting divisibility argument, the Seventh Circuit reviewed evidence that was presented at trial and held that a party may be able to prove that harm is divisible by showing the extent to which it contributed to the harm or contamination, and reversed the district court's rejection of divisibility, finding there was a reasonable basis for apportionment).

22. *See, e.g.*, United States v. Iron Mountain Mines, Inc., No. 91-0768, 2010 WL 1854118, at *3 (E.D. Cal. May 6, 2010) ("Plaintiffs are correct that *Burlington Northern* does not constitute a change in law as required for reconsideration. *Burlington Northern* simply reiterated the law as established in 1983 by *Chem-Dyne*, and then examined the record to resolve a factual question of whether the record supported apportionment. *Burlington Northern* did not add a new mandate that [courts] must apportion harm in CERCLA cost recovery cases. . . ."); Pakootas v. Teck Cominco Metals, Ltd., 868 F. Supp. 2d 1106 (E.D. Wash. 2012) (defendant held jointly and severally liable, despite request for apportionment, because the "harm was not shown to be "theoretically capable of apportionment").

In *Pakootas v. Teck Cominco*, the defendant's smelting operations were alleged to have contaminated the Upper Columbia River in Canada. Specifically, the defendant allegedly dumped smelting slag into the river, and the slag leached metals into the river, which caused contamination of tribal property in the United States. The Tribe and the United States sought recovery of cleanup costs from Teck Cominco. Prior to a bench trial, the plaintiffs moved for summary judgment and dismissal of the defendant's defense to limit its liability to apportionment based on divisible harm. Citing *Burlington Northern*, the court stated that "liability under CERCLA is generally joint and several," and that "apportionment is only proper when the evidence supports the divisibility of damages jointly caused by PRPs."[23]

Referring to *Burlington Northern*, the court stated that for apportionment to be available, the harm must be "theoretically capable of apportionment," which the court characterized as a question of law. Only if harm is theoretically capable of apportionment would evidence supporting "a reasonable basis for apportionment" be relevant. The court noted that although Teck Cominco presented three separate models for apportionment, Teck Cominco failed to prove that the harm was theoretically capable of apportionment:

> In this case, the harm is the entirety of the contamination of the [Upper Columbia River] Site and what the Plaintiffs seek are recovery of costs to investigate and clean up the entirety of that contamination. . . . None of Teck's apportionment theories address the entirety of the contamination. Instead, they begin with the assumption that the only harm at issue is whatever metals were released from Teck's slag and/or liquid effluent. . . . This is a fatal flaw. Because Teck has not addressed the relevant harm in the first instance, it has failed to establish as a matter of law that the relevant harm is a single harm divisible in terms of degree. Simply put, because it has failed to account for all of the harm at the UCR

23. 868 F. Supp. 2d at 1110.

> Site, it cannot prove the harm is divisible ("theoretically capable of apportionment").[24]

The result in *Pakootas* demonstrates that the presumption of joint and several liability under CERCLA is alive and well in the post-*Burlington Northern* era.[25]

Compare *Pakootas* with the decision in *Reichold*, where the plaintiff sought reimbursement for the cost of installing a cap as part of a remediation.[26] Both the defendant and a third party had caused the contamination necessitating the cap.[27] The court found that because each party's contaminants alone would have caused the agency to require the cap, "[t]ogether they constituted a single harm."[28] The court found that this single harm was divisible between the two parties:

> [T]he metals contamination . . . was a distinct or single harm that [Defendant] and a third party caused. There is a reasonable basis for division according to the contribution of each. The measurement is not the exact amount of metals contamination for which each was responsible; [Defendant] was undoubtedly the source of most of it. Rather, it is the circumstances that each was responsible for a sufficient amount of metals contamination that required the cap.[29]

Consequently, the court did not impose joint and several liability and instead split the damages. Plaintiff was only able to recover half of the costs related to the cap from the defendant, and had to recover the other half of the costs from the third party.[30]

---

24. *Id.* at 1117.

25. *See also* United States v. Saporito, 684 F. Supp. 2d 1043 (N.D. Ill. 2010) (another post-*Burlington Northern* decision in which the court imposed joint and several liability because it determined that the harm was not theoretically capable of apportionment).

26. 655 F. Supp. 2d at 448.

27. *Id.*

28. *Id.*

29. *Id.* at 448–49.

30. *Id.* at 449.

## III. PARTIES LIABLE UNDER SECTION 107

Section 107(a) of CERCLA identifies four classes of responsible parties who are subject to liability based on their nexus to a facility from which there has been a release or threat of release of a hazardous substance: (1) the current owner or operator of a facility from which there is a release of a hazardous substance; (2) the former owner or operator of a facility at the time of disposal of a hazardous substance; (3) any person who "arranged for disposal or treatment" of hazardous substances at the facility (commonly referred to as "generators" but also including "brokers"); and (4) any person who accepts hazardous substances for transport to a facility that that person selected (transporters).[31]

Before exploring the nuances of each class of responsible party, a brief word about the term "facility" is warranted. The CERCLA definition of "facility" embodies not only the traditional concept of a building or other structure ("any building, structure, installation, equipment, pipe or pipeline . . . pond, lagoon, impoundment, ditch, landfill, storage container, motor vehicle, rolling stock or aircraft"), but also includes "any site or area where a hazardous substance has been deposited, stored, disposed of, or placed, or otherwise come to be located."[32] Because the definition of facility is so broad, it has proven difficult to defeat liability on the premise that the hazardous substance was not released from a facility.[33] Note, however, that there are statutory limitations to the term "facility." The term does not include "any consumer products in

31. CERCLA § 107(a)(1)–(4).

32. CERCLA § 101(9).

33. *See, e.g.*, Pakootas v. Teck Cominco Metals, Ltd., 452 F.3d 1066 (9th Cir. 2006) (CERCLA liability applies to company's disposal of hazardous substances in Canadian part of Columbia River, which flows into the United States, because "facility" is broadly defined in CERCLA to include "every place where hazardous substances come to be located"); Westfarm Assocs. Ltd. P'ship v. Wash. Suburban Sanitary Comm'n, 66 F.3d 669 (4th Cir. 1995) (municipal sewer which leaked hazardous substances discharged by various industries is a "facility"); ITT Indus., Inc. v. Borgwarner, Inc., 700 F. Supp. 2d 848, 875 (W.D. Mich. 2010) (municipal sewer which leaked hazardous substances discharged by various industries is a "facility"); Adobe Lumber v. Hellman, 658 F. Supp. 1188 (E.D. Cal. 2009 (same); United States v. Conservation Chem. Co., 619 F. Supp. 162, 185 value (W.D. Mo. 1985) ("facility" is intentionally expansive and includes "every place where hazardous substances come to be located"); United States v. Saporito, 684 F. Supp. 2d 1043, 1056–57 (N.D. Ill. 2010) (owner of leased equipment at an industrial operator was the owner of a "facility").

consumer use." Like most of the CERCLA exclusions, the "consumer product in consumer use" exclusion has been narrowly construed.[34]

## A. Current Owners or Operators

The first, and sometimes only, PRPs to be identified at a Superfund site, are current owners and operators of the facility, who are potentially liable for response costs even though they may not have owned or operated the facility at the time of disposal. Liability attaches to current owners and operators even if they have no involvement with any hazardous substances, and the circular definition of "owner or operator" contained in the statute provides no relief from this burdensome result.[35] As discussed later in this chapter, courts have construed "owner or operator" broadly to not only the corporate entity directly owning and/or operating a facility, but also a corporation's parent, corporate officers, and shareholders.

The statute defines "owner or operator" collectively, rather than separately; it reads in pertinent part:

> The term "owner or operator" means (i) in the case of a vessel, any person owning, operating, or chartering by demise, such vessel, (ii) in the case of an onshore facility or an offshore facility, any person owning or operating such facility. . . .[36]

Most courts have distinguished the concept of owner liability from operator liability. The owner typically is the party with legal title to the facility. Thus, for example, a lessee does not hold title to leased property, and thus is not ordinarily considered an owner, although it may be considered an operator as discussed below. That said, at least one circuit has endorsed the notion of imposing "de facto" owner liability

34. *See, e.g.*, Nat'l R.R. Passenger Corp. v. N.Y.C. Hous. Auth., 819 F. Supp. 1271 (S.D.N.Y. 1993) (exception does not apply to asbestos-containing buildings and railroad tracks); Reading Co. v. City of Philadelphia, 823 F. Supp. 1218 (E.D. Pa. 1993) (electric cars do not fall within the exception). See discussion *infra* under "Defenses."

35. *See* CERCLA § 101(20)(A), which defines "owner or operator" to include, in the case of a facility, "any person owning or operating such facility."

36. *See* CERCLA § 101(20)(A).

on a lessee, where the lessee's connection to a site gives it attributes of ownership.[37]

The concept of operator liability is more elusive, especially in light of the statute's lack of any meaningful definition of operator. Accordingly, courts have turned to common law principles—examining whether a party exercises day-to-day control, hires employees, establishes production rates, and makes sales decisions—to determine whether a party meets the definition of operator.

In performing this analysis, two schools of thought emerged among courts to address the question of whether a party is an operator. The minority view was that authority to control the operations that released hazardous substances is sufficient for operator status to attach.[38] The majority view was that there must be evidence of "actual and substantial control" over day-to-day operations for operator liability to attach.[39] This split of authority was resolved by the Supreme Court's

37. *See* Commander Oil Corp. v. Barlo Equip. Corp., 215 F.3d 321 (2d Cir. 2000) (identifying attributes of ownership possessed by a lessee, but declining to impose owner liability); *but see* City of Los Angeles v. San Pedro Boat Works, 635 F.3d 440 (9th Cir. 2011) (rejecting Second Circuit de facto ownership rule, and relying instead on actual legal title to impose owner liability under CERCLA); United States v. S.C. Recycling & Disposal, Inc., 653 F. Supp. 984 (D.S.C. 1986), *aff'd in part, vacated in part sub nom.* United States v. Monsanto Co., 858 F.2d 160 (4th Cir. 1988).

38. *See, e.g.*, Nurad, Inc. v. William E. Hooper & Sons Co., 966 F.2d 837, 842 (4th Cir. 1992); Nw. Mut. Life Ins. Co. v. Atl. Res. Corp., 847 F. Supp. 389, 397 (E.D. Va. 1994) (tenant corporation liable as operator of leased facility because it had "authority to control" operators on property where releases occurred); Pierson Sand & Gravel, Inc. v. Pierson Twp., 851 F. Supp. 850 (W.D. Mich. 1994) *aff'd*, 89 F.3d 835 (6th Cir. 1996) (court denied "authority to control" standard in determining operator status); Hillsborough Cnty. v. A&E Rd. Oiling Serv., Inc., 853 F. Supp. 1402 (M.D. Fla. 1994) (same).

39. FMC Corp. v. Dep't of Commerce, 29 F.3d 833 (3d Cir. 1994) (en banc) (United States liable as operator for having exercised actual and substantial control over a manufacturing company's day-to-day operations and policy making decisions); Long Beach United Sch. Dist. v. Godwin Cal. Living Trust, 32 F.3d 1364 (9th Cir. 1994) (operator liability requires active role in running facility; mere allegation that defendant was in a position to prevent contamination is insufficient to establish operator liability); Nu-West, Inc. v. United States, 768 F. Supp. 2d 1082, 1091 (D. Idaho 2011) (U.S. Forest Service, which leased land to a mining company, held liable as operator for cleanup of property as "operator" because it was a "very active participant in designing and locating the waste dumps, in inspecting mining operations, and in ensuring compliance with all rules and plans" for the operation of those dumps); *cf.* United States v. Vertac Chem. Corp., 46 F.3d 803 (8th Cir. 1995) (United States not liable as operator because government was not sufficiently involved either directly or indirectly in operation of the facility).

1998 decision in *United States v. Bestfoods*,[40] where the Court held "an operator must manage, direct, or conduct operations specifically related to pollution, that is, operations having to do with the leakage or disposal of hazardous waste, or decisions about compliance with environmental regulations."[41] We discuss this split of authority and its resolution by *Bestfoods*, *infra*, in analyzing the potential liability of a parent corporation for releases at a facility owned by a subsidiary. This question has also arisen in the context of determining whether a lessee-sublessor should be deemed an operator. This was the question addressed by the Second Circuit in *Commander Oil Corp. v. Barlo Equipment Corp.*[42]

In this case, Commander owned two adjoining parcels, one housing an office and warehouse, the other housing above-ground petroleum storage tanks. Commander leased both parcels to Barlo, which, in turn, subleased the parcel with the above-ground tanks to Pasley Solvents. Barlo was responsible for basic maintenance and taxes on both parcels. The site became contaminated with hazardous substances that leaked from the above-ground tanks, and the question was whether Barlo could be held liable as an owner or operator. Citing the standard in *Bestfoods*, the court held that "under no version of those facts could Barlo be said to have 'manage[d], direct[ed], or conduct[ed] operations specifically related to the pollution.'"[43]

### B. Former Owners or Operators

As noted above, current owners or operators of a facility are strictly liable for response costs even if they had no interest in the facility at the time hazardous substances were placed there. The liability of *former* owners and operators, however, is more limited. Under section 107(a)(2) of CERCLA, only those who owned or operated a site at the time when hazardous substances were disposed of may be liable for the costs of

40. 524 U.S. 51 (1998).

41. *Id.* at 66–67.

42. 215 F.3d 321.

43. *Id.* at 332 n.3 (alterations in original). Note that the Second Circuit devoted most of its analysis to the question of whether Barlo could be considered a de facto "owner," based on five factors or attributes of ownership, *id.* at 330–31, and concluded that Barlo was not a de facto owner.

removal or remedial action if a release occurs or is threatened. Courts have rigorously enforced this limitation in evaluating allegations of former owner/operator liability.[44]

There is a split of authority, however, over the meaning of disposal. CERCLA defines "disposal" by incorporating the Resource Conservation and Recovery Act (RCRA) definition, namely "the discharge, deposit, injection, dumping, spilling, leaking, pouring or placing of solid waste or hazardous waste into or onto any land or water so that such solid or hazardous waste or any constituent thereof may enter the environment or be emitted into the air or discharged into the waters, including groundwaters."[45] The question is whether this definition of disposal encompasses passive migration of contaminants. A majority of courts have dismissed prior owner-operator claims brought against a party who had purchased a facility after a release had occurred but sold it prior to the suit being brought. In so ruling, these courts rejected the argument that continued passive migration or movement of the hazardous substances on the property constitutes disposal.[46] One court has taken

44. *See, e.g.*, G.J. Leasing Co. v. Union Elec. Co., 54 F.3d 379, 386 (7th Cir. 1995) (former owner of property that housed a decommissioned power plant containing asbestos was not liable as former owner/operator because plant was not leaking as of the time of sale of the property); Town of New Windsor v. Tesa Tuck, Inc., 935 F. Supp. 305 (S.D.N.Y. 1996) (no owner/operator liability where there was no evidence of disposal during period when defendant held the property); Joselyn Mfg. Co. v. Koppers Co., 40 F.3d 750 (5th Cir. 1994) (owner liability would apply only if disposal occurred while defendant owned the property); United States v. Cordova Chem. Co., 113 F.3d 572 (6th Cir. 1997), *vacated sub nom.* United States v. Bestfoods, 524 U.S. 51 (1998) (former owner not liable where there is no evidence of releases during brief period of ownership).

45. CERCLA § 101(29).

46. *See, e.g.*, Niagara Mohawk Power Corp. v. Jones Chem., Inc., 315 F.3d 171 (2d Cir. 2003) (interim owner not liable for passive migration because no disposal occurred); United States v. 150 Acres of Land, 204 F.3d 698, 706 (6th Cir. 2000) (interim landowner not liable for passive migration because disposal does not include passive migration); ABB Indus. Sys., Inc. v. Prime Tech., Inc., 120 F.3d 351 (2d Cir. 1997) (prior owner-operator not liable for passive migration of hazardous substances during period of ownership); United States v. CDMG Realty Co., 96 F.3d 706, 714 (3d Cir. 1996) (disposal requires human action); Carson Harbor Vill., Ltd. v. Unocal Corp., 270 F.3d 863 (9th Cir. 2001) (CERCLA definition of disposal does not encompass passive migration), *cert. denied*, 122 S. Ct. 1447 (2002); *In re* Diamond Reo Trucks, Inc. v. City of Lansing, 115 Bankr. 559 (Bankr. W.D. Mich. 1990) (mere ownership of property during a time period when hazardous substances are migrating is insufficient to confer liability under section 107(a)(2)); Ecodyne Corp. v. Shah, 718 F. Supp. 1454 (N.D. Cal. 1989), *abrogated by* Carson Harbor Vill., Ltd. v. Unocal Corp., 227 F.3d 1196 (9th Cir.

the contrary view, finding that passive migration can constitute disposal such that a prior owner is liable under CERCLA § 107(a)(2).[47]

## C. Party Who Arranged for Disposal

### 1. *Generators*

Generators of wastes disposed of at a site may be strictly liable for cleanup under section 107(a)(3) of CERCLA because they "arranged for the disposal or treatment" of hazardous substances. There is no threshold amount of hazardous substances necessary to trigger liability under CERCLA.[48] Thus, any quantity of hazardous substance is sufficient to trigger CERCLA liability.[49]

Generators who arrange for disposal of wastes may be held liable regardless of where the wastes are ultimately disposed. Indeed, liability has been imposed on generators even when their wastes were removed without their knowledge from the site of the original disposal.[50] In *Violet v. Picillo*, the generator was held liable for a cleanup at the second, or satellite, site, even though the generator had specified a different disposal site as part of the arrangement with the transporter and was not informed that the waste was sent to a different location.[51]

Since the passage of CERCLA, much litigation has focused on the meaning of "arranged for disposal," which is not defined in the statute. Many defendants have challenged the imposition of liability where they did not *intend* to arrange for disposal of hazardous substances. Up until the 2009 Supreme Court decision in *Burlington Northern & Santa*

2000) ("at time of disposal" does not refer to mere movement or migration of preexisting contamination).

47. *See Nurad, Inc.*, 966 F.2d at 846 (disposal includes not only "active human conduct," but also passive migration of contaminants).

48. *See, e.g.*, United States v. Alcan Aluminum Corp., 990 F.2d 711 (2d Cir. 1993); A&W Smelters & Refiners, Inc. v. Clinton, 146 F.3d 1107 (9th Cir. 1998).

49. Note, however, that under section 107(o) of CERCLA, a party may assert as an affirmative defense to liability that the total volume of waste containing the hazardous substance disposed of is less than 110 gallons or 200 lbs., subject to certain limitations. See discussion *infra* in this chapter.

50. *See* Missouri v. Indep. Petrochemical Corp., 610 F. Supp. 4 (E.D. Mo. 1985); Violet v. Picillo, 648 F. Supp. 1283 (D.R.I. 1986).

51. 648 F. Supp. 1283.

*Fe Railroad Co.*,[52] courts did not require a showing of actual intent to dispose of hazardous substances as a prerequisite to imposing liability under the phrase "arranged for disposal."

**Sale of Product or Disposal of Waste?** In general, the pre-*Burlington Northern* case law distinguished manufacturers and sellers of *virgin* materials, who have not been held liable for contamination caused by releases of those materials following their sale to third parties, from sellers of *used* (e.g., spent solvents, used oil) or *unusable* (e.g., off-specification or out-of-date) materials. The mere sale of a product that is a hazardous substance (or contains one) does not confer arranger liability.[53] On the other hand, one who is trying to get rid of a material may be held liable as an arranger, despite the fact that the materials were actually sold for valuable consideration and in the honest belief they would be reused or recycled, not disposed of or otherwise released into the environment.[54] Thus, pre-*Burlington Northern* courts uniformly held that sellers of spent lead-acid batteries, used industrial or electrical equipment or metal-containing by-products of manufacturing processes

52. 556 U.S. 599.

53. *See, e.g.*, Edward Hines Lumber Co. v. Vulcan Materials Co., 861 F.2d 155 (7th Cir. 1988) (supplier of wood-treating chemicals not liable under CERCLA for cleanup costs incurred by wood-treating company at its facility); Fla. Power & Light Co. v. Allis Chalmers Corp., 85 F.3d 1514 (11th Cir. 1996) (manufacturer of product containing hazardous substances not liable as arranger because plaintiff did not meet its burden of showing that manufacturer's transactions with plaintiff were anything other than "mere sales"); United States v. Wedzeb Enter., Inc., 844 F. Supp. 1328 (S.D. Ind. 1994) (no arranger liability for sale of PCB-containing capacitors because they were not "wastes" or "refuse" and because there was consumer demand for them).

54. *See, e.g.*, Gould Inc. v. A&M Battery & Tire Serv., 933 F. Supp. 431 (M.D. Pa. 1996) (sale of spent lead-acid batteries to recycling facility was an arrangement for disposal because only value of batteries was lead content); United States v. Mountain Metal Co., 137 F. Supp. 2d 1267 (N.D. Ala. 2001), *aff'd*, 91 F.3d 654 (11th Cir. 2004) (same); Pneumo Abex Corp. v. Bessemer & Lake Eerie R.R. Co., 921 F. Supp. 336 (E.D. Va. 1996), *rev'd*, 142 F.3d 769 (4th Cir. 1998) (sale of worn journal bearings was arrangement for disposal, rather than sale of useful product); Cal. Dep't of Toxic Substances Control v. Ako Pac., Inc., 508 F.3d 930 (9th Cir. 2007) (sale of lead slag and dross deemed arrangement for disposal because the material in question was not the principal product of seller; rather it was a by-product); United States v. Conservation Chem. Co., 619 F. Supp. 162, 237–41 (W.D. Mo. 1985) (sale of lime slurry and fly ash by-products as neutralizers for other hazardous substances construed as "arranging for disposal" of the lime slurry and fly ash), New York v. Gen. Elec. Co., 592 F. Supp. 291, 297 (N.D.N.Y. 1984) (sale of used transformer oil for dust control construed as an arrangement for disposal).

to a reclamation or recycling facility are liable under section 107(a)(3) for response costs incurred in responding to a release or threatened release of hazardous substances at that facility.

For example, in *United States v. Pesses*,[55] the defendants had sold scrap materials to a metal alloy manufacturing facility, which either processed them for use as raw material in the alloy manufacturing process or resold them to other reclamation facilities. The defendants contended that they sold valuable, useful commodities to the facility operator in order to make a profit. The court rejected this contention, holding that because the materials sold by defendants necessarily required processing in order to be put to productive use, defendants had implicitly arranged for the treatment and disposal of those materials at the processing site.[56]

**Constructive Intent to Dispose; Tolling Arrangements.** Two noteworthy cases expanded the scope of arranging for disposal to a category of commercial transactions that Congress probably never contemplated. Both cases involved pesticide manufacturers that used independent contractors to formulate their technical-grade pesticides into end-use products. In *United States v. Aceto Agricultural Chemical Corp.*,[57] the Eighth Circuit upheld the district court's ruling that pesticide manufacturers that hire a contractor to turn technical-grade pesticides into commercial products have arranged for disposal of the hazardous substances generated by the contractor, and, accordingly, can be liable under section 107(a)(3) of CERCLA for cleanup of environmental contamination at the contractor's plant. Central to the court's ruling was the fact that the contractor performed "a process on products owned by [the manufacturers of technical-grade pesticides] for [the manufacturers'] benefit and at their direction; waste is generated and disposed contemporaneously with the process."[58] The court, therefore, concluded that because the manufacturers owned the technical-grade pesticides provided to the contractor, the formulated pesticide product produced by

55. 794 F. Supp. 151 (W.D. Pa. 1992).
56. *Id.* at 156–67; *see also* Catellus Dev. Corp. v. United States, 34 F.3d 748 (9th Cir. 1994).
57. 872 F.2d 1373.
58. *Id.* at 1381.

the contractor and the work in progress, and because the generation of pesticide-containing wastes was inherent in the formulation process, the manufacturers could be deemed to have implicitly arranged for disposal of hazardous substances at the contractor's facility.[59] In *United States v. Velsicol Chemical Corp.*,[60] confronted with nearly identical facts, the court applied the same theory of liability to deny the defendant's motion to dismiss the government's claim for CERCLA response costs.[61]

Although the *Aceto* and *Velsicol* decisions only address the peculiar fact pattern of pesticide toll formulation, courts in the ensuing years applied the concept of arranger liability to a wide variety of circumstances in which there was evidence showing that the defendant knew or should have known of disposal of hazardous substances or had control over production processes.[62] The hallmark of all of these arranger cases is that liability could attach regardless of whether there was specific evidence that the defendant intended to dispose of hazardous substances, because evidence of intent to dispose could be inferred from the nature of the arrangement.

**Supreme Court Restores Intent as Requirement for Arranger Liability.** In 2009, the U.S. Supreme Court's decision in *Burlington Northern & Santa Fe Railroad Co. v. United States*[63] brought an abrupt halt to arranger liability without an intent to dispose. As noted earlier, the

59. *Id.*

60. 701 F. Supp. 140 (W.D. Tenn. 1987).

61. *See also* United States v. Vertac Chem. Corp., 966 F. Supp. 1491 (E.D. Ark. 1997); Jones-Hamilton Co. v. Beazer Materials & Servs, Inc., 973 F.2d 688 (9th Cir. 1992); Matthews v. Dow Chem. Co., 947 F. Supp. 1517 (D. Colo. 1996); Levin Metals Corp. v. Parr-Richmond Terminal Co., 781 F. Supp. 1448, 1451 (N.D. Cal. 1991).

62. *See, e.g.*, Cal. Dep't of Toxic Substances Control v. Payless Cleaners, 368 F. Supp. 2d 1069, 1080 (E.D. Cal. 2005) (court denied third-party-defendant dry cleaning equipment manufacturer's motion to dismiss, holding that it could be liable as arranger because it chose the locations of disposal drains, physically connected dry cleaning equipment to the drains and annually inspected the facility, and directed the dry cleaner to dispose of perchloroethylene (PCE) wastewater through the sewer system); United States v. Wash. State Dep't of Transp., 716 F. Supp. 2d 1009, 1015 (W.D. Wash. 2010) (WSDOT held liable as arranger because it designed, constructed and operated highway drainage system "whose sole function was to collect highway runoff and dispose of it into nearby water-bodies"); United States v. Shell Oil Co., 294 F.3d 1045, 1059 (9th Cir. 2002) (endorsing broad arranger liabilities standard wherein liability attaches if defendant had control over process that created the waste, regardless of intent to dispose).

63. 556 U.S. 599.

case involved cleanup of a parcel owned by two railroad companies, Burlington Northern and Union Pacific. They leased the land to B&B, which operated a pesticide blending operation on the property and had spilled chemicals on multiple occasions. Shell Oil supplied pesticides to B&B and was aware that some chemicals had been spilled. Ultimately, B&B went bankrupt, leaving the railroads and Shell as the only viable defendants in a cost recovery action by the United States.

The United States asserted that Shell was liable as an "arranger" because it delivered hazardous substances to the site and had knowledge of spills that occurred during and following such delivery. The Ninth Circuit agreed with the United States and held Shell liable as an arranger, reversing the district court's judgment. The Supreme Court reversed the Ninth Circuit, holding that arranger liability requires a showing of intent to dispose. Noting that CERCLA does not define or explain what is meant by "arrange for disposal," the phrase should be given its ordinary meaning.[64] "In common parlance, the word 'arrange' implies action directed to a specific purpose." Thus, "under [the] plain language [of the statute], an entity may qualify as an arranger . . . when it takes intentional steps to dispose of a hazardous substance."[65] The court added that "[w]hile it is true that in some instances an entity's knowledge that its product will be leaked, spilled, dumped, or otherwise discarded may provide evidence of the entity's intent to dispose of its hazardous wastes, knowledge alone is insufficient to prove that an entity 'planned for' the disposal, particularly when the disposal occurs as a peripheral result of the legitimate sale of an unused, useful product."[66]

**Post-*Burlington Northern* Case Law.** Post-*Burlington Northern* case law has focused on the evidence needed to establish the requisite intent to dispose in order to prove arranger liability. An instructive decision is *United States v. General Electric*.[67] The focus of this case was Pyranol, a material produced by GE in the 1950s for its dielectric properties. When GE produced off-spec Pyranol, it stored the material in 55-gallon

64. *Id.* at 600.
65. *Id.* at 600 (emphasis added).
66. *Id.* at 612.
67. 670 F.3d 377 (1st Cir. 2012).

drums labeled "scrap Pyranol" or "waste Pyranol." GE had a variety of arrangements for dealing with this material including landfill disposal, giving it to employees for use as weed killer, and selling it to local governments that used it for road dust suppression. GE also reached an agreement with Fred Fletcher, a paint manufacturer, who bought the material as a plasticizer at "bargain pric[ing]."[68]

In 1987, EPA found hundreds of drums at the Fletcher site and implemented response actions. In 2006, EPA sought recovery of its response costs from GE under an arranger theory of liability. The district court granted the United States' motion for summary judgment imposing arranger liability on GE, and GE appealed to the First Circuit. On appeal, GE argued that under *Burlington Northern* arranger liability only attaches where the defendant "has the distinctly apparent objective of disposing of its hazardous substances."[69] The First Circuit rejected this narrow reading of *Burlington Northern*, stating that the Supreme Court did not limit arranger liability solely to "circumstances in which the stated or facially-evident purpose of an arrangement is to dispose of hazardous substances. While the 'easy' cases may lie at one end of the spectrum, there necessarily remains a range of case law in which arranger liability is proper but the parties' intent will not be obvious."[70] Turning to the evidence in this case, the First Circuit affirmed the district court's imposition of liability, finding the facts sufficient to support an inference of an intent to dispose rather than a sale of a useful product. Among the more significant facts was that after ten years of purchases, Fletcher stopped paying GE, stating that the product had declined in quality: "[W]e could not dispose of [this product] as you had apparently added something and we could find nothing to neutralize the additive you were using. Secondly, you put the Pyranol in badly contaminated drums. . . . In other words, what we were buying as a reasonable scrap product, turned into a large percentage of what is known as garbage in the Chemical trade."[71] Addressing this and other evidence a GE former

68. *Id.* at 380.
69. *Id.* at 384.
70. *Id.*
71. *Id.* at 388.

employee testified "[w]e just hoped he was able to use some of it, and the balance of it he could dispose of it."[72] The First Circuit stated that such facts establish GE's "conscious and intentional step of leaving Fletcher to dispose of the materials."[73] Reviewing the total record, the appeals court said the evidence established GE purposefully entered into its arrangement with Fletcher to rid itself of the scrap Pyranol. Though the initial informal arrangement may not have *directed* Fletcher to dispose of GE's scrap Pyranol, the court said GE certainly understood this would be the result of its actions and took the conscious and intentional step of leaving Fletcher to dispose of the materials.

Another instructive post-*Burlington Northern* case is *Consolidated Coal Co. v. Georgia Power Co.*[74] At issue in this case was the sale of used transformers in the 1980s by Georgia Power, among others, to Ward Transformer Company, which reconditioned such transformers for resale. Georgia Power's practice was to test transformer oil for polychlorinated biphenyls (PCBs) before selling a used transformer. If transformer oil contained PCB concentrations over 50 ppm, Georgia would drain out the oil before sale to comply with the Toxic Substances Control Act. In this case, Georgia Power had sold 101 used transformers at auction to Ward, having drained oil from all but 14 of the transformers. The drained transformers contained small amounts of PCB-contaminated oil, according to Ward's records. Because Ward sometimes performed a significant amount of work to recondition transformers, some PCB-contaminated oil had spilled over the years, and EPA eventually listed the Ward site on the National Priorities List (NPL). The plaintiffs agreed to clean up the Ward site, and thereafter sought contribution from Georgia Power and others. The district court granted Georgia Power's motion for summary judgment. On appeal, the Fourth Circuit affirmed. After an in-depth analysis of the evidence, the Fourth Circuit concluded: "There is no direct evidence that Georgia Power intended, even in part, to arrange for disposal."[75] It reasoned that "[a]ny disposal of PCBs occurred only

72. *Id.* at 390.
73. *Id.* at 39.
74. 781 F.3d 129 (4th Cir. 2015).
75. *Id.* at 150.

as a result of Ward's later business judgments, not any implicit agreement or understanding between Ward and Georgia Power at the time of the auction."[76] Had this case arisen before the Supreme Court's ruling in *Burlington Northern*, one can easily imagine that the court would have imposed arranger liability such as happened in *United States v. Pesses*, *supra*.

### 2. *Brokers*

Liability has also been applied, under the "arranged for disposal" language of section 107, to parties that act as brokers of hazardous waste.[77] Generally, a broker will act as an outside consultant to a company that is seeking to arrange for the treatment or disposal of its waste.

If the broker's role in waste disposal is limited, arranger liability is unlikely to be imposed. For example, in *C.P. Systems, Inc. v. Recovery Corp.*, the court held that a disposal consultant was not liable as an arranger where it recommended a disposal site, found a transporter to take the waste to the site, and invoiced the waste generator for the marked up costs of transportation.[78] Similarly, in *United States v. Davis*,[79] the court did not impose arranger liability on a broker where its role in waste disposal was merely responding to a customer's call for waste removal, packaging it, labeling it, and delivering it to an appropriate disposal site.

By contrast, consider the ruling in *Emergency Technology Services Corp. v. Morton International*,[80] where the broker's role was much greater. The court held the broker liable as an arranger based on its having completed waste profile sheets, identifying and making arrangements with a disposal site, providing and completing shipping and manifest documents, and providing instructions for packaging of the waste.

76. *Id.* at 152.

77. *See, e.g.*, Am. Cyanamid Co. v. Capuano, 381 F.3d 6, 25 (1st Cir. 2004) (finding that "a broker can be liable as an arranger if the broker controls the disposal of the waste"); United States v. Bliss, 667 F. Supp. 1298, 1306 (E.D. Mo. 1987) (holding broker liable if it has authority to control the place and manner of the disposal).

78. No. 93-C-2905, 1994 WL 174162 (N.D. Ill. May 5, 1994).

79. 1 F. Supp. 2d 125 (D.R.I. 1998).

80. No. 92-C-3376, 1993 WL 210531 (N.D. Ill. June 11, 1993).

Generally, EPA does not pursue brokers for cost recovery. In a few instances, where extremely good records have been kept at disposal facilities and the broker's name appears on manifests, EPA has included such parties in the list of PRPs, but that has generally been a case of mistakenly believing that the broker was the original generator of the waste. In private cost recovery actions, however, brokers have been pursued more aggressively.

## D. Transporters

Section 107(a)(4) imposes liability on any party that transports hazardous substances for disposal or treatment to a facility of that party's selection. Historically, it was EPA's policy to issue notice letters regarding potential liability to all known transporters, regardless of whether the transporters had any involvement in selection of the site. This resulted in the institution of some enforcement actions against transporters prior to a determination that the transporter had selected the site.

In a 1985 guidance memorandum on transporter liability, EPA revised its practice with respect to enforcement actions against transporters. Although information requests under section 104(e) would continue to be sent to all transporters, notice letters would be sent to transporters only upon receipt of information indicating that the transporter selected the site. More important, EPA stated that enforcement actions would be brought against only those transporters that had selected the site for treatment or disposal.[81]

In the ensuing years, a virtually unbroken string of case law confirms that site selection by a transporter is an absolute prerequisite to imposition of CERCLA liability on such entities.[82] The site selection decision,

81. EPA, Policy for Enforcement Actions against Transporters under CERCLA (Dec. 23, 1985), http://www2.epa.gov/sites/production/files/2013-09/documents/enfact-transp-mem.pdf.

82. B.F. Goodrich v. Betkoski, 99 F.3d 505 (2d Cir. 1996) (only transporters who actively participated in selection of disposal site are subject to liability); United States v. USX Corp., 68 F.3d 811 (3d Cir. 1995) (shareholder of corporation would only be held liable as transporter if the shareholder actually participated in the decision to send wastes to the site); Tippins Inc. v. USX Corp., 37 F.3d 87 (3d Cir. 1994) (transporter liability will only be imposed if transporter actively and substantially participated in selection of disposal site); ACC Chem. Co. v. Halliburton Co., 932 F. Supp. 233, 239 (S.D. Iowa 1995); United States v. Aluminum Co. of

however, need not rest exclusively with the transporter for liability to be imposed.

The point is illustrated by *Tippins Inc. v. USX Corp.*[83] In this case, the plaintiff had contracted with Petroclean to remove and dispose of a hazardous waste (electric arc furnace dust). The agreement between Tippins and Petroclean required Petroclean to supply all labor, equipment, and material for the job, and to obtain an EPA identification number for the generation of the hazardous waste. Petroclean gathered information on a potential disposal site and submitted a cost proposal to Tippins. When subsequent developments rendered disposal at the original site prohibitively expensive, Petroclean surveyed substitute disposal sites and ultimately identified two landfills willing to accept the waste for disposal. Both of the landfills were presented to Tippins as possible disposal locations, and based on cost information gathered by Petroclean, Tippins directed Petroclean to dispose of the dust at the Four County Landfill. The question before the court was whether Petroclean's involvement with locating and gathering information on available landfill options was sufficient to support a finding that it selected the Four County Landfill for disposal of the dust.

Recognizing that one of CERCLA's primary objectives is to hold those who are actively involved in hazardous substance disposal financially accountable for the resulting harm, the Third Circuit asserted that the imposition of transporter liability must reflect that, by virtue of their expertise, "transporters often play an influential role in the decision to dispose waste at a given facility."[84] According to the court, a person is subject to transporter liability "not only if it ultimately selects the disposal facility, but also when it actively participates in the disposal

Am., Civ. A. No. 89-7421, 1995 WL 753862 (E.D. Pa. Dec. 12, 1995) (transporter defendant's motion for summary judgment granted because the plaintiff did not offer evidence showing that the transporter had selected the site for disposal); United States v. Atlas Minerals & Chems., Inc., Civ. A. No. 91-5118, 1995 WL 510304 (E.D. Pa. Aug. 22, 1995) (transporter held liable because its participation in site selection process was substantial).

83. 37 F.3d 87 (3d Cir. 1994).

84. *Id.* at 95.

decision to the extent of having had substantial input into which facility was ultimately chosen."[85] Such substantial input occurs:

> [w]hen a transporter with knowledge and understanding of the industry superior to its customer's investigates a number of potential disposal sites and suggests several to the customer from which it may pick, and the customer relies upon the transporter's knowledge and experience by choosing one of the winnowed sites.[86]

The court went on to explain that, in such a situation, while "the transporter has not made the ultimate decision, it has made the penultimate one; for all intents and purposes, the transporter has selected the facility by presenting it as one of a few disposal alternatives."[87] Based on the foregoing standard, the Third Circuit affirmed the district court's imposition of transporter liability on Petroclean.[88]

## E. Federal, State, and Local Government Liability

### *1. Federal Liability*

Section 120(a)(1) of CERCLA contains a limited waiver of sovereign immunity, which subjects the United States to CERCLA liability "in the same manner and to the same extent, both procedurally and substantively, as any non-governmental entity, including liability under Section [107]." Thus, the federal government may be held liable under any of the four categories identified above.[89]

Where EPA takes over a site to perform remedial activities, it arguably becomes an operator of the site. The waiver of sovereign immunity under

85. *Id.* at 94.

86. *Id.* at 95.

87. *Id.*

88. *Id.*; *see also* United States v. Aluminum Co. of Am., Civ. A. No. 89-7241, 1996 WL 27074, at *4 (E.D. Pa. Jan. 16, 1996) (simply acquiescing in customer's suggestion as to disposal location is insufficient to support finding that transporter selected the site; transporter must have taken "an affirmative step to exercise its formal discretion").

89. *See, e.g.*, FMC v. U.S. Dep't of Commerce, 29 F.3d at 849–50 (imposing owner/operator liability and arranger liability on the United States); United States v. Vertac Chem. Corp., 46 F.3d 803, 808–10 (8th Cir. 1995) (same).

the Federal Tort Claims Act (FTCA)[90] that could confer such liability, however, is subject to the "discretionary acts exception," under which the government or its agent cannot be held liable for a tort claim if the challenged government action involves an element of judgment or choice and such action is based on public policy considerations.[91] A number of courts have held that the discretionary acts exception applies when EPA officials making decisions in cleanups exercise discretion to weigh site-specific factors in determining the best way to achieve CERCLA's remedial goals. For example, in *United States v. Skipper*,[92] the court held that "administrative decisions involving cleanup and disposal of hazardous waste are grounded in environmental, economic and social considerations, and are just the sort of decisions the exception was designed to protect."[93] So long as EPA officials are selecting and implementing remedial actions based on their judgment as how best to protect public health, the exception would apply and the agency and its contractors would not be subject to liability under CERCLA as operators. On the other hand, operator liability could be imposed on EPA where response costs are incurred as a result of its own negligence.[94]

### 2. *State Liability*

Like the federal government, states and municipalities are not immune from liability under CERCLA. In this regard, the CERCLA definition of "owner or operator" encompasses any person owning or operating a facility, and the CERCLA definition of "person" includes "State, municipality, commission, political subdivision of a State." The U.S. Supreme Court held in *Pennsylvania v. Union Gas*[95] that these definitions evince Congress's intent to hold state and local governments liable for damages in federal court: "[S]tates are liable along with everyone else for cleanup

90. 28 U.S.C. §§ 2671–2680.
91. *See* Berkovitz v. United States, 486 U.S. 531, 536–37 (1988).
92. 781 F. Supp. 1106 (E.D.N.C. 1991).
93. *Id.* at 1114.
94. *See, e.g.*, United States v. Iron Mountain Mines, Inc., 881 F. Supp. 1432 (E.D. Cal. 1995).
95. 491 U.S. 1 (1989).

costs recoverable under CERCLA."[96] In so holding, the Court pointed out that in addition to expressly including "State" within the CERCLA definition of "person," Congress established exemptions from liability for states in certain circumstances from liability under section 101(20)(D) and section 107(d)(2) of CERCLA—a structure that is indicative of an intent to impose liability on states.

In this regard, section 101(20)(D) provides that the term "owner or operator" does not include a unit of state or local government that acquired ownership or control involuntarily through bankruptcy, tax delinquency, abandonment, or other circumstances in which the government involuntarily acquires title by virtue of its sovereign functions. This "exclusion," however, "shall not apply to any state or local government which has caused or contributed to the release or threatened release of a hazardous substance from the facility, and . . . shall be subject to the provisions of this chapter in the same manner and to the same extent, both procedurally and substantively, as any nongovernmental entity, including liability under [section 107] of this title."[97]

Section 107(d)(2) provides that states are not liable under section 107 for costs or damages that result from "actions taken in response to an emergency created by a release or threat of release of a hazardous substance generated by or from a facility owned by another person," except where the costs or damages result from "gross negligence or intentional misconduct by a State or local government." The Supreme Court held that this structure indicates that Congress intended states to be subject to liability as PRPs under section 107, unless exempted from liability under either section 101(20)(D) or section 107(d)(2).[98]

Notably, in *Pennsylvania v. Union Gas* the Supreme Court rejected the State of Pennsylvania's argument that CERCLA made states liable only to the United States, but not to private entities.[99] Three years later,

96. *Id.* at 10. As discussed below, *Pennsylvania v. Union Gas* was overruled on other grounds by the Supreme Court's decision in *Seminole Tribe of Florida v. Florida*, 517 U.S. 44 (1996).

97. CERCLA § 101(20)(D).

98. *Id.* at 7–10.

99. *Union Gas*, 491 U.S. at 11–13.

the Supreme Court overruled this holding in *Seminole Tribe of Florida v. Florida.*[100] In this case, the Supreme Court held that the Commerce Clause of the Constitution does not authorize Congress to abrogate states' sovereign immunity recognized by the Eleventh Amendment.[101] The Eleventh Amendment of the Constitution provides: "The Judicial power of the United States shall not be construed to extend to any suit in law or equity, commenced or prosecuted against any of the United States by citizens or subjects of any foreign state."[102] In *Pennsylvania v. Union Gas*, the Supreme Court held that Congress had the authority under the Commerce Clause to make states subject to suit by private parties, and therefore had authority to subject states to private party actions under section 107 of CERCLA.[103] In *Seminole Tribe*, the court overruled *Union Gas*, relying on earlier decisions that interpreted the Eleventh Amendment differently: "[E]ven when the Constitution vests in Congress complete law-making authority over a particular area, the Eleventh Amendment prevents Congressional authorization of suits by private parties, against unconsenting states."[104] The Supreme Court decision in *Seminole Tribe* thus precludes a private person from filing a CERCLA action against a state in federal district court.

## IV. INDIVIDUAL, PARENT, AND SUCCESSOR LIABILITY

Under general principles of corporate law, individuals and entities, such as stockholders, officers and directors, parent corporations, and successor corporations, are shielded from liability for the actions of their corporations, subsidiary corporations, and predecessor corporations, respectively, unless it is appropriate to ignore the corporate form and "pierce the corporate veil." In Superfund cases, however, these individuals and entities may be held directly liable, without having to pierce the

100. 517 U.S. 44 (1996).
101. *Id.* at 72.
102. U.S. CONST. amend. XI.
103. *Union Gas*, 491 U.S. at 19.
104. *Seminole Tribe*, 517 U.S. at 72.

corporate veil, if they operated the facility that released the hazardous substances.

### A. Individuals and Parent Corporations

One of the more enduring controversies under CERCLA is the circumstances under which a parent corporation and corporate executives, employees and shareholders may be held liable for a release of hazardous substances from a facility owned by a corporate subsidiary. Early on, EPA took the position that corporate shareholders may be held liable not only through application of corporate veil piercing but also directly to the extent a shareholder was an "operator" of the facility.[105] EPA's position spawned a split of authority on the circumstances under which a shareholder or parent could be held directly liable without the need for veil-piercing analysis.

***Veil-Piercing Test.*** One view was that CERCLA could not displace traditional veil-piercing analysis. For example, in *Joslyn Corp. v. T.L. James & Co.*, the court applied a traditional corporate veil analysis in determining whether to impose liability on a parent corporation.[106] The court expressly rejected analyses that chose "to ignore the corporate form without an express congressional directive."[107] The court stated:

> [T]he corporate form, including limited liability for shareholders, is a doctrine firmly entrenched in American jurisprudence that may not be disregarded absent a specific congressional directive. Neither the clear language of CERCLA nor its legislative history provides authority for imposing individual liability on corporate officers or direct liability on parent corporations.[108]

105. *See* EPA Memorandum, Liability of Corporate Shareholders and Successor Corporations (Office of Solid Waste and Emergency Response Directive No. 9832-10 (June 13, 1984)).

106. Joslyn Corp. v. T.L. James & Co., 696 F. Supp. 222 (W.D. La. 1988), *aff'd*, 893 F.2d 80 (5th Cir. 1990).

107. *Id.* at 224–25.

108. *Id.* at 226 (footnote omitted).

The court in *Joslyn* instead applied traditional principles of corporate law and considered, inter alia, whether (1) the parent and the subsidiary have common stock ownership, common directors or officers, or common business departments; (2) the parent and the subsidiary file consolidated financial statements and tax returns; (3) the parent finances the subsidiary; (4) the parent caused the incorporation of the subsidiary; (5) the subsidiary operates with grossly inadequate capital; and (6) the daily operations of the two corporations are not kept separate.[109] Based on the foregoing principles, the court declined to pierce the corporate veil and granted summary judgment in favor of the defendant parent company.[110]

***Authority-to-Control Test.*** Other courts interpreted CERCLA to allow for the direct imposition of liability on individuals and parent corporations, without piercing the corporate veil, based on such entities' authority to control the subsidiary. One of the foundational cases of this authority-to-control test is *United States v. Northeastern Pharmaceutical & Chemical Co.* (*NEPACCO*).[111] In this case, the plant manager, who was also the vice president, was held liable because he had the capacity to control the conduct that implicated CERCLA liability.

The decision in *Idaho v. Bunker Hill Co.*[112] extended the reasoning of *NEPACCO* to parent corporations. In this case, the State of Idaho sought recovery of CERCLA response costs from both the parent (Gulf) and its subsidiary (Bunker Hill). The parent contended that the subsidiary owned and operated the facility. Relying on *NEPACCO*, the court held that anyone who owns an interest in a facility and is actively participating in its management can be held liable under CERCLA as an operator. The court defined an operator as anyone who has the "power to direct the activities of persons who control the mechanisms causing the pollution."[113]

109. *Id.* at 227.

110. *Id.* at 233; *see also In re* Acushnet River & New Bedford Harbor Proceedings Re Alleged PCB Pollution, 675 F. Supp. 22, 31–35 (D. Mass. 1987).

111. 810 F.2d at 743; United States v. Ward, 618 F. Supp. 884 (E.D.N.C. 1985).

112. 635 F. Supp. 665 (D. Idaho 1986).

113. *Id.* at 672 (citation omitted).

In *United States v. Mottolo*,[114] the court found that "the absence of explicit statutory language addressing the effect of incorporation, the Act's strict liability scheme, and the broad and encompassing categories of potentially responsible parties ineluctably led the Court to the conclusion that CERCLA places no importance on the corporate form."

Other courts imposed CERCLA operator liability on parent corporations under the plain "own" or "operate" language of section 107(a)(1) or (2).[115] In another key case imposing personal liability without piercing the corporate veil, the court in *Kelley v. Arco Industries Corp.* held that when a corporate officer's liability is based upon his personal participation in the creation of the hazardous waste site, it is not necessary to pierce the corporate veil to impose personal liability.[116]

Some courts imposing operator liability under the authority-to-control test used more roundabout reasoning to interpret CERCLA as *implicitly* authorizing the imposition of liability on parent corporations and individuals. Section 101(20)(A) of CERCLA defines "owner or operator" to exclude "a person, who, without participating in the management of a . . . facility, holds indicia of ownership primarily to protect his security interest in the . . . facility." In light of this exclusion, some courts have construed the definition as implicitly authorizing the imposition of liability on a person who participates in the management of a facility and who does not hold indicia of ownership solely to protect a security interest.[117]

***Actual Control Test.*** Other courts rejected the authority-to-control test as overbroad, holding that a parent must have actual control over the operations of a facility that released a hazardous substance into the environment. In *United States v. Kayser-Roth Corp.*,[118] the United States brought a CERCLA cost recovery action against the parent Kayser-Roth

114. 695 F. Supp. 615, 624 (D.N.H. 1988).

115. 721 F. Supp. 873 (W.D. Mich. 1989).

116. United States v. Ne. Pharm. & Chem. Co., 579 F. Supp. 823, 848 n.29 (W.D. Mo. 1984), *aff'd in part, rev'd in part*, 810 F.2d 7267 (8th Cir. 1986).

117. *See, e.g.*, New York v. Shore Realty Corp., 759 F.2d 1032 (2d Cir. 1985); Vermont v. Staco, Inc., 684 F. Supp. 822 (D. Vt. 1988), *vacated in part*, Civ. No. 86-190, 1989 WL 225428 (D. Vt. Apr. 20, 1989); Idaho v. Bunker Hill Co., 635 F. Supp. 665.

118. 910 F.2d 24 (1st Cir. 1990).

for cleanup of a mill that was owned by its subsidiary, Stamina Mills, Inc. The court imposed operator liability on the parent, but in so doing it first rejected the authority-to-control test, stating: "To be an operator requires more than merely complete ownership and concomitant general authority or ability to control that comes with ownership."[119] Rather, to impose liability on a parent "requires active involvement in the activities of the subsidiary," which existed in this case because the parent controlled the finances of the subsidiary, including funding of environmental matters.[120]

***Supreme Court Resolves Split of Authority.*** In the 1998 decision *United States v. Bestfoods*,[121] the U.S. Supreme Court resolved this split of authority. The case involved an effort by the United States to recover the costs of cleaning up a chemical manufacturing facility in Michigan that had been owned by Ott Chemical Company. In 1965, the Bestfoods Company (then known as CPC International Inc.) created a subsidiary corporation that purchased Ott Chemical. The subsidiary (also called Ott Chemical) continued to produce chemicals at the Michigan facility until 1972, when Bestfoods sold the subsidiary to another company. The district court held that Bestfoods was liable under CERCLA as an operator, noting that Bestfoods had selected the subsidiary's board and that Bestfoods executives served as the subsidiary's executives.[122] The Sixth Circuit reversed, holding that a parent could be held liable for the actions of a subsidiary only if the state common law rules for piercing the corporate veil were met or if the parent directly participated in the subsidiary's operations.[123]

The Supreme Court vacated the Sixth Circuit decision, ruling that a parent company can be liable under CERCLA for the acts of its subsidiaries under two alternate theories. A parent may be vicariously liable when a basis exists for piercing the corporate veil under traditional

119. *Id.* at 27.

120. *Id.* Other "actual control" cases include *Lansford-Coaldale Joint Water Authority v. Tonolli Corp.*, 4 F.3d 1209 (3d Cir. 1993); *Shiavone v. Pierce*, 79 F.3d 248 (2d Cir. 1996); and *United States v. Cordova Chemical Co.*, 113 F.3d 572 (6th Cir. 1997) (en banc).

121. 524 U.S. 51.

122. CPC Int'l, Inc. v. Aerojet-Gen Corp., 777 F. Supp. 549, 572 (W.D. Mich. 1991).

123. *Cordova Chem. Co.*, 113 F.3d 572.

corporate principles. Alternatively, a parent may be directly liable under CERCLA as an operator of the subsidiary's polluting facility if the parent actually "manage[s], direct[s], or conduct[s] the operations specifically related to pollution, that is, operations having to do with the leakage or disposal of hazardous waste, or decisions about compliance with environmental regulations."[124] Thus, liability could no longer be imposed based solely on a parent's mere capacity to control or authority to control a subsidiary's facility. Nor could even actual control of a facility be sufficient, unless the parent's control extends to the polluting operation. The determinative question is "not whether the parent operates the subsidiary, but rather whether it operates the facility, and that operation is evidence by participation in the activities of the facility, not the subsidiary."[125]

The Court observed that when evaluating the activities of a parent corporation or its officers for evidence that the parent operates a subsidiary's facility, the "critical question is whether, in degree and detail, actions directed to the facility by an agent of the parent alone are eccentric under accepted norms of parental oversight of a subsidiary's facility."[126] The Court provided three examples of what might qualify as "eccentric" parental oversight:

> [T]he parent operates the facility in the stead of its subsidiary or alongside the subsidiary in some sort of a joint venture.
>
> [A] dual officer or director might depart so far from the norms of parental influence exercised through dual officeholding as to serve the parent, even when ostensibly acting on behalf of the subsidiary in operating the facility.
>
> [A]n agent of the parent with no hat to wear but the parent's hat might manage or direct activities at the facility.[127]

124. *Id.* at 66–67.
125. *Id.* at 68 (emphasis added).
126. *Id.* at 72.
127. *Id.* at 71.

In sum, under the Supreme Court's decision in *Bestfoods*, there are two ways that a parent may be held liable as an operator under CERCLA for the cleanup of a facility owned by its subsidiary: (1) the direct liability test wherein it must be shown that the parent actually manages, directs, or conducts operations that released hazardous substances from the subsidiary's facility; and (2) vicarious liability under traditional veil-piercing analysis. The contours of these tests, however, remain open to interpretation, as a survey of post-*Bestfoods* decisions illustrates.

As to the direct liability test, a number of courts have wrestled with the Supreme Court's three examples of parental behavior that would confer operator liability on the parent for managing, directing, or controlling the subsidiary's facility. For example, when does a parent's involvement in a subsidiary facility's operations become eccentric when compared to corporate norms? Post-*Bestfoods* decisions examining this question are all over the lot.[128]

As to the veil-piercing test, a question left unresolved by the Supreme Court is whether state law or federal common law applies to this analysis.[129] Some post-*Bestfoods* courts have applied the law of the forum when analyzing this issue, reasoning that *Bestfoods* "left little room for the creation of a federal rule of liability" and that state law principles of

128. *Compare* BP Amoco Chem. Co. v. Sun Oil Co., 316 F. Supp. 2d 166 (D. Del. 2004) (parental involvement in design of pollution control systems does not demonstrate that parent's actions were unusual under the accepted norms of parental oversight of a subsidiary's facility), *with* Pinal Creek Grp. v. Newmont Mining Corp., 352 F. Supp. 2d 1037, 1042 (D. Ariz. 2005) (evidence of parent's involvement in design of pollution control systems was relevant to the question of "whether the actions of Anaconda's sole agents . . . were consistent with investor oversight, or if they were beyond corporate norms and were eccentric"). *See also* United States v. Friedland, 173 F. Supp. 2d 1077, 1094–95 (D. Colo. 2001) (parental involvement in facility operations, including "pollution and cleanup issues," did not establish that parent managed, directed or controlled operations specifically related to pollution); Schiavone v. Pearce, 77 F. Supp. 2d 284, 290–92 (D. Conn. 1999) (parental involvement in subsidiary facility operations, including environmental operations, deemed typical of what one would expect of a parent, and was insufficient to establish operator liability); Browning-Ferris Indus. of Ill. Inc. v. Ter Maat, 195 F.3d 953, 956 (7th Cir. 1999) (president and principle shareholder could be held directly liable as landfill operator for "negotiating waste-dumping contracts with the owners of the wastes or directing where the wastes were to be dumped or designing or directing measures for preventing toxic substances in the wastes from leeching into the ground and thence into the groundwater").

129. 524 U.S. at 63 n.9.

corporate veil piercing should be followed.[130] Other courts have applied federal common law, emphasizing the importance of having uniform principles for adjudicating federal causes of action under a statute like CERCLA that has nationwide application.[131]

## B. Successors

A bedrock principle of corporate law is that after a consolidation or merger of two corporate entities, the surviving corporation succeeds to the liabilities of the former corporations. Courts have referenced that principle in imposing CERCLA liability on successor corporations for the acts of predecessors.[132]

A closely related corporate law principle is that if a corporation acquires assets of another corporation, the corporation acquiring the assets does not succeed to the liabilities of the selling corporation except where (1) the purchasing corporation expressly or implicitly agreed to assume the seller's liability; (2) when the transaction amounts to a [de facto] consolidation or merger of the purchaser and seller corporations; (3) when the purchaser corporation is merely a continuation of the seller corporation; or (4) when the transaction is entered into fraudulently to escape liability. Courts have cited this same analytical

130. United States v. Davis, 261 F.3d 1, 54 (1st Cir. 2001); *see also* Marsh v. Rosenbloom, 499 F.3d 165, 177 (2d Cir. 2007) ("While *Bestfoods* stops short of expressly instructing courts to apply state law, it clearly admonishes courts to refrain from creating CERCLA-specific rules in the face of applicable longstanding common law principles."); New York v. Nat'l Serv. Indus., Inc., 352 F.3d 682, 685 (2d Cir. 2003) (same); Opinion and Order, Cyprus Amax Minerals Co. v. TCI Pac. Commc'ns, Inc., No. 11-CV-0252-CVE-PJC, (N.D. Okla. Feb. 2, 2015), ECF No. 183 (ruling that under *Bestfoods*, "CERCLA does not replace or supersede state corporation law, and piercing the corporate veil is decided under state law").

131. *See, e.g.*, United States v. Gen. Battery Corp., 423 F.3d 294, 298–305 (3d Cir. 2005) (interpreting *Bestfoods* as being "in favor of a uniform federal standard"); N. Shore Gas Co. v. Salomon, 152 F.3d 642, 650 (7th Cir. 1998) ("resort to federal common law is warranted because of the need for national uniformity with respect to CERCLA").

132. *See, e.g.*, Smith Land & Improvement Corp. v. Celotek Corp., 851 F.2d 86 (3d Cir. 1988), *cert. denied*, 488 U.S. 1029 (1989); B.F. Goodrich v. Betkoski, 99 F.3d 505 (1996), *reh'g denied*, 112 F.3d 88 (2d Cir. 1997), *overruled on other grounds*, New York v. Nat'l Serv. Indus., Inc. 352 F.3d 682, 685 (2d Cir. 2003); United States v. Carolina Transformer Co., 978 F.2d 832 (4th Cir. 1992); United States v. Mexico Feed & Seed Co., 980 F.2d 478 (8th Cir. 1992); La.-Pac. Corp. v. Asarco, Inc., 909 F.2d 1260 (9th Cir. 1990); *but see* Anspec Co. v. Johnson Controls, Inc., 922 F.2d 1240 (6th Cir. 1991) (directing the district court to "follow Michigan law in it application of successor liability").

framework in deciding whether an acquiring corporation is subject to CERCLA liability for the acts of the selling corporation, usually based on federal common law.[133]

In a few states, courts have recognized a fifth exception to the general principle that a corporation acquiring assets does not acquire liabilities: the "substantial continuity test." The substantial continuity test broadens the "mere continuation" exception by making a purchaser of assets liable based on continuity of the business without continuity of ownership. It focuses on whether the successor maintained the same employees, supervisory personnel, production facilities, products, name, and overall operations. Prior to *Bestfoods*, several federal courts adopted this test as a special standard for determining successor liability under CERCLA. All of the federal circuit courts that have addressed the issue since *Bestfoods*, however, have rejected the substantial continuity test as untenable as a basis for successor liability under CERCLA.[134]

## V. LENDER LIABILITY

A party who has loaned money to the owner of a facility can itself, under certain circumstances, be deemed the owner or operator of the facility. As defined by section 101(20)(A) of CERCLA, the term "owner or operator . . . does not include a person, who, without participating in the management of a facility, holds indicia of ownership primarily to protect his security interest in the vessel or facility." While this language clearly exempts a secured lender who does nothing other than receive and hold a security interest from the definition of "owner" or "operator," it provides little guidance on what constitutes "participating in the management of a facility," the ramifications of foreclosure on a contaminated property, or the steps that a secured lender can take to "protect his security interest."

133. *In re* Acushnet River & New Bedford Harbor Proceedings Re Alleged PCB Pollution, 712 F. Supp. 1010, 1014 (D. Mass. 1989) (quoting Dayton v. Peck, Stow & Wilcox Co., 739 F.2d 690, 692 (1st Cir. 1984)); United States v. Vertac Chem. Corp., 671 F. Supp. 595, 614–15 (E.D. Ark. 1987), *vacated on other grounds*, 855 F.2d 856 (8th Cir. 1988).

134. Action Mfg. Co. v. Simon Wrecking Co., 387 F. Supp. 2d 439, 452 (E.D. Pa. 2005) (collecting cases). *See, e.g.*, United States v. Gen. Battery Corp., 423 F.3d 294, 309 (3d Cir. 2005); New York v. Nat'l Servs. Indus., Inc., 352 F.3d 682, 687 (2d Cir. 2003).

This ambiguity was the subject of extensive litigation, which led to a split of authority before Congress amended the statute in 1996 as discussed below. The first significant case, *United States v. Mirabile*,[135] determined that a secured lender faced potential CERCLA liability as an operator if it immersed itself in the day-to-day operational management of the borrower's facility. Participation in the purely financial aspects of the borrower's facility—such as monitoring the borrower's cash collateral accounts, ensuring that the borrower's receivables were sent to the proper account, and establishing a reporting system between the secured lender and the borrower—was viewed by the *Mirabile* court as being insufficient for establishing liability as an operator under CERCLA. Furthermore, the *Mirabile* court declined to impose CERCLA owner or operator liability on a secured lender that foreclosed on a real property security interest when the secured lender's activities were undertaken for the sole purpose of protecting its security interest.[136]

Other courts, however, held that lenders become owners upon foreclosure.[137] In 1990, the Eleventh Circuit, however, dropped a bomb on the lender community when it ruled in *United States v. Fleet Factors Corp.* that mere ability to exercise control, regardless of whether such control was in fact exercised, constituted "participating in management" of a facility, and subjected lenders to owner/operator liability.[138]

In 1992, EPA promulgated regulations that effectively reversed the *Fleet Factors* decision by declaring that mere ability to control management would not subject a lender to CERCLA liability.[139] EPA's lender liability rule, however, was vacated by the D.C. Circuit in *Kelley v. EPA*.[140]

In 1996, Congress stepped in, passing the Asset Conservation, Lender Liability, and Deposit Insurance Protection Act amending CERCLA.[141]

---

135. Civ. A. No. 84-2280, 1985 WL 97 (E.D. Pa. Sept. 6, 1985).

136. *Id.*

137. *See, e.g.*, Guidice v. BFG Electroplating & Mfg. Co., 732 F. Supp. 556 (W.D. Pa. 1989) (holding bank liable under CERCLA as owner after foreclosing on contaminated property).

138. *See* United States v. Fleet Factors Corp., 901 F.2d 1550 (11th Cir. 1990), *cert. denied*, 498 U.S. 1046 (1991).

139. Nat'l Oil & Hazardous Substances Pollution Contingency Plan; Lender Liab. under CERCLA, 57 Fed. Reg. 18,344 (Apr. 29, 1997).

140. 15 F.3d 1100 (D.C. Cir. 1994).

141. Pub. L. No. 104-208, 110 Stat. 3009-462 (1996).

The 1996 amendments essentially codified EPA's vacated lender liability rules. Under the 1996 amendments, the exemption from the term "owner or operator" extends to any party "that is a lender" that does not "participate in management."[142] The amendments then clarify what constitutes participating in management: mere capacity to influence a facility or an "unexercised right to control" a facility does not constitute "participating in management." Rather, a lender will be deemed to have participated in management if it (1) undertakes decision-making control or responsibility for the facility's hazardous substance handling or disposal practices; or (2) exercises control at the level of a manager over (a) day-to-day decision making with respect to environmental matters; or (b) the "operational functions" of the facility other than the functions of environmental compliance.[143] In addition, as amended, CERCLA expressly identifies certain actions that will not qualify as "participation in management," including

- holding, abandoning or releasing a security interest;
- including environmental covenants or warranties in a credit or security agreement;
- monitoring or enforcing, or altering, restructuring, or renegotiating, the terms and conditions of the extension credit or security agreement;
- conducting one or more inspections of the facility;
- requiring a response action "or other lawful means of addressing the release or threatened release of a hazardous substance";
- providing financial or other advice or counseling seeking to mitigate, prevent or cure default or diminution in value; and
- exercising other legal remedies for the breach of a credit or security agreement.

Not surprisingly, Congress's attempt to eliminate the ambiguities over lender liability did not entirely eliminate litigation over the issue.[144]

142. CERCLA § 101(20)(E)(i).
143. CERCLA § 101(20)(E)(ii).
144. *See, e.g.*, United States v. Pesses, No. Civ. A. 90-0654, 1998 WL 937235 (W.D. Pa. May 6, 1998) (small savings and loan defendant entitled to secured creditor exemption because

But having clarified the conditions for qualifying as an exempt secured creditor under CERCLA, such litigation is diminishing.

## VI. CAUSATION

To establish that a defendant is liable under CERCLA, the plaintiff need not demonstrate that a defendant caused the release that resulted in the plaintiff's incurrence of response costs.[145] Rather, under the express language of section 107(a) of CERCLA the plaintiff need only prove that (1) there has been a release or threatened release covered by CERCLA, (2) the release or threatened release "cause[d] the incurrence of response costs," (3) the response costs were "not inconsistent" with the National Contingency Plan (NCP) (or, in the case of a private plaintiff, "necessary" and "consistent" with the NCP) and (4) the defendant is a PRP as defined by CERCLA.[146] If the plaintiff succeeds in proving these elements by a preponderance of the evidence, a rebuttable presumption of liability arises, and the burden shifts to the defendant to assert an affirmative defense under section 107(b) of CERCLA by disproving causation.[147]

As the foregoing makes clear, CERCLA does not require a plaintiff to prove causation in the traditional sense—e.g., that the defendant's actions somehow caused the plaintiff to incur cleanup costs. Rather, a plaintiff need only (1) link the defendant to the site (e.g., based on a defendant's status as current or prior owner-operator or transporter or its having arranged for disposal of wastes that were disposed of at the site, and (2) demonstrate that its response costs were "caused" by the

---

it did not participate in management of the facility, had no influence over tenant's waste disposal practices and made reasonable efforts to sell, re-lease, or otherwise divest itself of the facility).

145. *See, e.g.*, United States v. Alcan Aluminum Corp., 990 F.2d 711 (2d Cir. 1993), *aff'd*, 315 F.3d 179 (2d Cir. 2003); Elf Atochem N. Am., Inc. v. United States, 833 F. Supp. 488 (E.D. Pa. 1993) (plaintiff need not link a specific defendant with a specific release to establish CERCLA liability); Premium Plastics v. LaSalle Nat'l Bank, 904 F. Supp. 809 (N.D. Ill. 1995) (private plaintiff need not prove that defendant caused the particular releases that caused plaintiff to incur response costs).

146. *See* 42 U.S.C. § 9607(a); *see also Alcan Aluminum Corp.*, 990 F.2d 711 at 721; Amoco Oil Co. v. Borden, Inc., 889 F.2d 664, 668 (5th Cir. 1989); Dedham Water Co. v. Cumberland Farms Dairy, Inc., 889 F.2d 1146, 1150 (1st Cir. 1989), *aff'd*, 972 F.2d 453 (1st Cir. 1992).

147. *See Amoco*, 889 F.2d at 668; United States v. Monsanto Co., 858 F.2d 160, 170 (4th Cir. 1988).

release or threatened release of hazardous substances (not necessarily the defendant's specific hazardous substances).[148]

One of the first cases to address causation under CERCLA is *United States v. Wade*.[149] In this case, the court held that it was not necessary for the government to identify a generator's specific waste at a site. Rather, it was sufficient for the government to establish that the generator had sent waste to the site, that wastes of the same type were found there, and that hazardous wastes of *any* type had been released from the site. The defendant's wastes need not have been released or even threatened to be released, need not comprise any substantial portion of the total volume of wastes at the site, and need not require any response action for the generator to be held liable for all or a portion of the response costs. As the court noted, "to require a plaintiff under CERCLA to 'fingerprint' wastes is to eviscerate the statute."[150]

Indeed, a generator may be held liable even if the plaintiff cannot prove that a specific substance generated by the defendant and sent to a site was present there at the time of release; a showing of chemical similarity between that substance and released substances is sufficient.[151]

The issue of causation can become more complex in the context of two-facility sites, where a release or threat of release at one facility gives rise to response costs at another facility, typically an adjoining property. In these situations, the only causation that need be proven is that the response costs of the plaintiff (e.g., adjoining property owner) were incurred as a reasonable reaction to the release or threatened release from the defendant's facility; there is no requirement to prove actual

148. *See, e.g.*, Acushnet v. Mohasco Corp., 191 F.3d 69, 77 (1st Cir. 1999) ("To satisfy causal element, it is usually enough to show that a defendant was a responsible party within the meaning of [section 107(a)]; that cleanup efforts were undertaken . . . and that reasonable costs were expended during the operation."); United States v. Rohm & Haas Co., 939 F. Supp. 1142, 1150 (D.N.J. 1996) ("[F]inding a causal connection between a [defendant's] wastes and USA-EPA's costs is not required by statute"). *See also* United States v. P.H. Glatfelter Co. & NCR Corp., 768 F.3d 662, 674–75 (7th Cir. 2014) (upstream discharger of PCBs held liable for cleanup of entire river, including downstream operable units, as "nothing in the statute limits the defendant's liability to the response costs its release caused").

149. 577 F. Supp. 1326 (E.D. Pa. 1983).

150. *Id.* at 1332.

151. *Monsanto Co.*, 858 F.2d at 169.

contribution to contamination at the plaintiff's site.[152] As the court in *Dedham Water Co. v. Cumberland Farms Dairy Inc.* noted:

> To our knowledge, every court that has addressed this issue, with the exception of the district court in the instant case, has held that it is not necessary to prove actual contamination of plaintiff's property by defendant's waste in order to establish liability under CERCLA. There is nothing in the statute, its legislative history, or the case law, which requires proof that the defendant's hazardous waste actually have migrated to plaintiff's property . . . before CERCLA liability is triggered.[153]

Thus, the *Dedham* court held that the plaintiff in a CERCLA cost recovery action need only demonstrate the possibility that hazardous substances migrated from one site to the other to complete the site nexus link.[154]

## VII. DEFENSES TO LIABILITY

### A. Statutory Defenses

It is a common misconception that pursuant to section 107(b) of CERCLA there are but three exclusive statutory defenses to liability, namely, that the release was caused solely by (1) an act of God; (2) an act of war; or (3) an act or omission of a third party other than an employee or agent or one in a contractual relationship with the defendant.[155] This misconception is understandable since the text of section 107(a) begins as follows: "Notwithstanding any other provision of law, and subject

152. *See Dedham Water Co.*, 889 F.2d 1146; Artesian Water Co. v. New Castle Cnty., 851 F.2d 643 (3d Cir. 1988).

153. 889 F.2d at 1154.

154. *See also* Castaic Lake Water Agency v. Whittaker Corp., 272 F. Supp. 1053 (C.D. Cal. 2003) (plaintiff, who alleged that its wells were contaminated by perchlorate migrating from another contaminated site, could sustain its burden by merely providing "evidence of a plausible migration pathway by which the contaminant could have travelled from the defendant's facility to the plaintiff's site").

155. CERCLA § 107(b)(1)–(3).

only to the defenses set forth in [section 107(b)]. . . ." In fact there are far more defenses to liability than those three defenses, as discussed below.

### 1. *Act of God*

CERCLA defines an "act of God" as follows:

> [A]n unanticipated grave natural disaster or other natural phenomenon of an exceptional, inevitable, and irresistible character, the effects of which could not have been prevented or avoided by the exercise of due care or foresight.[156]

The "act of God" defense will rarely present a viable defense to liability under section 107. The only case substantively addressing the "act of God" defense held that heavy rainfalls were not the "kind of exceptional natural phenomena" to which the defense applies because the rains were foreseeable, the harm could have been prevented through design of proper drainage channels, and the rains were not the sole cause of the release.[157]

The act of God defense was raised but not ruled on in *Wagner Seed Co. v. Bush*.[158] In this case, a defendant, after conducting a cleanup pursuant to EPA orders, made a claim against the Fund under section 106(b)(2). The defendant claimed, under the act of God defense, that it was not liable for the release because lightning struck the defendant's warehouse, which contained pesticides, burning it down and releasing various hazardous substances into the environment. The court, however, never addressed this defense because it held that section 106(b)(2) did not apply retroactively to defendant's cleanup and, therefore, defendant could not make a claim against the Fund.

156. CERCLA § 101(1).
157. United States v. Stringfellow, 661 F. Supp. 1053, 1061 (C.D. Cal. 1987).
158. 709 F. Supp. 249 (D.D.C. 1989).

### 2. *Act of War*

Unlike an act of God, CERCLA does not define "act of war." But like the act of God defense, courts have mostly rejected the act of war defense, typically with little explanation. For example, in *United States v. Shell Oil Co.*,[159] the government sought to recover response costs from defendants who had dumped wastes associated with products of aviation fuel during World War II. The defendants argued that aviation fuel had been deemed "critical to the war effort" and that all aspects of production, including waste disposal, were subject to significant control and that such disposal should be covered by the act of war defense. The Ninth Circuit rejected the act of war defense.[160] Given the lack of a CERCLA definition of "act of war," the court looked to "act of war" definitions from international law, as well as cases outside the CERCLA context, all of which narrowly construed the phrase as forceful actions by one state against another featuring "massive violence" or "catastrophies" beyond a party's control. Thus, the court concluded that the defendant's disposal could not be construed as resulting from or caused by an act of war.[161]

The lone case in which the act of war defense was sustained is *In re September 11 Litigation*.[162] In that case, a real estate developer sued the owners and lessees of the World Trade Center under CERCLA for costs incurred in remediating a building contaminated with dust containing hazardous substances, following the September 11 attack. In sustaining the defense, the court noted that CERCLA's remedial goals were "not advanced here by imposing CERCLA liability on the airlines and the owners (and lessors) of the real estate."[163] The court concluded that the "sole responsibility" for the event and its environmental consequences

159. 294 F.3d 1045.

160. *Id.* at 1061.

161. *See also* Coeur d'Alene Tribe v. Asarco Inc., 280 F. Supp. 2d 1094, 1135 (D. Idaho 2003) (court rejected PRP act of war defense, stating that it was not relevant and in any event was not the "sole cause" of releases).

162. 751 F.3d 86 (2d Cir. 2014).

163. *Id.* at 91.

rested with the terrorists "whose acts the defendants were not bound by CERCLA to anticipate or prevent."[164]

### 3. *Third-Party Defense, Innocent Landowners, and Innocent Purchasers*

The third-party defense in section 107(b)(3) is the most heavily used—and litigated—statutory defense to liability. This section provides that a party will not be liable for the release or threat of a release of a hazardous substance, and any resulting damages, caused solely by the act or omission of a third party other than an employee, agent, or party with whom the defendant has a contractual relationship, if the defendant establishes by a preponderance of the evidence that:

> (A) he exercised due care with respect to the hazardous substance concerned, taking into consideration the characteristics of such hazardous substance, in light of all relevant facts and circumstances, and
>
> (B) he took precautions against foreseeable acts or omissions of any such third party and the consequences that could foreseeably result from such acts or omissions. . . .[165]

To prevail on this defense, a defendant must prove that a totally unrelated "third party was the sole cause of the release."[166] For example, a landlord could not assert the third-party defense for contamination caused by a tenant because the lease signed by the landlord and tenant is a contractual relationship within the meaning of section 107(b)(3) of CERCLA and thus the parties are related by that contract.[167] The classic situation for which the third-party defense was intended is a landowner whose property becomes contaminated because of the actions

---

164. *Id.*

165. CERCLA § 107(b)(3).

166. *See* G.J. Leasing Co. v. Union Elec. Co., 854 F. Supp. 539, 566 (S.D. Ill. 1994), *aff'd*, 54 F.3d 379 (7th Cir. 1995); New York v. Lashins Arcade Co., 856 F. Supp. 153 (S.D.N.Y. 1994), *aff'd*, 91 F.3d 353 (2d Cir. 1996).

167. *See, e.g.*, United States v. Monsanto Co., 858 F.2d 160 (4th Cir. 1988); Int'l Clinical Labs., Inc. v. Stevens, 710 F. Supp. 466 (E.D.N.Y. 1989); United States v. Northernaire Plating Co., 670 F. Supp. 742 (W.D. Mich. 1987).

of a trespasser—the "midnight dumper" who disposes of hazardous substances on the landowner's property without permission. If the landowner, after learning of the contamination, took reasonable measures to address the contamination it should in theory be eligible for the third-party defense. Notably, however, the author is unaware of any case in which this fact pattern has arisen and the third-party defense was successfully asserted.

Rather, since the passage of CERCLA, the most frequent fact pattern in which PRPs have sought to apply the third-party defense is where a party purchases property without knowing that it was contaminated. It was unclear whether Congress intended the third party to apply in this situation. Several district courts said yes.[168] One circuit court, however, said no.[169] Faced with confusion over whether the third-party defense could apply to an innocent purchaser, Congress in the 1986 amendments clarified that a contract for sale of land or a deed establishes a contractual relationship and, therefore, that persons who purchase contaminated land cannot qualify for the third-party defense.[170] At the same time, however, Congress established an "innocent purchaser" defense, pursuant to which a party that acquires contaminated property without knowing of its contamination—and without reason to know of such contamination—may have a defense to liability.[171]

Specifically, a defendant will be considered an innocent purchaser only if (1) the real property was acquired after the disposal or placement of the hazardous substance at the facility; and (2) at the time the facility was acquired, the defendant did not know and had no reason to know that any hazardous substances had been disposed of at the facility.[172] To meet the "no reason to know" condition, a defendant must show that it

168. United States v. Mirable, Civ. A. No. 84-2280, 1985 WL 97, at *16–17 (E.D. Pa. Sept. 6, 1985); United States v. Md. Bank & Trust Co., 632 F. Supp. 573, 581 (D. Md. 1986); United States. v. Tyson, Civ. A. No. 84-2663, 1986 WL 9250 at *9, *11–12 (E.D. Pa. Aug. 22, 1986).

169. New York v. Shore Realty Corp., 759 F.2d 1032, 1048 (2d Cir. 1985) ("the acts or omissions referred to in the statute are doubtless those occurring *during* the ownership or operation of the defendant") (emphasis added).

170. CERCLA § 101(35)(A).

171. *Id.*

172. *Id.*

had "undertaken, at the time of acquisition, all appropriate inquiry into the previous ownership and uses of the property consistent with good commercial or customary practice in an effort to minimize liability."[173] As discussed below, in 2005, EPA promulgated a regulatory interpretation of actions that constitute "all appropriate inquiries," which created a "safe harbor" for real estate investors. Prior to that point, however, parties were left to guess as to the level of investigation that constituted "all appropriate inquiries," and, not surprisingly, they often guessed wrong.

The cases rejecting the innocent purchaser defense are legion.[174] It has been accepted, however, on several occasions.[175]

---

173. CERCLA § 101(35)(B) (emphasis added). Under this section, lack of knowledge may also be established if (1) the defendant is a government entity and acquired the facility by escheat, involuntary transfer, or eminent domain; or (2) the facility was acquired by inheritance or bequest.

174. *See, e.g.*, Hidden Lakes Dev., LP v. Allina Health Sys., No. 02-406, 2004 WL 2203406 (D. Minn. Sept. 27, 2004) (current property owner ineligible for innocent purchaser defense unless it can show it did not know of contamination at time of purchase); Franklin Cnty. Convention Facilities Auth. v. Am. Premier Underwriters, Inc., 240 F.3d 534, 548 (6th Cir. 2001) (court rejected innocent purchaser defense, even though the contamination predated the owner's acquisition of the land, the owner had no way of knowing about the contamination, and the release was accidental and not the owner's fault, because the owner "failed to exercise due care after discovering" the contamination); Kaladish v. Uniroyal Holding, Inc., No. Civ. A. 300-CV-854, 2005 WL 2001174, at *4 (D. Conn. Aug. 9, 2005) (innocent purchaser defense denied where purchaser "admitted that he made no inquiries regarding the property before purchasing it; did not examine any court records pertaining to the property before purchase; and either failed to perform a title search or cannot remember conducting one before the purchase"); New York State Elec. & Gas Corp. v. FirstEnergy Corp., 808 F. Supp. 2d 417, 518 (N.D.N.Y. 2011), *aff'd in part, vacated in part*, 766 F.3d 212 (2d Cir. 2014) (innocent purchaser defense rejected for failure to exercise due care because it unnecessarily delayed remediation); Containerport Grp., Inc. v. Am. Fin. Grp., Inc., 128 F. Supp. 2d 470, 480 (S.D. Ohio 2001) (innocent purchaser defense rejected where proponent had "done nothing to secure the site or made any effort to clean up the allegedly hazardous substances" after learning of the contamination); United States v. A&N Cleaners & Launderers, Inc., 854 F. Supp. 229, 243 (S.D.N.Y. 1994) (innocent purchaser defense rejected because news reports of groundwater contamination gave defendants constructive knowledge of contamination); BCW Assocs. Ltd. v. Occidental Chem. Corp., Civ. No. 86-5947, 1988 WL 102641 (E.D. Pa. Sept. 29, 1988) (court rejected the defense despite the fact the defendants had hired consultants to conduct site evaluations and that these consultants had concluded that the facility did not pose environmental concerns).

175. *See* Redwing Carriers, Inc. v. Saraland Apartments, Ltd., 875 F. Supp. 1545 (S.D. Ala. 1995), *aff'd in part and rev'd in part*, 94 F.3d 1489 (11th Cir. 1996) (defendants that owned property identified on the NPL were found to have exercised due care when tar began to rise to the surface of the property and they notified EPA and the state environmental agency); State of New York v. Lashins Arcade Co., 91 F.3d 353, 360 (2d Cir. 1996) (affirming innocent purchaser defense, focusing on "contractual relationship" element, concluding that

#### 4. *Bona Fide Prospective Purchaser Defense*

The innocent purchaser defense does not provide protection to parties who acquire lightly contaminated property (colloquially referred to as "brownfields") who are prepared to invest in cleanup and redevelopment. This had the unintended effect of spurring the development of greenfields while lightly contaminated brownfields would be "mothballed" by owners and thereby exacerbating urban blight.

In 2002, Congress sought to eliminate this impediment to the acquisition and development of contaminated property by passing the Small Business Liability Relief and Brownfields Revitalization Act.[176] This law established the bona fide prospective purchaser defense to CERCLA liability. A bona fide prospective purchaser is a person who acquires ownership of a facility after January 11, 2002, and who establishes the following by a preponderance of the evidence:

- The hazardous substances disposal occurred before the purchaser acquired the property;
- The purchaser made "all appropriate inquiries" (described below) into the previous ownership and uses of the facility;
- The purchaser complies with all continuing obligations after acquiring the property; and
- The purchaser has no affiliation with any potentially responsible party.

A party may purchase property with knowledge of the contamination after performing all appropriate inquiries and still qualify for the bona fide prospective purchaser defense, provided that it meets the other criteria set forth in section 101(40) of CERCLA.

---

"straightforward sale" did not "relate to hazardous substances" or give purchaser ability to "exert some element of control" over third party's hazardous substance activities); United States v. Serafini, 706 F. Supp. 346, 353 (M.D. Pa. 1988) (partners who in 1969 acquired property that had 55-gallon drums in plain sight granted innocent purchaser defense, notwithstanding their failure to inspect property, because "government has presented no evidence from which the court can conclude that the defendants' failure to inspect or inquire was inconsistent with good commercial or customary practices").

176. Pub. L. No. 107-118, 115 Stat. 2356 (2002).

"All appropriate inquiries" refers to the process of evaluating a property's environmental conditions and assessing potential liability for contamination. The 2002 Brownfields Amendments require the EPA to promulgate regulations establishing standards and practices for conducting all appropriate inquiries into the previous ownership and uses of a property necessary to qualify for certain landowner liability protections. In 2005, EPA promulgated regulations establishing a standard for innocent landowners and bona fide prospective purchasers to demonstrate that they have conducted all appropriate inquiries.[177] Under the 2005 regulations, a party seeking to demonstrate that it has made all appropriate inquiries must comply with the terms of the American Society for Testing and Materials (ASTM) E1527-05 standard (titled Standard Practice for Environmental Site Assessments: Phase I Environmental Site Assessment Process). In 2013, EPA revised the "all appropriate inquiry" regulation by replacing ASTM E1527-05 with an updated version, namely, ASTM E-1527-13.[178]

Even with extensive guidance from EPA, the bona fide prospective purchaser defense can be elusive. Consider the fate of a developer who purchased a brownfield property in Charleston, South Carolina in 2003.[179] The property had been used for phosphate fertilizer production for decades, and consequently was contaminated with arsenic, lead, and other heavy metals. To perfect the bona fide perspective purchaser defense, the developer conducted an ASTM Phase I due diligence review prior to acquiring the property. The developer also retained an environmental consultant to assure compliance with its "continuing obligations," and worked with EPA to develop a remedial plan for the contamination. The developer's efforts, however, fell short in several respects. The developer had allowed a capped area of the site to deteriorate. In addition, the developer's demolition of certain buildings inadvertently exposed sumps that had been identified as potential sources of contamination during the Phase I review. In a subsequent lawsuit over cleanup costs,

177. 70 Fed. Reg. 66,070 (Nov. 1, 2005) (codified at 40 C.F.R. pt. 312).
178. 78 Fed. Reg. 79,319 (Dec. 30, 2013).
179. *Ashley II of Charleston*, LLC, 791 F. Supp. 2d 431.

these shortcomings were sufficient to deprive the developer of a bona fide prospective purchaser defense.[180]

### *5. Exempt Releases of Hazardous Substances*

CERCLA liability does not attach to the following types of releases of hazardous substances.

**Pesticide Application.** The application of pesticide products is exempt from CERCLA liability under section 107(1), which provides:

> No person (including the United States or any State or Indian tribe) may recover under the authority of [section 107] for response costs or damages resulting from the application of a pesticide product registered under the Federal Insecticide, Fungicide, and Rodenticide Act [FIFRA].

The legislative history of this provision clarifies that Congress intended the pesticide application exemption to apply to the "use of a pesticide generally in accordance with its purpose."[181] Congress added, however, that failure to comply with FIFRA requirements should not "be a basis for invoking [CERCLA] liability for pesticide application." *Id.* Thus, it appears that strict conformance to FIFRA pesticide labeling is *not* a prerequisite to qualifying for the pesticide application exemption.[182]

---

180. Other cases denying the bona fide prospective purchaser defense include *Voggenthaler v. Maryland Square LLC*, 724 F.3d 1050 (9th Cir. 2013) (buyer's hiring of an advisor to review state files and/or environmental consultant to demolish on-site structures is not sufficient to qualify for bona fide purchaser defense because buyer took no steps to remove contaminated soil); *Saline River Props. LLC v. Johnson Controls, Inc.*, 823 F. Supp. 2d 670 (E.D. Mich. 2011) (denying summary judgment to plaintiff because he failed to establish all elements of the defense and facts were in dispute). One reported case in which the bona fide prospective purchaser defense succeeded is *3000 E. Imperial, LLC. v. Robertshaw Controls Co.*, No. CV 08-3985, 2010 WL 5464296 (C.D. Cal. Dec. 29, 2010) (holding it was not unreasonable for Imperial to wait two years before excavating underground storage tanks because it had taken reasonable steps to stop any continuing leaks and prevent future releases from the tanks).

181. S. REP. No. 848, 96th Cong., 2d Sess. 45 (July 13, 1980).

182. See chapter 11 for discussion of the pesticide application exemption in the release reporting context.

Nevertheless, this exemption is extremely narrow, and one of the few reported cases involving an attempt to invoke the defense failed.[183]

**Federally Permitted Releases.** Also exempt from the reach of CERCLA's liability provisions are federally permitted releases. Section 107(j) provides:

> Recovery by any person (including the United States or any State or Indian tribe) for response costs or damages resulting from a federally permitted release shall be pursuant to existing law in lieu of [section 107].

Section 101(10) defines "federally permitted release(s)" to include releases subject to permit and other controls under specified federal environmental laws.[184]

**Petroleum Exclusion.** The CERCLA definition of hazardous substance expressly excludes "petroleum, crude oil or any fraction thereof which is not otherwise specifically listed or designated as a hazardous substance . . . natural gas, natural gas liquids or synthetic gas usable for fuel (or mixtures of natural gas and such synthetic gas)."[185] Thus, releases of petroleum, natural gas, or synthetic gas are not subject to liability under CERCLA. The author is unaware of any judicial decision addressing the exemption from liability for releases of natural gas, natural gas liquids, or synthetic gas usable for fuel. By contrast, there has been an enormous amount of litigation over the contours of the petroleum exclusion, such as the question of whether the exemption applies to used petroleum products, and if so under what circumstances. This case law is discussed in detail in chapter 1, and will not be repeated here.

**Consumer Product in Consumer Use.** As noted in chapter 1, EPA's response authority extends to a release of a hazardous substance from

183. *See, e.g.*, United States v. La.-Pac. Corp., No. Civ. S-89-871, 1994 U.S. Dist. LEXIS 12303, at *27 (E.D. Cal. Oct. 7, 1994) (spills of a registered pesticide incidental to its lawful application are not excluded releases).

184. See chapter 11 for a discussion of the contours of this exemption. See also chapter 10 for a discussion of the application of the federally permitted release exemption to natural resource damages.

185. *See* CERCLA § 101(14).

a facility into the environment. The statutory definition of facility "does not include any consumer product in consumer use."[186] Thus, a release of a hazardous substance from a consumer product in consumer use in theory should not create liability under CERCLA. Case law, however, has diverged widely on the applicability of this defense.[187]

### 6. *Recycling Exemption*

A new statutory exemption from liability was established in 1999 when Congress passed and the president signed into law the Superfund Recycling Equity Act (SREA).[188] The act amended CERCLA by adding a limited exemption from liability for recyclers of hazardous materials that meet certain requirements.

The recycling exemption, which is set forth in new section 127 of CERCLA, exempts certain persons who "arranged for recycling of recyclable materials" from liability under section 107(a)(3) and 107(a)(4). Owners and operators of facilities from which there have been releases are not eligible for the exemption. "Recyclable materials" is defined as

186. *See* CERCLA § 101(9).

187. Cases supporting application of the defense include *Knox v. AC & S, Inc.*, 690 F. Supp. 752, 756 (S.D. Ind. 1988) (asbestos-containing insulation, sold between businesses, could be considered a consumer product); *Electric Power Board of Chattanooga*, 716 F. Supp. 1069, 1080 (E.D. Tenn. 1988), *aff'd*, 879 F.2d 1368 (6th Cir. 1989) (holding that electrical transformers that leaked dielectric cooling fluid containing polychlorinated biphenyl are consumer products in consumer use); *Vernon Village, Inc. v. Gottier*, 755 F. Supp. 1142, 1150 (D. Conn. 1990) (holding that contaminated drinking water is a consumer product in consumer use based on the apparent reasoning that water is a useful product).
*Contra Uniroyal Chemical Co. v. Deltech Corp.*, 160 F.3d 238 (5th Cir. 1998) (tanker truck and trucking terminal are not covered by the "consumer product in consumer use" exemption based on commonsense interpretation of CERCLA § 101(9)); *United States v. M/V Santa Clara I*, 887 F. Supp. 825, 842 (D.S.C. 1995) (holding that consumer product exception not applicable in case where shipping containers carrying barrels of arsenic trioxide were lost from vessel in heavy seas); *KN Energy, Inc. v. Rockwell Int'l Corp.*, 840 F. Supp. 95, 99 (D. Colo. 1993) (holding that pipelines sealed with substance containing PCBs were commercial facilities, not consumer products in consumer use); *Reading Co. v. City of Philadelphia*, 823 F. Supp. 1218, 1232–34 (E.D. Pa. 1993) (holding that railcars that leaked PCBs while used in commuter train service not consumer products in consumer use); *CP Holdings, Inc. v. Goldberg-Zoino & Assocs., Inc.*, 769 F. Supp. 432, 438 (D.N.H. 1991) (holding that commercial hotel built with asbestos-containing materials not consumer product in consumer use); *see also R.R. Passenger Corp. v. New York City Housing Authority*, 819 F. Supp. 1271, 1276 (S.D.N.Y. 1993) (support pillars and undersides of building coated with asbestos are not consumer products in consumer use).

188. Pub. L. No. 106-113, 113 Stat. 1501 (1999).

"scrap paper, scrap plastic, scrap glass, scrap textiles, scrap rubber (other than whole tires), scrap metal, or spent lead-acid, spent nickel-cadmium and other spent batteries, as well as minor amounts of material incident to or adhering to the scrap material as a result of its normal and customary uses prior to becoming scrap."[189] To qualify for the exemption, the person seeking exempt status must demonstrate, by a preponderance of evidence, that the following conditions existed at the time of the recycling transaction:

- The recyclable material met a commercial specification grade;
- A market existed for the material;
- A substantial portion of the recyclable material was made available for use as feedstock for the manufacture of a new saleable product; and
- For transactions occurring after February 27, 2000, the person exercised "reasonable care" to determine that the recycling facility was in compliance with applicable law.[190]

"Reasonable care" is to be determined by the price paid in the recycling transaction, the ability of the person to detect the nature of the recycling facility's operations concerning its management of the recyclable material, and the result of inquiries made to appropriate federal, state, or local environmental agencies regarding the facility's environmental compliance history.[191] Transactions involving scrap metal or spent batteries are subject to additional criteria.[192]

189. CERCLA § 127(b).
190. CERCLA § 127(c).
191. CERCLA § 127(c)(6).
192. The additional conditions applicable to scrap metal transactions are set forth in CERCLA § 127(d). In brief, the person claiming the exemption must also show that it was in compliance with applicable provisions of RCRA, and that the person did not melt the metal before the transaction. The additional conditions for spent battery recycling are found in CERCLA § 127(e). The person claiming the exemption for spent batteries must show that it "did not recover the valuable components of such batteries." For lead-acid batteries, the person must also demonstrate compliance with applicable federal environmental standards for recycling lead-acid batteries. For all other types of batteries, the person must demonstrate both the existence of recycling standards and compliance with them. *See* CERCLA § 127(e)(1), (2).

The recycling exemption does not apply where there is an "objectively reasonable basis" to believe at the time of the transaction that:

- Recyclable material would not be recycled;
- Recyclable materials would be burned as fuel, or for energy recovery or incineration;
- For transactions occurring before February 2000, the consuming (recycling) facility was not in compliance with substantive environmental requirements;
- Hazardous substances had been added to the recyclable material for purposes other than recycling; or
- The person failed to exercise "reasonable care" with respect to the handling of the recyclable material.[193]

One of the more unusual features of the recycling exemption is the fee shifting provision under which a person who seeks contribution from a party asserting the exemption shall be liable for the party's attorney's fees if the party is adjudged not liable under section 127. At least one court has awarded attorney's fees to a prevailing defendant invoking this provision.[194]

Finally, section 127 includes a transition rule that provides that the recycling exemption is not intended to affect concluded cases or pending judicial actions initiated by the United States prior to November 29, 1999.[195] This transition rule has been the subject of considerable litigation.[196]

193. CERCLA § 127(f).

194. Evansville Greenway & Remediation Trust v. S. Ind. Gas & Elec. Co., No. 3:07-CV-66, 2012 WL 602638 (S.D. Ind. Feb. 25, 2012).

195. CERCLA § 127(i).

196. United States v. Atlas Lederer Co., 97 F. Supp. 2d 830 (S.D. Ohio 2000) (SREA does not apply to pending actions initiated by the United States prior to enactment, including cross-claims and contribution claims brought in the same action); United States v. NL Indus., Inc., No. 91-CV-578, 2005 WL 1267419 (S.D. Ill. May 4, 2005) (same); Dep't of Toxic Substances Control v. Interstate Non-Ferrous Corp., 99 F. Supp. 2d 1123 (E.D. Cal. 2000) (SREA is applicable to pending claims of state; a state and the United States are not equivalent for purposes of interpreting SREA); United States v. Mountain Metal Co., 137 F. Supp. 2d 1267 (N.D. Ala. 2001) (the SREA exemption is applicable to a private cost recovery action pending

Among the first cases to test the new recycling exemption was *Gould Inc. v. A & M Battery & Tire Service*,[197] which concerned spent lead-acid batteries. The court ruled that spent lead-acid batteries were recyclable under section 127 of CERCLA, and thus scrap metal dealers that sold such batteries to battery breakers were not liable under CERCLA for cleanup of the battery breaking facility, even though not all parts of the batteries were in fact recyclable. Defendants have asserted the recycling exemption in connection with numerous other materials and arrangements, with varying degrees of success.[198]

A noteworthy rejection of the recycling exemption came in *Del-Ray Battery Co. v. Douglas Battery Co.*,[199] in which the Fifth Circuit ruled that it does not provide a defense to claims arising under the Texas Solid Waste Disposal Act. In this case a group of battery recyclers incurred costs in connection with cleanup of a battery recycling facility at the direction of the Texas Commission on Environmental Quality. Those companies then sought contribution in state court from other battery recyclers under the Texas Solid Waste Disposal Act (SWDA). The defendants asserted that they should be exempted from liability under the SWDA based on application of the CERCLA recycling exemption. The state court apparently rejected the defense and granted partial summary judgment to the plaintiff, but the case was later dismissed for reasons that are not apparent in the record.

at the time of enactment but not to a pending United States cost recovery action, even though the two actions were consolidated).

197. 232 F.3d 162 (3d Cir. 2000).

198. *See, e.g.*, Cal. Dep't of Toxic Substances Control v. Ako Pac., Inc., 508 F.3d 930 (9th Cir. 2007) (court rejected defendants' claim that recycling exemption applied to various lead sources, including dross and slag, because they failed to prove that the recyclable material met a commercial specification at the time of the transaction); United States v. Mallinckrodt, Inc., 343 F. Supp. 2d 809, 819 (E.D. Mo. 2004) (court denied Shell's claim for application of the recycling exemption to surface of reconditioned 55-gallon drums because Shell "failed to present any evidence to support its claim that the [lead] paint on the outside of its drums was an integral part of the drums, rather than merely adhering to the drums"); Cal. Dep't of Toxic Substances Control v. Interstate Non-Ferrous Corp., 298 F. Supp. 2d 930 (E.D. Cal. 2003) (court held that insulated copper wire and scrap aluminum qualified for the recycling exemption, but denied summary judgment as to bottom ash, dross, and battery parts because of genuine issue of material facts).

199. 635 F.3d 725 (5th Cir. 2011).

Del-Ray Battery and Golden Eagle Battery, the defendants in the state court case, then filed a claim in federal court against Douglas Battery and Interstate Battery, the plaintiffs in the state action, seeking recovery of attorneys' fees and expert witness fees incurred in defending the state court action. Del-Ray based the claim on SREA, asserting that Congress's purpose in passing SREA was to "'promote the reuse and recycling of scrap material'" and therefore must have intended that SREA apply to the SWDA. The district court rejected this argument and the Fifth Circuit affirmed the district court, holding that "the defense contained in the SREA that may have been available to Plaintiffs, had they been sued under CERCLA, is not available to them in a suit brought pursuant to the SWDA."[200]

### *7. Service Station Dealers Exemption*

Under the 1986 amendments, Congress established an exemption from liability for service station dealers who accept used oil for recycling. The exemption was intended to encourage service station dealers to accept used oil from "do-it-yourself recyclers," that is, people who change the oil in their own cars, trucks, and appliances. Under section 114(c) of CERCLA, service station dealers that accept or handle used oil may qualify for an exemption from liability if they do not mix the used oil with hazardous substances and otherwise comply with federal used oil management standards and accept do-it-yourself-generated oil.[201] A service station dealer is defined as a person who owns or operates a retail establishment that sells, repairs, or services motor vehicles and who accepts for collection, accumulation and delivery to an oil recycling facility recycled oil that has been removed from light duty motor vehicles or household appliances.[202] EPA's used oil management standards, which are codified at 40 C.F.R. pt. 279, specify procedures for service station dealers to qualify for this exemption. The exemption also applies to government agencies that establish a facility solely for

200. *Id.* at 733.
201. *See* CERCLA § 114(c).
202. CERCLA § 101(37)(A).

purposes of accepting recycled oil from do-it-yourself generators, as well as owners and operators of refuse collection services who are required by state law to collect, accumulate, and deliver do-it-yourself oil to an oil recycling facility.[203]

### 8. *De Micromis Exemption*

One of the enduring legends of Superfund is that a party who tosses a penny into a landfill could be held liable—jointly and severally liable, no less—for cleanup of that landfill. This fiction has endured in part because the original text of Superfund did not exempt a party from liability even if the party was responsible for disposing a tiny amount of waste in an ocean of debris.

Congress put that legend to rest in 2002, as part of the Small Business Liability Relief and Brownfields Revitalization Act, which amended several provisions of CERCLA among other things.[204] One of the changes was the addition of a "de micromis" exemption from liability. Specifically, under new section 107(o), if the total amount of material that a party arranges for disposal or treatment, or accepts for transport for disposal or treatment, is less than 110 gallons of liquid materials or 200 pounds of solid materials, then such party "shall not be liable with respect to response costs at a facility on the National Priorities List."

There are several limitation to this exemption. First, this exemption is applicable only if all or part of the disposal, treatment, or transport occurred before April 1, 2001.[205] Second, the exemption shall not apply if EPA determines that the materials at issue "contributed significantly, either individually, or in the aggregate, to the cost of the response action or natural resource damage restoration."[206] Third, the de micromis exemption is not available to convicted criminals.[207]

203. CERCLA § 101(37)(B).
204. Pub. L. No. 107-118, 115 Stat. 2356 (2002).
205. CERCLA § 107(o)(1).
206. CERCLA § 107(o)(2)(A)(i).
207. CERCLA § 107(o)(2)(B).

### 9. *Exemption for Municipal Solid Waste*

In addition to establishing the de micromis exemption, the Small Business Liability Relief and Brownfields Revitalization Act exempted from liability certain PRPs that generate municipal solid waste (MSW). Specifically, the MSW exemption applies only to (1) business entities (including a parent, subsidiary or affiliate of the entity) that generate household waste and employ 100 or fewer workers and are nonprofit organizations under § 501(c)(3) of the Internal Revenue Code or small business concerns as defined in the Small Business Act; and (2) owners, operators, and lessees of residential property.[208] MSW is broadly defined as waste material generated by a household (or multiple households), as well as waste material generated by business if it is of the same character as household waste.[209] The statute enumerates examples of MSW, such as food and yard waste, paper, clothing, and the like.[210]

As with the de micromis exemption, there are a number of limits on the MSW exemption. First, the MSW exemption does not provide relief to owners or operators of municipal landfills or transporters of municipal waste. Nor did Congress provide relief to small companies that are in any way owned or affiliated with large generators of hazardous waste.[211]

Second, if EPA determines that the MSW has contributed significantly either individually or in the aggregate to the cost of the response action or natural resource damages restoration the exemption would not apply.[212]

Third, the MSW exemption does not apply if the person seeking the exemption has either failed to comply with an EPA information request or has otherwise impeded, through action or inaction, the performance of the cleanup.[213]

208. CERCLA § 107(p)(1).
209. CERCLA § 107(p)(4)(A).
210. CERCLA § 107(p)(4)(B).
211. CERCLA § 107(p)(1).
212. CERCLA § 107(p)(2)(A).
213. CERCLA § 107(p)(2)(B) & (C).

In contribution suits, the plaintiff has the burden of showing that each defendant asserting an MSW defense does not qualify for the exemption. If the plaintiff fails to meet this burden, it is liable for the defendant's attorney fees.[214]

### *10. Contiguous Property Owner Defense*

Another defense established by Congress in 2002 under the Small Business Liability Relief and Brownfields Revitalization Act is a defense for owners of land adjacent to the contaminated site, which itself becomes contaminated by passive migration of hazardous substances from such site. Under the so-called contiguous property defense, a current owner is not liable for contamination that migrates onto his or her property from a contiguous site that is owned by another party.[215] Congress established this defense to protect "parties that are essentially victims of pollution incidents caused by their neighbor's actions."[216] To be eligible for this defense, the landowner must not have caused the contamination on the adjacent property, nor be affiliated with those who created the contamination; the landowner must have exercised due diligence at the time of purchase; the landowner must not have been aware of the contamination; and the landowner must cooperate with any investigation and cleanup.[217]

### *11. Cleanup Contractors' Exemption*

After the enactment of CERCLA in 1980, it became clear that a private contractor hired to clean up a Superfund site may be exposed to potential liability as an operator of that site, or as a person who arranged for disposal or transport of waste in instances in which the contractor removed waste as part of the cleanup. This potential liability could deter private contractors from performing cleanup work. In response to this concern, Congress in 1986 added section 119 to CERCLA to limit

214. CERCLA § 107(p)(5) & (7).
215. CERCLA § 107(q).
216. S. REP. NO. 107-2, at 10 (2001).
217. CERCLA § 107(q)(1)(A)(i)–(vii).

the liability of "response action" contractors who are hired to perform cleanup actions.[218]

Section 119(a)(1) states that cleanup contractors shall not be liable under CERCLA, or any other federal law, to any person for "injuries, costs, damages, expenses or other liability" resulting from the release or threatened release of a hazardous substance, pollutant, or contaminant.[219] (However, no immunity from liability under state law is conferred under CERCLA.) section 119(a)(2) states that a cleanup contractor shall not be exempt from federal liability for a release caused by that contractor as a result of conduct that is "negligent, grossly negligent, or which constitutes intentional misconduct."[220] Under certain circumstances, section 119(c) authorizes the president the discretion to indemnify a cleanup contractor for negligent conduct, but not grossly negligent conduct or intentional misconduct.[221] Such indemnification is intended to cover a contractors' liability that cannot be covered by insurance at a fair and reasonable price.

### B. Equitable Defenses

In addition to the statutory defenses described above, defendants in cost recovery actions have attempted to assert a variety of equitable defenses including restitution, laches, unclean hands, estoppel, and waiver. In government cost recovery actions, courts generally decline to recognize equitable defenses.[222] Equitable defenses fare only slightly better in actions between private parties. (See discussion in chapter 9.)

In private cost recovery actions, several courts have adopted the view that the action is essentially one for contribution, which provides for

218. 42 U.S.C. § 9619.
219. 42 U.S.C. § 9619(a)(1).
220. 42 U.S.C. § 9619(a)(2).
221. 42 U.S.C. § 9619(c).
222. United States v. Mottolo, 695 F. Supp. 615, 626 (D.N.H. 1988) (doctrine of laches inapplicable in suit by sovereign); Violet v. Picillo, 648 F. Supp. 1283, 1294 (D.R.I. 1986) (same); United States v. Stringfellow, 661 F. Supp. 1053, 1062 (C.D. Cal 1987) ("plain meaning" of language used in section 107(a) precludes defenses other than those enumerated in the statute); *but see* United States v. Conservation Chem. Co., 619 F. Supp. 162, 204 (W.D. Mo. 1985) ("Equitable defenses are proper under CERCLA in determining liability, the nature of the remedy and the amount of damage.").

allocation of response costs based on equitable considerations, and, therefore, allowed these arguments to be raised.[223] Among those courts allowing the assertion of equitable defenses, there is a dispute as to whether such defenses act as a complete bar to liability or are relevant only in apportioning damages.[224] The majority of cases, however, hold that an equitable defense is merely one of the factors to be considered in apportioning liability for response costs.[225]

### C. Indemnification and "As Is" Sales

Section 107(e) of CERCLA provides:

> No indemnification, hold harmless, or similar agreement or conveyance shall be effective to transfer from the owner or operator of any vessel or facility or from any person who may be liable for a release or threat of release under this section, to any other person the liability imposed under this section. Nothing in this subsection shall bar any agreement to insure, hold harmless, or indemnify a party to such agreement for any liability under this section.

Under the first sentence of section 107(e), a party may not avoid liability in a government cost recovery action by asserting the existence of an indemnification or "hold harmless" agreement in either the transfer

---

223. *See, e.g.*, Smith Land & Improvement Corp. v. Celotek Corp., 851 F.2d 86 (3d Cir. 1988); PVO Int'l, Inc. v. Drew Chem. Corp., 16 Chem. Waste Litig. Rep. 669 (D.N.J. June 27, 1988).

224. *Compare* Versatile Metals, Inc. v. Union Corp., 693 F. Supp. 1563, 1572 (E.D. Pa. 1988) ("Versatile Metals' assertions as to relative fault will be considered in this court's apportionment of necessary response costs, if any"), *with* Mardan Corp. v. C.G.C. Music, Ltd., 600 F. Supp. 1049 (D. Ariz. 1984), *aff'd*, 804 F.2d 1454 (9th Cir. 1986) (plaintiff's claim for response costs barred by, inter alia, the doctrine of unclean hands).

225. *See, e.g.*, *Smith Land & Improvement Corp.*, 851 F.2d at 90 (court rejected the defense of caveat emptor as a complete bar to recovery and held that the doctrine may only be considered in mitigation of the amount due as determined in a contribution action); Southland Corp. v. Ashland Oil, Inc., 696 F. Supp. 994, 1003 (D.N.J. 1988), *rev'd in part*, Civ. No. 88-0700, 1988 WL 125855 (D.N.J. Nov. 23, 1988) ("equitable considerations are applicable in allocating the amount of contribution between parties under CERCLA").

of title to real property or in contracts between generators, transporters, and owners or operators of the facility.[226]

The second sentence of section 107(e), however, has always been understood to preserve the rights of private parties to contractually transfer financial responsibility arising out of CERCLA liability. Such warranty or indemnification must explicitly allocate the risk to one of the parties to be binding in a private cost recovery or contribution action between those parties.[227]

Notably, such private agreements do not affect the CERCLA contribution rights of third parties. For example, in *Cyprus Amax Minerals Co. v. CBS Operations, Inc.*,[228] the plaintiff had incurred response costs in cleaning up contamination from its smelting operations. The plaintiff sought contribution from two potential successors of a neighboring smelter. One of the defendants, TCI, offered to stipulate that it was successor to the operator of the second smelter. The other defendant, CBS Operations, had agreed to indemnify TCI against potential CERCLA liability. Cyprus argued that the agreement between TCI and CBS Operations meant that they should both be considered operators, based on the language in the second sentence of §107(e). The court rejected that argument:

> Whatever liability CBS Ops. may have assumed as TCI's indemnitor, the underlying CERCLA liability could not have been transferred from [TCI] to CBS Ops. by way of the [indemnification agreement].[229]

---

226. *See, e.g.*, United States v. Hardage, 985 F.2d 1427, 1433 (10th Cir.1993); AM Int'l, Inc. v. Int'l Forging Equip. Corp., 982 F.2d 989, 994 (6th Cir.1993); Mardan Corp. v. C.G.C. Music, Ltd., 804 F.2d 1454, 1459 (9th Cir.1986).

227. *See, e.g.*, Harley-Davidson, Inc. v. Minstar, Inc., 41 F.3d 341 (7th Cir. 1994) (indemnification agreement between plaintiff and defendant held enforceable and barred plaintiff's contribution action against defendant under section 107(e) of CERCLA); Lyncott Corp. v. Chem. Waste Mgmt., Inc., 690 F. Supp. 1409 (E.D. Pa. 1988) (granting summary judgment on contribution claim based on indemnification agreement); *cf.* Fischer Dev. Co. v. Boise Cascade Corp., 37 F.3d 104, 108 (3d Cir. 1994) (tenant held not liable to property owner for contamination based on owner's contractual release of all claims that it "ever had, or now has . . . [and] hereafter can, shall or may have").

228. No. 11-CV-252, 2012 WL 4857924 (N.D. Okla. Oct. 12, 2012).

229. *Id.* at *3.

In sum, the second sentence of § 107(e) of CERCLA only preserves the rights of parties to contract among themselves; it does not affect any third party—government or private—asserting a CERCLA cost recovery or contribution claim. Those claims are determined purely by reference to the liability provisions of section 107(a) of CERCLA.

A closely related issue is the effect of an "as is" clause used in the transfer of title to real estate. Courts generally have held that these provisions are merely warranty disclaimers that preclude claims based on breach of warranty. They do not shift liability from one party to another and, therefore, are unavailable as a defense in a cause of action not based on breach of warranty, that is, a cost recovery action under section 107.[230]

## VIII. STATUTE OF LIMITATIONS

Through the 1986 amendments to CERCLA, Congress provided explicit statutes of limitations for both cost recovery and natural resource damage actions. Actions to recover removal costs generally must be commenced within three years of the completion of the removal action.[231] For remedial costs, the action must be commenced within six years of initiation of "physical on-site construction of the remedial action."[232] The difference in the limitations periods applicable to removal and remedial actions, as well as the somewhat ambiguous boundaries between removal and remedial actions, has spawned an immense amount of litigation. Typically, the defendant moves for dismissal based on a lapsed limitation period, which often is premised on characterizing the cleanup as a removal with the shorter limitation period, whereas the government argues the cleanup is a remedial action with the longer limitation period that has yet to lapse.[233] As with numerous other statutory limitations on liability, courts have demonstrated an inclination to narrowly construe

230. *See, e.g., Southland Corp.*, 696 F. Supp. 994.

231. CERCLA § 113(g)(2)(A).

232. *Id.* § 113(g)(2)(B). Note, however, that the time within which removal cost recovery actions may be brought becomes coextensive with the time for commencing remedial cost recovery actions if remedial action is commenced within three years following completion of the removal action. *Id.*

233. See discussion of this case law in chapter 1, section III.

the statute of limitations.[234] Moreover, courts tend to give deference to EPA's characterization of the response action.[235]

Another statute of limitations provision that Congress added with the 1986 amendments is section 158 of CERCLA.[236] Section 158 provides that in "any action brought under State law for personal injury, or property damages, which are caused or contributed to by exposure to any hazardous substance, or pollutant or contaminate . . . if the applicable limitations period for such action . . . provides a commencement date which is earlier than the federally required commencement date," the federal commencement date applies in lieu of the state date.[237] This provision was adopted by Congress in response to a 1982 report that was commissioned when CERCLA was enacted.[238] The report recommended that states "remove unreasonable procedural and other barriers to recovery in court action for personal injuries resulting from exposure to hazardous waste."[239] This measure was needed because "many of the hazardous wastes are carcinogens" with lengthy latency periods.[240]

234. *See, e.g.*, United States v. Moore, 698 F. Supp. 622, 625–27 (E.D. Va. 1988) (declining to apply statute of limitations retroactively where to do so would bar United States' cost recovery claim and, thereby, frustrate the broad remedial intent of CERCLA).

235. *See, e.g.*, Colorado v. Sunoco, Inc., 337 F.3d 1233 (10th Cir. 2003) (EPA characterization of cleanup as "remedial action" entitled to some deference based on extent to which EPA's consideration was thorough, and its reasoning valid).

236. 42 U.S.C. § 9658.

237. 42 U.S.C. § 9658(a)(1).

238. The "Study Group" report was commissioned under section 301 of CERCLA, a provision with an interesting history. It was one of the compromises that made passage of CERCLA possible. One group of senators had wanted the statute to include a federal toxic tort cause of action; another group of senators opposed that provision. The compromise was to commission a study group to research the need for legislation in the area, under section 301(e). The Study Group was composed of 12 lawyers drawn equally from the following four legal organizations: American Bar Association, American Law Institute, Association of Trial Lawyers of America, and National Association of Attorneys General. In 1982, the Study Group published its report. *See* Committee Print, Serial 97-12, Injuries and Damages from Hazardous Wastes—Analysis and Improvement of Legal Remedies, A Report to Congress in 1980 (P.L. 96-510) by the "Superfund Section 301(e) Study Group" (Senate Comm. on Environment and Public Works, Sept. 1982).

239. Study Group Report at 255.

240. *Id.*

The Supreme Court was called on to interpret section 158 of CERCLA in *CTS Corp. v. Waldburger*.[241] The question at issue was whether section 158 applies not only to state statutes of limitation but also to state statutes of repose. The case arose in North Carolina where CTS Corporation had operated an electronics plant until 1985. The plant operations involved use of a solvent, trichloroethylene (TCE), among other chemicals. In 1987, CTS sold the property to a buyer who eventually sold a portion of the property to individuals who brought a nuisance action in 2011 against CTS alleging property damage from TCE contamination. CTS moved for dismissal of the action, citing a North Carolina statute of repose, which provides that tort action may not be brought more than 10 years after the last culpable act of the defendant.[242] Because CTS's last culpable act was the sale in 1987, the district court granted the motion to dismiss.

The Fourth Circuit reversed on the ground that the state's statute of repose was preempted under section 158 of CERCLA. The Supreme Court reversed the Fourth Circuit, noting that the state statutes of repose are different from state statutes of limitation, which section 158 preempts: "[A] statute of limitation creates 'a time limit for suing in a civil case, based on the date when the claim [has] accrued,'" whereas a statute of repose "puts an outer limit on the right to bring a civil action" that is "measured not from the date on which the claim accrues but instead from the date of the last culpable act or omission of the defendant."[243] In light of the different purposes of statutes of limitations and statutes of repose, the court held that section 158 does not preempt North Carolina's statutes of repose.[244]

## IX. TYPES OF RECOVERABLE COSTS

Section 107(a) of CERCLA outlines four categories of recoverable costs:

> (A) all costs of removal or remedial action incurred by the United States Government or a State or an Indian tribe not inconsistent

241. 134 S. Ct. 2175 (2014).
242. *Id.* at 2181.
243. *Id.* at 2182.
244. *Id.*

> with the national contingency plan; (B) any other necessary costs of response incurred by any other person consistent with the national contingency plan; (C) damages for injury to, destruction of, or loss of natural resources, including the reasonable costs of assessing such injury, destruction, or loss resulting from such a release; and (D) the costs of any health assessment or health effects study carried out under [section 104(i)].[245]

In this section, we elaborate on the types of response costs that government entities may recover under section 107(a)(4)(A). In chapter 9 (Private Party Actions), we discuss in detail the differences between response costs that government entities may recover under section 107(a)(4)(B)—i.e., response costs that are "not inconsistent with the NCP"—and those that private parties may recover—i.e., response costs that are "necessary" and "consistent with the NCP." In chapter 10, we discuss natural resource damage claims that government trustees of damages resources may assert.

## A. Response Costs

Numerous cases address the types of costs that qualify as recoverable costs of removal or remedial action. The majority of these cases were decided on the basis of summary judgment motions wherein the court held that specific response costs may be recoverable, but that the plaintiff must meet its burden at trial in substantiating its claims before those costs could be awarded. Following is a discussion of the relevant case law in this area.

### 1. *Cleanup Activities*

There is little controversy surrounding recovery of direct costs of cleaning up environmental media, that is, the cost of removing hazardous substances in soil and groundwater. Moreover, there is little in the way of controversy over the recovery of costs of activities that may seem ancillary to cleanup are expressly included in the CERCLA definitions of "remove," "removal," "remedy," and "remedial action," such as security

245. CERCLA § 107(a)(4)(A)–(D).

fencing to limit access, provision of alternative water supplies, temporary evacuation for threatened individuals, storage, perimeter protection, collection of leachate, repair or replacement of leaking containers, and on-site treatment or incineration. Based on the express language of the statute, courts have approved recovery for costs of (1) evacuation and relocation;[246] (2) destroying contaminated personal property;[247] (3) experts to remove PCB contamination;[248] and (4) security measures.[249]

Similarly noncontroversial are claims for recovery of costs associated with evaluating the effects of a release and determining which response activities are most appropriate are recoverable under CERCLA. Thus, the costs of monitoring, assessing, and evaluating sites where there has been a release of hazardous substances are recoverable.[250] Likewise, the cost of investigating the extent of the release is generally held to be a recoverable response cost.[251]

Courts are split on the issue of whether plaintiffs may recover medical testing costs as "response costs."[252] This disagreement stems in part from the statutory definitions of "remove," "removal," "remedy," and "remedial action," which contain no references to medical expenses. Additionally, the legislative history of CERCLA indicates that Congress considered including medical monitoring, but the statute as passed contained no language reflecting this intention.[253] Finally, the split of opinion also appears to result from a case law distinction between medical

246. T & E Indus., Inc. v. Safety Light Corp., 680 F. Supp. 696, 707 (D.N.J. 1988).

247. Wehner v. Syntex Corp., 681 F. Supp. 651, 653 (N.D. Cal. 1987).

248. Emhart Indus., Inc. v. Duracell Int'l Inc., 665 F. Supp. 549, 574 (M.D. Tenn. 1987).

249. Cadillac Fairview/Cal., Inc. v. Dow Chem. Co., 840 F.2d 691, 695 (9th Cir. 1988).

250. Lykins v. Westinghouse Elec., Civ. A. No. 85-508, 1988 WL 114522, at *4 (E.D. Ky. Feb. 29, 1988); Brewer v. Ravan, 680 F. Supp. 1176, 1179 (M.D. Tenn. 1988).

251. Tanglewood E. Homeowners v. Charles-Thomas, Inc., 849 F.2d 1568, 1575 (5th Cir. 1988); *Emhart Indus.*, 665 F. Supp. at 574.

252. *See, e.g.*, Coburn v. Sun Chem. Corp., Civ. A. No. 88-D120, 1988 WL 120739 (E.D. Pa. Nov. 9, 1988) (holding that costs of medical screening and monitoring are clearly not "necessary costs of response" under section 107); *contra Brewer*, 680 F. Supp. at 1179 (holding that costs of medical testing and screening to assess the effect of the release or discharge on public health or to identify potential public health problems presented by the release are necessary to monitor, assess, or evaluate a release and, therefore, constitute "removal").

253. *See* S. REP. NO. 848 (1978), *reprinted in* 1 LEGIS. HIST. 308, 360–61.

monitoring as a personal injury remedy and medical monitoring as a means to assess the extent of contamination at the site.[254]

One issue on which the courts are *not* split is that response costs do not include traditional tort damages. For example, neither lost profits nor other compensatory damages are recoverable.[255] Similarly, lost income, diminution of property value, and personal injury, are not available under CERCLA.[256]

### 2. *Indirect Costs*

Unlike the direct costs of response activity, there has been considerable controversy over the recovery of indirect costs of response. The government's indirect costs are costs of administering the Superfund program that cannot be directly linked to the cleanup of any site. They include costs such as compensation and benefits of EPA staff, travel, rent, utilities, communications equipment, contract services, and supplies necessary for EPA administrative activity. Since the inception of the Superfund program, EPA always sought to recover a portion of these costs from PRPs, with varying degrees of success.

For many years, EPA based its allocation on the percentage of labor hours that were devoted to a given site. Early on, courts were split on whether such a showing justified recovery of indirect costs. In *United States v. Ottati & Goss*, the court declined to award EPA $336,922 in indirect costs, which included overhead expenses such as rent, utilities, supplies, and clerical staff, reasoning that "[t]hese indirect costs necessary

254. *Compare Brewer*, 680 F. Supp. at 1179 (holding that medical expenses incurred in the treatment of personal injuries or disease are not recoverable, but the costs of medical testing to assess the effect of a release on public health are), *with* Chaplin v. Exxon Corp., Civ. A. No. H-84-2524, WL 13130 (S.D. Tex. June 10, 1986) (finding that, although the legislative history of CERCLA contained language imposing liability for personal injury, as well as medical expenses and economic loss, these specific liability provisions were deleted from CERCLA).

255. *See, e.g.*, Artesian Water Co. v. Gov't. of New Castle Cnty., 659 F. Supp. 1269, 1285–86 (D. Del. 1987), *aff'd in part and remanded in part*, 851 F.2d 643 (3d Cir. 1988) (Congress "clearly manifested an intent not to provide compensation for economic losses or personal injury resulting from the release of hazardous substances"); Jones v. Inmont Corp., 584 F. Supp. 1425 (S.D. Ohio 1984).

256. *Wehner*, 681 F. Supp. at 653 ; Piccolini v. Simon's Wrecking, 686 F. Supp. 1063, 1067–68 (M.D. Pa. 1988); Allied Towing Corp. v. Great E. Petrol. Corp., 642 F. Supp. 1339, 1348 (E.D. Va. 1986).

to operate the Superfund program cannot be attributed directly to the . . . sites, and are therefore disallowed."[257] By contrast, in *United States v. R.W. Meyer, Inc.*,[258] the Sixth Circuit affirmed the award of the government's indirect costs. The court reasoned that CERCLA's broad remedial purpose "supports a liberal interpretation of recoverable costs" and found that "the challenged indirect costs are part and parcel of all costs of the removal action, which are recoverable under CERCLA."[259] Over time, however, courts have increasingly accepted and awarded EPA's indirect costs as recoverable costs of response actions.[260]

In 2000, EPA published a revised methodology for calculating indirect costs allocable to a site.[261] Under the new methodology, EPA identifies a pool of indirect costs to allocate to Superfund sites. The pool includes (1) agency-wide costs such as facilities and human resources management; (2) regional administrative, support and management costs for all regions combined; and (3) the regional and nationwide costs of running the Superfund program. EPA then calculates the percentage of nationwide, site-specific direct costs attributable to each EPA region and assigns an indirect rate to each region. For an individual site, EPA multiplies the total direct costs incurred in remediating a site by the rate applicable to the region in which the site is located. The new methodology has been accepted in at least one case.[262]

### 3. *Oversight Costs*

Another category of response costs over which there has been considerable controversy is oversight costs, which conceptually are the

257. 694 F. Supp. 977, 995 (D.N.H. 1988), *aff'd in part and vacated in part*, 900 F.2d 429 (1st Cir. 1990).

258. 889 F.2d 1497.

259. *Id.* at 1503 (citation omitted).

260. *See, e.g.*, United States v. Dico, Inc., 266 F.3d 864 (8th Cir. 2001), *cert. denied*, 535 U.S. 1095 (2002); United States v. Domenic Lombardi Realty, Inc., 290 F. Supp. 2d 198 (D.R.I. 2003); United States v. Kramer, 913 F. Supp. 848 (D.N.J. 1995).

261. Guidance on Exercising CERCLA Enforcement Discretion in Anticipation of Full Cost Accounting Consistent with "Statement of Federal Financial Accounting Standards No. 4," 65 Fed. Reg. 35,339 (June 2, 2000).

262. *See* United States v. W.R. Grace & Co., 280 F. Supp. 2d 1149 (D. Mont. 2003) (rejecting defendant's argument that the new indirect cost calculation methodology was flawed and awarding indirect costs in full), *aff'd*, 429 F.3d 1224 (9th Cir. 2005).

costs incurred by a government agency in overseeing or supervising a private party who is performing a response action. Although courts now unanimously agree that such costs are recoverable, it was not always so.

The first significant case concerning the recoverability of oversight costs is *New York v. Shore Realty Corp.*[263] In this case, the State of New York brought suit against Shore Realty to clean up a hazardous waste disposal site that the defendant had purchased for development. The New York State Department of Environmental Conservation had undertaken to evaluate site conditions, but the actual cleanup—removal of leaking drums—had been performed by the defendant. The district court held the defendant liable for the State's response costs, including oversight costs it incurred supervising the defendant's drum removal work. On appeal, the Second Circuit affirmed, stating "[t]he State's costs in assessing the conditions of the site and supervising the removal of drums of hazardous waste squarely fall within CERCLA's definition of response costs, even though the State is not undertaking to do the removal."[264] Other courts followed the reasoning of *Shore Realty*.[265]

Then came the 1993 decision in *United States v. Rohm & Haas Co.*,[266] in which the Third Circuit rejected EPA's claim for recovery of oversight costs. In this case, Rohm & Haas had entered into an administrative consent order under section 3008(h) of RCRA to clean up a facility where it had disposed of hazardous wastes. The order required Rohm & Haas to perform certain cleanup activities at the site, but did not require it to compensate EPA for the agency's costs of implementing the order including oversight of Rohm & Haas's work. In 1990 EPA brought a cost recovery action under section 107 of CERCLA to recover such costs. The district court agreed with EPA that Rohm & Haas was liable under CERCLA for oversight costs. On

263. 759 F.2d 1032.
264. *Id.* at 1042–43.
265. *See, e.g.*, Lykins v. Westinghouse Elec. Corp., 1988 WL 114522 (E.D. Ky. 1988); *Ottati & Goss*, 694 F. Supp. at 1002; United States v. R.W. Meyer, Inc., 889 F.2d 1497.
266. 2 F.3d 1265 (3d Cir. 1993).

appeal, the Third Circuit reversed the award of oversight costs. In so holding the court relied on the Supreme Court decision in *Skinner v. Mid-America Pipeline Co.*,[267] for the proposition that "Congress must indicate clearly its intention to delegate to the Executive the discretionary authority to recover administrative costs not inuring directly to the benefit of the regulated parties by imposing additional financial burdens, whether characterized as fees or taxes, on those parties."[268] Applying this analysis to the facts, the Third Circuit concluded that EPA oversight costs were administrative costs inuring to the benefit of the public at large rather than private parties. After further finding the lack of any express statutory language authorizing the recovery of such administrative fees, the Third Circuit reversed the award of oversight costs.[269]

Notably, however, the Fifth, Eighth, Ninth and Tenth Circuits criticized the Third Circuit's reasoning in *Rohm & Haas*, and held that EPA's oversight costs are recoverable response costs under section 107 of CERCLA.[270] Then, in 2005, the Third Circuit reversed its 1993 *Rohm & Haas* ruling, holding in *United States v. E.I. du Pont de Nemours & Co., Inc.* that EPA may recover oversight costs under section 107 of CERCLA.[271] All of these courts, including the Third Circuit, noted that under section 107(a)(4)(A) of CERCLA a responsible party is liable for "*all costs* or removal or remedial action incurred by the United States" (emphasis added), and thus statutory interpretation was not required to find the costs recoverable. Thus, at this point there no longer appears to be any controversy over recovery of oversight costs. The circuits are unanimous.

267. 490 U.S. 212 (1989).
268. *Id.* at 1272 (citing *Skinner*, 490 U.S. at 224).
269. *Id.* at 1275.
270. *See* United States v. Lowe, 118 F.3d 399 (5th Cir. 1997); *Dico, Inc.*, 266 F.3d 864; United States v. Hyundai Merch. Marine Co., 172 F.3d 1187 (9th Cir. 1999); Atl. Richfield Co. v. Am. Airlines, Inc., 98 F.3d 564 (10th Cir. 1996).
271. 432 F.3d 161 (3d Cir. 2005) (en banc).

### *4. Attorney's Fees*

Courts have uniformly held that under section 104(b)(1) the government may recover litigation costs incurred in connection with a cleanup.[272] However, this is one of the areas where there is a distinction between recoverable costs for the government and those for private parties.

Neither section 107 nor section 113 expressly authorize the recovery of attorney's fees by a private plaintiff as part of the response costs it incurs. During the first two decades of CERCLA litigation, courts split on the question of whether attorney's fees were recoverable in private cost recovery and/or contribution actions.[273]

The U.S. Supreme Court resolved this split in *Key Tronic v. United States*.[274] The Court held that a private party may not recover attorney's fees associated with bringing a cost recovery action. The Court, however, distinguished fees paid to an attorney for other services related to the cleanup:

> The conclusion we reach with respect to litigation-related fees does not signify that all payments that happen to be made to a lawyer are unrecoverable expenses under CERCLA. On the contrary, some lawyers' work that is closely tied to the actual cleanup may constitute a necessary cost of response in and of itself under the terms of § 107(a)(4)(B).[275]

The Court then held that attorney's fees incurred to identify other PRPs are recoverable response costs; by contrast, the Court deemed attorney's fees for negotiating a consent decree nonrecoverable.[276] On remand, the

272. *See, e.g.*, United States v. S.C. Recycling & Disposal, Inc., 653 F. Supp. 984, 1009 (D.S.C. 1986); United States v. Ne. Pharm. & Chem. Co., 579 F. Supp. 823, 851 (W.D. Mo. 1984), *aff'd in part and rev'd in part*, 810 F.2d 726 (8th Cir. 1986).

273. *See, e.g.*, Gen. Elec. Co. v. Litton Indus. Automation Sys., Inc., 920 F.2d 1415 (8th Cir. 1990) (fees recoverable); Donahey v. Bogle, 987 F.2d 1250, 1256 (6th Cir. 1993) (same); Juniper Dev. Grp. v. Kahn, 993 F.2d 915, 933 (1st Cir. 1993) (litigation fees not recoverable); FMC Corp. v. Aero Indus., Inc., 998 F.2d 842 (10th Cir. 1993) (only nonlitigation fees recoverable).

274. 511 U.S. 809 (1994).

275. *Id.* at 820–21.

276. *Id.*

Ninth Circuit reversed the award of legal fees for negotiating the consent decree and litigating the cost recovery action, and affirmed that portion of the award of attorney's fees for services performed in searching for other PRPs.[277] A number of cases since *Key Tronic* have reached similar rulings on the award of attorney's fees.[278]

### 5. *Consistency with the NCP*

Section 107(a)(4)(A) provides that the United States may recover all costs of removal or remedial action incurred that are *not inconsistent* with the NCP. Accordingly, the burden is on defendants to show that the government's response costs are inconsistent with the NCP. All costs incurred by the United States are "conclusively presumed to be reasonable" so long as they "are not inconsistent with the NCP."[279] In determining whether the government's response costs were not inconsistent with the NCP, courts look to whether EPA's remedial decision and associated activities and their costs are arbitrary and capricious.[280]

This has proven to be onerous, but not insurmountable. In *United States v. Newmont USA, Ltd.*,[281] the government brought a motion to establish the amount of costs for which the responsible parties were liable. The defendants challenged EPA's sampling-related activities during the remedial investigation as duplicative and unnecessary, and therefore inconsistent with the NCP. The defendants contended that the sampling conducted by the EPA had previously been conducted by the defendants and were available for the EPA's use during the remedial investigation. The EPA tried to undermine the defendants' argument by pointing out

277. Key Tronic Corp. v. United States, 30 F.3d 1105 (9th Cir. 1994).

278. *See, e.g.*, United States v. Gurley, 43 F.3d 1188 (8th Cir. 1994) (government can recover the attorney's fees it incurred identifying other PRPs, even though it ultimately dismissed its actions against those other PRPs); Hillsborough Cnty. v. A&E Rd. Oiling Serv., Inc., 853 F. Supp. 1402 (N.D. Fla. 1994) (attorney's fees recoverable in cost recovery actions); Hatco Corp. v. W.R. Grace & Co., 849 F. Supp. 931 (D.N.J. 1994) (attorney's fees that are not related to litigation may be recoverable if necessary to the containment and cleanup of hazardous releases).

279. *Northernaire Plating Co.*, 685 F. Supp. at 1417; United States v. Vertac Chem. Corp., 671 F. Supp. 595, 614 (E.D. Ark. 1987).

280. *NEPACCO*, 810 F.2d at 748.

281. 504 F. Supp. 2d 1050 (E.D. Wash. 2007).

that defendants' own expert recognized that the EPA may legitimately take additional samples to verify prior sampling results. The court concluded that, viewing the evidence in the light most favorable to defendants, a genuine issue on inconsistency with the NCP existed as to the sampling alleged to be unnecessary and duplicative and therefore denied the government's motion.[282]

Note that a different standard applies to private cost recovery actions. Specifically, section 107(a)(4)(B) provides that any other person may recover "any other necessary costs of response [incurred] . . . consistent with the [national contingency plan]." Thus, the presumption of reasonableness does not apply to private parties; instead, private parties must demonstrate that their response costs are *consistent* with the NCP.[283] (In chapter 9, we discuss EPA's interpretation of the phrase "consistent with the NCP" as applied to private cost recovery actions.)

Although courts have added consistency with the NCP as an element to a private plaintiff's prima facie case, courts hold that the fact that response costs are inconsistent affects the amount, not the existence, of defendant's liability.[284]

## B. Natural Resource Damages

In addition to response costs for site cleanup, parties liable under section 107 are liable for "damages for injury to, destruction of, or loss of" publicly owned or managed natural resources, and "the reasonable costs of assessing such injury."[285] This liability runs to the natural resource trustee—the official designated to act on behalf of the public to protect

282. *See also* United States v. JG-24, Inc., 478 F.3d 28, 32 (1st Cir. 2007) (court rejected defendants' argument that a delay of 21 months between EPA's inspection of a site and EPA's initiation of a removal action was inconsistent with the NCP requirement that a removal action begin "'as soon as possible' after discovery of contamination"); United States v. Broderick Inv. Co., 963 F. Supp. 951 (D. Colo. 1997), *aff'd in part and rev'd in part*, 200 F.3d 679 (10th Cir. 1999) (EPA did not violate the NCP by failing to run every possible test when characterizing the site).

283. *See NEPACCO*, 810 F.2d at 747.

284. United States v. Mottolo, 695 F. Supp. 615, 630 (D.N.H. 1988); O'Neil v. Picillo, 682 F. Supp. 706, 729 (D.R.I. 1988).

285. CERCLA § 107(a)(4)(C).

the natural resource. Chapter 10 discusses the elements of an action for recovery of natural resource damages, as well as case law in this area.

## C. Health Effects Studies

With the enactment of CERCLA, Congress created the Agency for Toxic Substances Disease Registry (ATSDR) within the Public Health Service, but limited its authority to maintaining a registry of serious diseases and persons exposed to toxic substances and an inventory of literature, research, and studies on health effects, and to providing medical care and testing to exposed individuals.

SARA amended section 104(i) to require ATSDR to prepare toxicological profiles of hazardous substances and to perform health assessments and health effects studies at hazardous waste sites.[286] These studies include preliminary assessments of the potential risk to human health posed by a facility.[287] Under section 107(a)(4)(D), the costs incurred in conducting these studies may be recovered from responsible parties in cost recovery actions.

286. CERCLA § 104(i)(3).

287. CERCLA § 104(i)(6)(A). If the results of these studies indicate that exposure presents a significant risk to human health or the environment, ATSDR may, inter alia, notify EPA so that it may evaluate the site to determine whether it should be placed on the NPL or, if already on the NPL, accorded a higher priority. *Id.* § 104(i)(6)(H).

# CHAPTER 5

# Settlement

## I. INTRODUCTION

Faced with a steady stream of precedent supporting the imposition of strict, joint, and several liability under CERCLA, a fairly limited set of affirmative defenses, and the high costs of litigation, the litigation reflex that predominated potentially responsible party (PRP) strategy during the first decade of the Superfund program has gradually shifted toward consideration of early settlement evaluation.

As originally enacted in 1980, CERCLA provided little, if any, illumination on settlements with PRPs. In the absence of such settlement guidance, litigation ruled the day, and the Superfund program fell short of expectations. In the first five years of the program, the federal government spent $1.6 billion, but only ten sites were completely remediated.[1]

In 1985, EPA formulated its Interim CERCLA Settlement Policy.[2] Nearly two years later, much of the Interim CERCLA Settlement Policy was incorporated by the Superfund Amendments and Reauthorization Act of 1986 (SARA) as newly added section 122 of CERCLA. Section 122 established a statutory framework for settlement that had been sorely lacking. Of particular significance, section 122 specifically sets forth a process by which PRPs may be allowed to assume control over the conduct of the remedial investigation/feasibility study (RI/FS) and of

1. GAO, Comptroller Report to the Congress of the United States, Cleaning Up Hazardous Wastes: An Overview of Superfund Reauthorization Issues 28 (Mar. 29, 1985), http://www.gao.gov/assets/150/142614.pdf.

2. 50 Fed. Reg. 5034 (Feb. 5, 1985).

the remedy. Section 122 also provides some guidance for PRPs embroiled in a pending section 106 and/or section 107 action.

Section 122 was initially viewed by some as a settlement panacea for a program mired in litigation. Although section 122 did significantly improve the process for (and the prospects of) settling Superfund cases, its full potential has not been realized. Indeed, some of the settlement authorities established under section 122 have rarely been used.[3] Because the use of section 122 lies entirely within the discretion of EPA, and because EPA's decision to use or not to use the section 122 procedures is not subject to judicial review,[4] PRPs seeking innovative settlements will be largely powerless to change this state of affairs.

This chapter will explore the prospects, process, and nature of settlement between the government and a private PRP both before and after the commencement of litigation, and the statutory provisions that create the framework for such settlements.[5]

## II. PRELITIGATION SETTLEMENT EVALUATION

A prelitigation settlement evaluation should be undertaken by any company identified by EPA as a PRP at a given site. Each case, and each PRP, will present a unique set of factors that must be carefully evaluated in conjunction with a number of general factors that will be common to any case and to any PRP. As discussed in turn below, foremost of these general factors are the PRP's possible liability at the site, the position of other PRPs at the site, the strength of the government's case against the PRP, the PRP's projected total costs at the site, and the prospects for later recovering those costs.

3. GAO, Report to Congressional Requestors, GAO/RCED-94-90, Superfund: Further EPA Management Action is Needed to Reduce Legal Expenses (Jan. 1994), http://gao.gov/assets/220/219044.pdf.

4. CERCLA § 122(a).

5. This chapter does not address settlement in the context of private cost recovery actions. For a discussion of that topic, see chapter 9.

## A. Possible Liability

A PRP's initial inquiry in assessing the merits of pursuing settlement is to evaluate the evidence connecting it to the contaminated site. This inquiry should enable a PRP to determine whether it falls within one or more of the four categories of persons subject to liability under section 107 of CERCLA.[6] It should be noted that cases of mistaken corporate identity are not uncommon. At the same time, the obligations of an unfamiliar corporate entity may have been assumed by the PRP in a long-forgotten transaction, thereby exposing the PRP to possible liability as a successor in interest. Thereafter, the extent to which the remaining prima facie elements of liability under CERCLA sections 106[7] and/or 107[8] appear to have been met must be evaluated. Along those lines, the availability of any affirmative defenses to liability, such as the third-party defense provided by section 107(b)(3) of CERCLA, should also be closely examined.

A PRP's initial inquiry should also focus on a review of the site's history and management practices at the site. Inquiry should also be directed toward ascertaining the volume, toxicity, and type of materials at the site that are allegedly attributable to the PRP, and the extent to which materials allegedly attributable to the PRP are commingled with other materials at the site. Such information may provide grounds for asserting that a PRP's liability is divisible or, alternatively "reasonably capable of apportionment," the test articulated by the Supreme Court in *Burlington Northern & Santa Fe Railway Co. v. United States* as justifying apportionment rather than joint and several liability.[9]

## B. Position of Other PRPs at the Site

Superfund cases, unlike the vast majority of cases involving the enforcement of federal environmental laws, often will involve dozens, if not hundreds, of PRPs. Frequently, the spectrum of PRPs at a site will range from solvent, large-volume national corporations to small-volume, "mom

6. See chapter 4.
7. See chapter 3.
8. See chapter 4.
9. 556 U.S. 599 (2009). See discussion in chapter 4.

and pop" operations on the verge of insolvency. EPA is far more likely to pursue the former type of PRP in a CERCLA section 106 and/or section 107 action. Consequently, the major players at a site often assume active roles in attempting to settle their exposure without litigation, while minor players typically assume a much lower profile.[10]

A PRP should, therefore, attempt to characterize the other PRPs at the site by focusing on factors such as the volume and percentage of materials attributable to each of the other PRPs, the toxicity of such materials, and the apparent solvency of each of the other PRPs. For reasons that will become apparent later in this chapter, this inquiry should also attempt to identify the existence of government PRPs at the site and the extent to which materials at the site are apparently attributable to PRPs that are insolvent, dissolved, or otherwise unreachable ("orphans").

## C. Strength of the Government's Case

Establishing the liability of those PRPs that are later named as defendants in a CERCLA action will, of course, depend on the government's ability to prove each of the requisite elements for liability. Information that EPA believes to be adequate for the purpose of adding a company's name to a PRP list may, in fact, be insufficient for later sustaining the government's burden of proof in a lawsuit. Thus, in close conjunction with its own internal assessment of possible liability, a PRP should attempt to determine whether information already in the government's possession—or nonprivileged information that *could* become available to the government through later discovery procedures—will be sufficient for establishing CERCLA liability. If possible, the same type of inquiry should be directed toward other PRPs at the site. Inquiry should also be directed toward the government's conduct at the site; some courts have been responsive to PRP efforts to establish that EPA itself exacerbated the conditions of contamination at the site.[11]

10. See related discussion in chapter 7.

11. *See, e.g.*, United States v. Iron Mountain Mines, Inc., 881 F. Supp. 1432 (E.D. Cal. 1995) (operator liability can be imposed on EPA where response costs increase as a result of EPA's own negligence); United States v. Ottati & Goss, Inc., 694 F. Supp. 977 (D.N.H. 1988), *aff'd in part and vacated in part on other grounds*, 949 F.2d 429 (1st Cir. 1990).

Particularly in those cases in which liability is a forgone conclusion, the technical component of the government's case should also be carefully scrutinized. Bearing in mind that section 113(j) of CERCLA, the "record review" provision,[12] will usually limit a PRP's ability to challenge certain aspects of the government's technical case, an evaluation of the strength of this part of the government's case might include the following areas of inquiry:

- If an RI/FS has been completed, did it conform to the National Contingency Plan (NCP)?
- If a remedial option has been chosen, does it conform to the NCP?
- Have all the prerequisites for proceeding with the response action been met?
- If remediation has actually been implemented, did it conform to the NCP?
- For a cost recovery action for work deemed by EPA to constitute a removal action, did the work exceed the one year/$2 million limits of section 104(c)(1) of CERCLA and, if not, was such work actually remedial in nature?

### D. Cost Outlays and Cost Recovery Alternatives

Total response costs and natural resource damages must be identified, estimated, and compared under either a settlement or a litigation scenario. A company's financial status should be analyzed to determine whether it can afford the possibly unrecoverable costs of protracted litigation and/or a lengthy cleanup effort. Fiscally sound companies that can afford such expenditures should nevertheless note that transaction costs involved in litigating a CERCLA case (attorneys' fees, consultants' fees, internal and PRP group administrative costs, etc.) can approach, or even exceed, a PRP's likely share of the actual costs of cleanup. Cleanup cost estimates provided by the government should be reviewed for reliability and compared to the costs of privately implementing such activities.

12. See chapter 6.

Also influential in a PRP's decision to settle or litigate will be the ability of the PRP to later recover its costs through insurance, third-party actions for contribution and/or indemnification, or even, perhaps, a claim against the Fund. Each of these avenues, of course, could entail additional transaction costs.

## E. Additional Considerations

### 1. *PRP Organization*

In cases with large numbers of PRPs, EPA typically will negotiate only with a single organization representing a number of PRPs; a PRP choosing to stay outside of such a group, and hence outside the mainstream of negotiations, could risk considerable exposure as an outsider.[13] That said, some circumstances may justify going it alone. For instance, a PRP who has an ironclad defense to liability, which differentiates it from other PRPs, may believe it has a better chance of settling for a nominal amount or avoiding liability entirely by distancing itself from the other PRPs.

### 2. *Publicity*

If adverse publicity is a significant concern to a PRP, then it must be factored into any decision regarding whether to litigate or settle a CERCLA claim. In most cases, litigating a CERCLA case is likely to create a greater degree of adverse publicity for the PRP than would be created by settling. Unfortunately, any stigma that may occur as a result of a company being named as a PRP will not be fully alleviated simply by settling.

### 3. *Consequential Risks of Litigation*

No evaluation of the option to litigate would be complete without an understanding of that option's consequential risks. Discovery could reveal activities by the PRP at another Superfund site, unearth embarrassing facts at the instant site, and disrupt the PRP's business with time-intensive searches for documents and depositions. Facts developed

13. See chapter 7.

through litigation could be used to the PRP's detriment in pending or future third-party litigation, including toxic tort suits.

## III. PRELITIGATION SETTLEMENT NEGOTIATION PROCEDURES WITH THE GOVERNMENT

Section 122(e) of CERCLA establishes formal prelitigation settlement negotiation procedures with the government, including issuance of a "special notice" that triggers a period of settlement negotiations during which a moratorium on certain EPA activities is statutorily imposed. In October 1987, EPA published its Interim Guidance on Notice Letters, Negotiations, and Information Exchange, which established the general process by which EPA implements the special notice and moratorium procedures.[14]

Notably, EPA is not precluded by statute from engaging in informal settlement negotiations prior to or in place of the formal CERCLA section 122(e) procedures discussed at length below. Indeed, EPA has made clear that issuance of the general notice letter (notifying a party that EPA regards it as a PRP) "should be viewed as a mechanism for initiating negotiations" whereas issuance of a special notice letter under section 122(e) "should be viewed as a mechanism for concluding negotiations."[15]

Whether negotiating informally or through the formal section 122(e) mechanism, PRPs engaging in prelitigation settlement negotiations with the government would be well advised to create a record documenting all good-faith settlement efforts.

### A. Special Notice and Moratorium Procedures

Prior to the issuance of a special notice, the Interim Guidance on Notice Letters directs EPA's regional personnel to issue a general notice letter to PRPs at sites that are proposed for or listed on the National Priorities

14. OSWER Directive No. 9834.10-1a (Sept. 26, 1988); 53 Fed. Reg. 5298 (Feb. 23, 1988) [hereinafter Interim Guidance]. Note that EPA reissued or revised model documents appended to the Interim Guidance, such as Appendix C, the "Model Notice Letters," on multiple occasions. The text referred to herein, however, has not been changed.

15. Interim Guidance, *supra* note 14, at 7.

List.[16] The general notice, which does not impose any type of moratorium on EPA action, should contain the following information: (1) "notification of potential liability for response costs"; (2) "a discussion about future notices and the possible future use of special notice procedures"; (3) a discussion of "site response activities"; (4) "a request for information about the site"; (5) "the release of certain site-specific information" (if it is available); (6) "a discussion about the merits of forming" a PRP group; (7) "a notice regarding the development of an administrative record"; and (8) a deadline for responding to the letter.[17]

Thereafter, if EPA "determines that a period of negotiation . . . would facilitate [settlement] . . . and would expedite remedial action," a special notice is sent to all PRPs at the site.[18] There are two types of special notice letters: (1) a special notice letter that requests PRPs to agree to perform a remedial investigation/feasibility study (RI/FS); and (2) a special notice letter that requests PRPs to perform a remedial design/remedial action (RD/RA).

Although, as discussed below, each type of special notice letter has its own peculiar set of timing considerations, both letters should contain the following information: (1) a notification of potential liability; (2) a discussion about the special notice and subsequent negotiation moratorium; (3) a discussion about the response activities to be conducted; (4) a copy of a statement of work or workplan and a draft administrative order on consent for the RI/FS; (5) a copy of a draft consent decree for the RD/RA; (6) a release of certain site-specific information (names and addresses of all PRPs, the volume and nature of the substances disposed of at the site and the ranking by volume of such substances); (7) a demand for payment of EPA costs incurred to date; (8) a notification about the administrative record; (9) a deadline for response; and (10) a discussion about EPA's conception of what constitutes a "good-faith offer."[19]

---

16. 53 Fed. Reg. at 5301.
17. *Id.* at 5301–02.
18. CERCLA § 122(e)(1).
19. 53 Fed. Reg. at 5305–06.

### 1. *RI/FS Special Notice Timing and Moratorium Procedures*

EPA's Interim Guidance on Notice Letters instructs regional personnel to issue an RI/FS special notice no later than 90 days before the date for initiating the RI/FS.[20] Upon receipt, PRPs have 60 days within which to respond with a "good-faith offer."[21] If no offer is received, or if EPA determines that a PRP proposal is not a good-faith offer, EPA may then proceed to conduct its own RI/FS. If a good-faith offer is made, the agency may not commence action under section 104(b) of CERCLA for a total of 90 days after issuance of the RI/FS special notice (i.e., an additional 30 days), and negotiations between the parties may proceed during that period.[22]

Certain things are immediately obvious in analyzing these procedures. First, it may not be possible to successfully conclude negotiations with the government within the allotted time frames (except, of course, on the government's terms). Second and consequently, both EPA and the PRPs must make a considerable effort in order to achieve success during the limited period of negotiation before the formal notice provision is triggered. As discussed in chapter 1, Congress provided EPA with substantially enhanced information-gathering authority through section 104(e) of CERCLA. EPA must use that authority to develop an adequate database—regarding the site, the waste, and the PRPs—to enable the PRPs to make informed judgments regarding their RI/FS offer. For their part, PRPs wishing to pursue a prelitigation settlement dialogue with the government should use the general notice letter as a springboard for preparing a strategy for responding to the RI/FS special notice well in advance of its issuance. PRPs cannot wait for the special notice to be issued before considering individual strategies for the site, collectively communicating and organizing with the other PRPs, and opening up lines of communication with EPA.

20. *Id.* at 5303.
21. CERCLA § 122(e)(2)(B).
22. *See* CERCLA § 122(e)(2)(A).

### 2. *RD/RA Special Notice Timing and Moratorium Procedures*

The second type of special notice, the remedial design/remedial action or RD/RA special notice, places a moratorium on EPA activity when and if EPA determines to allow the PRPs to implement the selected remedy.[23] Pursuant to EPA's Interim Guidance on Notice Letters, an RD/RA special notice can be issued at three possible junctures: (1) when the draft FS and proposed plan are released for public comments; (2) prior to the release of the FS; or (3) when the Record of Decision (ROD) is signed.[24] Like the RI/FS special notice, the RD/RA special notice commences a 60-day period during which the PRPs may make a good-faith offer.[25] Upon receipt of a good-faith offer, EPA and the PRPs have 120 days from the issuance of the RD/RA special notice—as few as 60 days from the PRP offer—to negotiate an agreement.

Once again, experience demonstrates that this is an extremely tight schedule. As with the RI/FS special notice, all efforts should be made by the PRPs to anticipate the RD/RA special notice and to informally negotiate and resolve any issues that may arise relating to the agreement to implement the remedy. Such efforts will be greatly facilitated if the PRPs are conducting the RI/FS.

Agreements reached between PRPs and the government at the conclusion of the RI/FS and RD/RA special notice periods may take different forms. An agreement to perform the RI/FS may take the form of either an administrative consent order or a consent decree.[26] An agreement under which the PRPs agree to implement the remedy must be in the form of a judicial consent decree (which is subject to public comment).[27]

## B. Nonbinding Preliminary Allocation of Responsibility

Reacting to the widespread criticism of early EPA efforts to provide meaningful allocation of PRP responsibility, Congress directed EPA in section 122(e)(3)(A) of CERCLA to develop guidelines for such

23. CERCLA § 122(e)(2)(A).
24. 53 Fed. Reg. at 5303–04.
25. CERCLA § 122(e)(2)(B).
26. CERCLA § 122(d)(3).
27. CERCLA § 122(d)(1)–(2).

allocation. The system, called the nonbinding preliminary allocation of responsibility (NBAR), is to factor in "volume, toxicity, mobility, strength of evidence, ability to pay, litigative risks, public interest considerations, precedential value, and inequities and aggravating factors."[28] These factors are the same settlement criteria contained in EPA's Interim CERCLA Settlement Policy, which EPA has used for some time to evaluate PRP liability.

EPA is to conduct the NBAR only after the RI/FS has been completed, and "[w]hen it would expedite settlements . . . and remedial action. . . ."[29] An EPA NBAR is not admissible in litigation and is not subject to judicial review.[30] Oddly, a substantial offer made by the PRPs that is based on EPA's NBAR may be rejected by EPA, with explanation, and that action by EPA is also not subject to judicial review.[31]

In order to provide EPA with the data and information necessary to conduct the NBAR, Congress has created new subpoena power by which the agency may "require the attendance and testimony of witnesses and the production of reports, papers, documents, answers to questions, and other information that the [EPA] deems necessary."[32] EPA is directed to allocate the entire response costs among all PRPs, not just those PRPs with whom it is negotiating.[33] The costs incurred by the agency in conducting the NBAR must be included in the PRP's offer and, in any event, are added to the ever-growing pot of recoverable response costs.[34]

EPA's "Interim Guidelines for Preparing Nonbinding Preliminary Allocations of Responsibility" indicate that the agency will not routinely prepare NBARs.[35] More likely, EPA will wait until "a significant percentage of PRPs at a site request one."[36] This may be good news for PRPs,

28. CERCLA § 122(e)(3)(A).
29. *Id.*
30. CERCLA § 122(e)(3)(C).
31. CERCLA § 122(e)(3)(E).
32. CERCLA § 122(e)(3)(B).
33. H.R. Conf. Rep. No. 99-962 at 253 (1986).
34. CERCLA § 122(e)(3)(D).
35. 52 Fed. Reg. 19,919 (May 28, 1987).
36. *Id.*

given the allocation factors EPA emphasized in its guidelines. Although EPA will initially allocate responsibility based on the volume of waste contributed by each PRP, EPA will adjust this allocation based on criteria set forth in its Interim Settlement Policy. These include, among other things, ability of the settling parties to pay, and inequities and aggravating factors.[37]

These are clear danger signals. By allowing EPA to consider (indeed, to develop, by information gained through subpoena) the "ability to pay" criterion, the PRPs negotiating with EPA, who are often the deepest pockets, may be leading with their economic chins in asking EPA to conduct an NBAR procedure. Inviting EPA to factor in ability to pay and "inequities and aggravating factors" is risky, at best. Although there may be circumstances in which requesting an NBAR would be appropriate, PRPs should carefully evaluate their respective situations before requesting EPA to allocate liability.

## IV. SETTLEMENT EVALUATION AND NEGOTIATION PROCEDURES WITH THE GOVERNMENT DURING PENDING LITIGATION

The formal commencement of a CERCLA section 106 and/or section 107 action does not, of course, preclude the possibility that the PRP's exposure at the site might still be settled without trial. To that end, section IV of this chapter considers factors pertinent to a settlement evaluation of pending CERCLA litigation before trial.

Guidance on formal settlement negotiation procedures during a pending CERCLA lawsuit does not exist. Section 122(e) and its elaborate special notice procedures appear to be confined to use in the prelitigation stage.[38] Similarly, EPA's Interim Settlement Guidance is largely geared toward settlement at the prelitigation stage. EPA personnel will rely heavily on the ten settlement criteria contained in the guidance for evaluating PRP settlement offers made during the course of litigation and, perhaps, to a lesser extent on the admonition of the guidance "to

37. 50 Fed. Reg. at 5037–38.
38. *See* CERCLA § 122(c)(2), (e)(2).

commence negotiations if responsible parties make an initial offer for a *substantial proportion* of the cleanup costs."[39]

Once named as defendants in a CERCLA action, PRPs should constantly assess and monitor the prospects of settling the case; a rough settlement position should always be developed in advance of any actual negotiations with the government. Although many of the same considerations that were relevant to the initial settle-or-litigate decision will still be relevant after litigation actually begins, the onset of litigation creates additional points that must be considered.

### A. Reevaluation of Major Prelitigation Settlement Considerations

Turning first to the major considerations that were discussed in the prelitigation context, the PRP's possible liability at the site should be reassessed in light of whatever discovery and motion practice has been or will be taken in the case. For instance:

- Has information obtained to date through discovery enhanced or weakened the government's liability case against the PRP?
- What will be the likely impact of additional discovery?
- Has the government already moved for summary judgment?
- If not, what are the prospects for surviving such a motion?

Information obtained through discovery can also be used to reevaluate the strength of the government's liability case against the other named defendants in the case and to reassess the possibility of government liability at the site. Technical information generated through discovery, as well as by the PRP's own experts, may help to cast new light on the strength of the government's technical case at the site.

The costs of litigation must be continually monitored and analyzed. The total costs of litigation to date should be determined and compared against the likely costs that will be incurred during the remaining preparation and trial of the case. At the same time, of course, the latest

39. 50 Fed. Reg. at 5041 (emphasis added).

prospects for obtaining recovery for such costs and/or the actual cost of cleanup from third parties must be reassessed.

### B. Additional Considerations

Each CERCLA lawsuit will present a host of additional considerations to be factored into the decision to continue litigation or seek settlement. The following are some of the more important considerations:

- What are the chances that favorable (or unfavorable) precedent will be created if litigation proceeds?
- Is there a possibility that some discrete phase of the case (e.g., a separate operable unit) can be settled?
- If the PRP is jointly litigating the case through a PRP group, are the prospects for liability the same for each defendant or is it now clear that some defendants face a greater chance of liability than others? As a consequence, have the dynamics of the PRP group changed? Should the PRP try to split off from the group and—either alone or in conjunction with similarly situated defendants—devise a separate settlement/litigation strategy?
- In that vein, has the group been effective in developing and implementing an innovative settlement strategy?
- Has there been a change in personnel on either side of the lawsuit that may create an opportunity to settle? For example, intransigence over a major sticking point by a departing player may be replaced by moderation and a willingness to compromise with the entry of a less-seasoned (or hardened) player.
- What is the status of negotiations between any remaining de minimis players and the government?

Information generated through litigation can help to clarify liability and technical and cost issues, thereby providing a more comprehensive, probing view of the case. Clarification, however, may not make the settlement decision-making process any easier; rather, the development of new facts and/or issues underscores the need for a continuing reevaluation of assumptions and strategies.

# V. SETTLEMENT AGREEMENTS

The culmination of successful negotiations with the government, either before or during a pending CERCLA section 106 and/or section 107 action, is a settlement agreement that, depending on a number of factors, will take the form of either a consent decree or an administrative order on consent. EPA has published "model consent decrees" to be used as the basis for these settlement agreements.[40] Section V of this chapter focuses on a few of the more significant provisions found in such agreements, special types of settlement agreements provided for under section 122 of CERCLA, and procedural considerations connected with CERCLA settlement agreements.

## A. Settlement Provisions

### 1. *Covenant Not to Sue*

Besides the obligation of the settling PRPs to perform or pay for some agreed-upon undertaking at the site, one of the most important aspects of the settlement agreement with the government will be the extent to which the PRP is protected against present and future liability to the United States. With the amendment of CERCLA in 1986, Congress made it clear that *any* agreement that limits the liability of a party to the United States must be in the form of a covenant not to sue, rather than a release.[41]

---

40. EPA's website publishes 13 separate settlement models, including "Cashout Consent Decree for Ability to Pay Peripheral Parties (CERCLA § 107)," "Cashout Consent Decree for Peripheral Party Settlements Not Based on Ability to Pay (CERCLA § 107)," "Cashout Settlement Agreement for Ability to Pay Peripheral Parties (CERCLA § 122(b) and AG Authority)," "Cashout Settlement Agreement for Peripheral Parties Not Based Upon Ability to Pay (CERCLA § 122(h) and AG Authority)," "Consent Decree for Recovery of Past Response Costs (CERCLA § 107)," "De Minimis Contributor Administrative Settlement Agreement and Order on Consent (CERCLA § 122(g)(4))," "De Minimis Contributor Consent Decree (CERCLA § 122(g)(4))," "De Minimis Landowner Administrative Settlement Agreement and Order on Consent (CERCLA § 122(g)(4))," "Municipal Solid Waste Generator/Transporter Consent Decree," "Non-Exempt De Micromis Party Consent Decree (CERCLA § 122(g)(4))," "Remedial Design/Remedial Action Consent Decree," and "Settlement Agreement for Recovery of Past Response Costs (CERCLA § 122(h))." In addition, EPA has in the past used a model Administrative Order on Consent for CERCLA Remedial Investigation/Feasibility Study OSWER Directive No. 9835.3-1A (issued Jan. 30, 1990).

41. CERCLA § 122(c)(1).

A covenant not to sue under CERCLA can be either discretionary or special. The former type of covenant, which is within the discretion of EPA to grant, may or may not include future liability; the latter unconditionally limits future liability and must be granted to a PRP if one of two statutory requirements is met.[42] Whether discretionary or special, a covenant not to sue for future liability cannot take effect until EPA has certified that the required remedial action has been completed.[43]

Most PRPs can expect to receive the discretionary covenant not to sue, which will be available if (1) the response action has been approved by EPA; (2) the PRP is in full compliance with a consent decree; (3) the covenant would expedite response action consistent with the NCP; and (4) the covenant is in the "public interest."[44] Special factors to be considered in determining whether the consent decree would be in the public interest are provided for by section 122(f)(4)(A)-(G) of CERCLA.

If granted, the discretionary covenant not to sue will usually contain a "reopener," a provision reserving EPA's right to require a settling party to take further response action notwithstanding EPA's covenant not to sue. EPA must include a reopener to cover future liability at the site that "arises out of conditions which are unknown at the time" the covenant becomes effective.[45] In vague, but extremely limited situations ("extraordinary circumstances"), EPA may decide not to include the "unknown conditions" reopener provision.[46] EPA's guidance does little to define such extraordinary circumstances, but confirms that they will rarely arise. Significantly, EPA will include a "reopener for remedy failure" in any covenant not to sue.[47] EPA will invoke this reopener to address "any situation in the future at the site which is judged to present a threat to public health and the environment."[48] In effect, PRPs become guarantors of the effectiveness of the remedy.

42. In its guidance on covenants not to sue, EPA narrowly construes these requirements. *See* 52 Fed. Reg. 28,038, 28,042 (July 27, 1987).
43. CERCLA § 122(f)(3).
44. CERCLA § 122(f)(1)(A)–(D).
45. CERCLA § 122(f)(6)(A).
46. CERCLA § 122(f)(6)(B).
47. 52 Fed. Reg. at 28,041.
48. *Id.*

For a price, some PRP may be able to eliminate the usual reopener requirements. Building upon a concept that was originally discussed within the context of de minimis settlement agreements, EPA's Guidance on Premium Payments in CERCLA Settlements (Nov. 17, 1988) allows a PRP to obtain a covenant not to sue without the usual reopeners in exchange for a "premium payment." In effect, a premium payment is "a risk taken by the government for providing PRPs with a release from liability not usually available (e.g., a covenant not to sue without the usual reopeners . . .) offset by a payment in excess of the cost projected to complete the remedy."[49] The guidance does not provide a formula for computing the amount of the premium payment. Instead, EPA will broadly consider the likelihood that future remediation will be required and the cost of such remediation.[50]

### 2. *Contribution Protection*

In section 113(f)(1) of CERCLA, Congress expressly created a federal right of contribution among parties that are liable under section 107. At the same time, Congress provided contribution protection (albeit limited) to those parties entering into settlements at Superfund sites.[51]

Specifically, section 113(f)(2) of CERCLA provides that a responsible party who has resolved its CERCLA liability through a judicially approved consent decree "shall not be liable [to other responsible parties] regarding matters addressed in the settlement." Contribution protection, which protects a settling PRP from contribution claims by a nonsettling PRP, provides an important incentive to settle, namely, a degree of finality. This can be a significant benefit where the matters addressed in the settlement are not being fully funded by the settling PRPs, and EPA intends to recover the difference from the nonsettlors.

There are limitations to this benefit, however. The statute makes it clear that the protection only applies to "matters addressed in the

49. EPA, OSWER Directive No. 9835.6, Guidance on Premium Payments in CERCLA Settlements (issued Nov. 17, 1988) , at 11.

50. *Id.*

51. CERCLA §§ 113(f)(2), 122(g)(5), 122(h)(4).

settlement."[52] Thus, for example, in a settlement that addresses only the surface cleanup requirements, the settling PRPs have no protection from other PRPs for remedial requirements for groundwater contamination.

Moreover, while contribution protection may extinguish the rights of other PRPs to obtain contribution from settling PRPs, it does not affect the rights of other PRPs to recover their own response costs in a section 107 cost recovery action.[53] Thus, for example, where PRPs voluntarily incur response costs, the United States cannot take away such PRPs' rights to recover such costs from another PRP. The statute makes contribution protection equally applicable to "an administrative or judicially approved settlement. . . ."[54] However, Congress created considerable doubt regarding the wisdom of relying on administrative settlements for contribution protection by adding the following language to section 308 of CERCLA:

> If an administrative settlement under Section 122 has the effect of limiting any person's right to obtain contribution from any party to such settlement, and if the effect of such limitation would constitute a taking without just compensation in violation of the fifth amendment of the Constitution of the United States, such person shall not be entitled, under other laws of the United States, to recover compensation from the United States for such taking, but in any such case, such limitation on the right to obtain contribution shall be treated as having no force and effect.

In those limited circumstances where a choice exists between an administrative or judicial settlement, this provision should weigh heavily in support of the latter.

The contribution protection provision has consequences for nonsettlors. In addition to barring contribution claims against settlors for matters addressed in the settlement, section 113(f)(2) of CERCLA provides

52. *Id.*

53. *See* United States v. Atl. Res. Corp., 551 U.S. 128 (2007) (distinguishing private cost recovery actions under section 107(a)(4)(B) from contribution actions under section 113(f)).

54. CERCLA § 113(f)(2).

that such settlement "reduces the potential liability of the [nonsettlors] by the amount of the settlement." In effect, this language embodies the "pro tanto" rule of the Uniform Contribution among Tortfeasors Act (UCATA).[55] Under the UCATA pro tanto rule, settlement in good faith between the claimant and settling tortfeasor reduces the liability of the other tortfeasors to the claimant on a pro tanto or dollar-for-dollar basis.[56] This approach places on the claimant the risk of settling for less than tortfeasor's pro rata share of liability. This differs from the alternative "pro rata" approach embodied by the Uniform Comparative Fault Act (UCFA).[57] Under the UCFA pro rata approach, a settlement between the claimant and the settling tortfeasor reduces the liability of the other tortfeasors to the claimant by the amount of the settling tortfeasor's equitable or proportionate share of the common liability.[58]

A number of courts have confirmed that section 113(f)(2) of CERCLA embodies the UCATA pro tanto rule, and reflects Congress's "conscious choice in 1986 not to adopt UCFA principles for CERCLA purposes."[59] In the private party settlement context, however, courts are split over which doctrine applies. See chapter 9.

---

55. 12 U.L.A. 201. The National Conference of Commissioners on Uniform State Laws promulgated UCATA as a model law in 1955.

56. *Id.* § 4 at 284.

57. 12 U.L.A. 124. The National Conference of Commissioners on Uniform State Laws promulgated UCFA in 1977 as a model law for use in states that follow the principle of comparative fault.

58. *See id.* § 6 at 147.

59. *See* United States v. Cannons Eng'g Corp., 720 F. Supp. 1027, 1048 (D. Mass. 1989), *aff'd*, 899 F.2d 79 (1st Cir. 1990); Smith Land & Improvement Corp. v. Celotex Corp., 851 F.2d 86, 89 (3d Cir. 1988) (settlements with the government give rise to pro tanto reduction of claims against nonsettlors); Kelley v. Thomas Solvent Co., 790 F. Supp. 731, 735–36 (W.D. Mich. 1991) ("According to the express terms of the statute, a settlement 'reduces the potential liability of others by the amount of the settlement'") (citation omitted); United States v. Rohm & Haas Co., 721 F. Supp. 666, 675–79 (D.N.J. 1989) (same); *In re* Acushnet River & New Bedford Harbor, 712 F. Supp. 1019, 1027 (D. Mass. 1989) ("the words of the statute are clear: the potential liability of the others is reduced 'by the amount of the settlement,' not by the settlor's proportionate share of any damages ultimately determined to have been caused"). *But see* United States v. Laskin, No. C84-2035Y, 1989 U.S. Dist. LEXIS 4900, at *19 (N.D. Ohio Feb. 27, 1989) (nonsettlors' liability to the United States should be governed by UCFA, and therefore any claims against nonsettlors would "be reduced by the greater of the amount of the settling defendants' combined equitable share of the obligation or the amount of the settlement).

### 3. *Stipulated Penalties and Dispute Resolution Procedures*

Pursuant to section 121(e)(2) of CERCLA, a settlement agreement entered as a consent decree must "contain stipulated penalties for violations of the decree in an amount not to exceed $25,000 per day." To limit the potential harshness of this provision, PRPs often attempt to negotiate the amount of the stipulated penalty and narrow the scope of events or circumstances that will constitute a "violation of the decree."[60] Particular attention should be directed to the drafting of a favorable force majeure clause and, if possible, a notice provision designed to provide an opportunity to rectify any problems before penalties begin to actually accrue.

A dispute resolution procedure must also be contained in a consent decree for the purposes of requiring "the parties to attempt expeditiously to resolve disagreements concerning implementation of the remedial action informally. . . ."[61] Typically, such dispute resolution clauses require referral to, and ultimate resolution of the dispute by, EPA, not a court.

As an alternative to this type of dispute resolution, PRPs may wish to seek inclusion in the consent decree of an alternative dispute resolution (ADR) technique that makes use, at a minimum, of a third-party mediator to provide independent, advisory fact-finding. In this regard, EPA's Final Guidance on Use of Alternative Dispute Resolution Techniques in Enforcement Actions (August 14, 1987), a document of general applicability to CERCLA as well as other federal environmental laws, approvingly discusses the use of mediation as a possible ADR technique.

## B. Special CERCLA Section 122 Settlements

### 1. *De Minimis Settlements*

At multiple-party sites, it is possible for a de minimis PRP—i.e., a PRP whose contribution of wastes to the site is minimal—to bear the same weight of transaction costs as a large volume PRP, such as tracking the progress of the case, attending PRP group meetings, reviewing documents, and on occasion, becoming entangled in full-scale, costly litigation

60. CERCLA § 121(e)(2).
61. CERCLA § 121(e)(2).

with either the government, their fellow PRPs, or both. As noted earlier, under this worst-case scenario, a de minimis PRP's transaction costs could exceed its share of cleanup costs. In response to this problem, in 1986 Congress added section 122(g) of CERCLA, which specifically authorizes EPA to enter into de minimis settlements with both generators and landowners. A summary of each type of de minimis settlement is set forth below.

**a. Section 122(g)(1)(A) Settlements (De Minimis Generators)**

Under section 122(g)(1)(A) of CERCLA, EPA may enter into a de minimis settlement with generator PRPs who meets the following criteria: (1) such settlement involves a "minor portion of the response costs" at the site;[62] (2) the amount and toxicity of the hazardous substances of a PRP are minimal in comparison to other hazardous substances;[63] and (3) EPA determines that settlement is "practicable and in the public interest."[64] Beyond the preceding instructions, neither section 122(g) of CERCLA nor its legislative history provides any general or specific criteria as to who or what may qualify for a de minimis settlement. Similarly, the appropriate time for EPA to enter into such settlements is left to be interpreted from vague phrases such as "[w]henever practicable," "as promptly as possible," and "as soon as possible after [EPA] has available the information necessary. . . ."[65] These and other issues led EPA to adopt its Interim Guidance on Settlements with De Minimis Waste Contributors under Section 122(g) of SARA.[66] The guidance attempts to deal with the key issues of eligibility, timing, and the terms of settlement.

*i. Eligibility*

In determining de minimis eligibility, EPA's guidance stresses that a "de minimis contributor will be defined on a site-specific basis."[67] For a PRP

62. CERCLA § 122(g)(1).
63. CERCLA § 122(g)(1)(A)(i)–(ii).
64. CERCLA § 122(g)(1).
65. CERCLA § 122(g)(1), (3).
66. 52 Fed. Reg. 24,333 (June 30, 1987).
67. *Id.* at 24,335.

to be eligible for a de minimis settlement, the guidance requires that the amount of waste contributed must be minimal in comparison to the total universe of waste, but does not set a definite eligibility cutoff per PRP, nor a cutoff for the aggregate of all waste contributed by a group of de minimis PRPs.

Toxicity is likewise treated broadly. The guidance requires that the hazardous substances contributed by a potential de minimis PRP must not be "significantly more toxic and not of significantly greater hazardous effect than other hazardous substances at the facility. . . ."[68] Further, the hazardous substance must not require "disproportionately high treatment and disposal costs" or require "a different or more costly remedial technique than that which otherwise would be technically adequate for the site. . . ."[69]

EPA's thoughts are clearest in the hypothetical cases posed in the guidance. For example, where all PRPs are relatively small, a de minimis settlement would not be appropriate, both because individually their hazardous substances are not *relatively* minimal, *and* because cumulatively they represent more than a minor portion.[70] Where the major PRP(s) "are bankrupt or otherwise non-viable," EPA concludes that "it may not be in the public interest to 'cash out' smaller contributors before reaching a settlement with the remaining parties."[71] Because of joint and several liability, EPA could require each of the remaining PRPs, even if individually and collectively responsible for only a minor portion of the hazardous substances at the site, to pay the orphan share, as well as their own share, of the current and future remedial costs.

*ii. Timing Considerations*

The statute directs EPA to reach de minimis settlements only after the agency "has available the information necessary. . . ."[72] Under the

68. *Id.* at 24,336.
69. *Id.*
70. *Id.*
71. *Id.* at 24,336.
72. CERCLA § 122(g)(3).

guidance, before EPA can enter a de minimis settlement, the agency must have "adequate" information regarding (1) "the identity, waste contributions and viability of PRPs"; and (2) the cost of remediating the site.[73] EPA may consider early settlements, but will impose expansive reopeners or substantial premiums.[74] However, the guidance cautions that, generally, de minimis settlements should not be entered into before

> the Agency is able to estimate, with a reasonable degree of confidence, the total response costs. . . . The Agency usually will arrive at this level of confidence only after a . . . ("RI/FS") and a Record of Decision ("ROD") have been (or are close to being) completed at the site.[75]

Thus, as a practical matter, EPA will likely defer de minimis settlements until after the RI/FS is completed and the ROD is issued. This practice effectively frustrates the fundamental purpose of de minimis settlement—allowing small generators to settle out early.

The agency suggested several approaches that could make pre-ROD de minimis settlement possible, including the cap mechanism, under which de minimis PRPs pay their projected share of all future costs, but remain subject to additional cost recovery action if "actual costs exceed EPA's estimate by an agreed-upon dollar amount or percentage."[76] Under this approach, the de minimis parties would be protected (by the covenant not to sue) from future participation in remedial activities unless and until the remedial costs exceed a specific level—the cap.

*iii. Terms*

The extent to which the de minimis settlor is released from future liability arising out of the cleanup and the amount that the settlor must pay to account for early settlement uncertainties are key components

73. 52 Fed. Reg. at 24,336.
74. *Id.*
75. *Id.*
76. *Id.* at 24,337.

to any de minimis settlement. CERCLA provides no standards on these issues other than to authorize EPA to use covenants not to sue in de minimis settlements.[77]

Under EPA's guidance, the terms of the release will depend, to a great extent, on the timing of the settlement. The earlier the settlement, the greater the uncertainty and, therefore, the narrower the scope of the release. Irrespective of the timing of the de minimis settlement, EPA will always require at least two reopeners:

1. claims for natural resource damages will be "expressly reserved unless the Federal natural resource trustee has agreed in writing" to the settlement; and
2. claims for further relief "if information not known to the Government at the time of the settlement is discovered which indicates that the volume or toxicity criteria for the site's de minimis parties are no longer satisfied with respect to that party."[78]

The cash offer submitted by the de minimis parties "must be at least equal to their volumetric share of the total past and projected response costs at the site."[79] The agency's projection of future response costs generally should be based on the most probable costs of the response action. EPA will generally require payment of a premium in exchange for granting a covenant not to sue that does not include reopeners for cost overruns and future response action. EPA does not elaborate on the calculation of premiums in its guidance; however, the degree of uncertainty as to the total cleanup costs bears directly on the amount of premium payment, as well as on the scope of the release.

**b. Section 122(g)(1)(B) Settlements (De Minimis Landowners)**

Under section 122(g)(1)(B) of CERCLA, EPA may enter into a de minimis settlement with a landowner who meets the following criteria: (1) the PRP "is the owner of the real property on or in which the facility is

77. *See* CERCLA § 122(g)(2).
78. 52 Fed. Reg. at 24,337.
79. *Id.* at 24,338.

located"; (2) "did not conduct or permit the generation, transportation, storage, treatment or disposal of any hazardous substance at the facility"; (3) "did not contribute to the release or threat of release of a hazardous substance at the facility through any action or omission"; and (4) had no "actual or constructive knowledge that the property was used for the generation, transportation, storage, treatment, or disposal of any hazardous substance."[80] As EPA notes in its Guidance on Landowner Liability under Section 107(a)(1) of CERCLA, De Minimis Settlements under Section 122(g)(1)(B) of CERCLA, and Settlements with Prospective Purchasers of Contaminated Property,[81] these criteria are the same as the elements that the defendant must prove to establish the affirmative "innocent landowner" defense under section 107(b)(3). However, inasmuch as section 122(g)(1)(B) comes into play in the settlement context, whereas section 107(b)(3) comes into play in the litigation context, the quality and quantity of evidence provided by a landowner in support of his eligibility for a de minimis settlement may differ from that necessary for establishment of the third-party defense at trial. To that end, the guidance instructs EPA's regional personnel to "make an effort in the early stages of a case to determine whether a landowner satisfies the elements necessary to establish [the innocent landowner] defense under Section 107(b)(3) of CERCLA."[82] For nongovernment entities that acquire the property in question through means other than inheritance or bequest, qualification as an "innocent landowner" will depend in large part upon whether "all appropriate inquiry" was made by the private landowner at the time of acquisition.

As described in chapter 4, "all appropriate inquiry" within the context of a commercial transaction will frequently entail a number of investigative steps. CERCLA does not require landowners to submit the results of such investigations to EPA, but it seems unlikely that the

80. CERCLA § 122(g)(1)(B).

81. 54 Fed. Reg. 34,235 (Aug. 18, 1989). EPA has modified its guidance on several occasions. *See, e.g.*, EPA Memorandum on Revised Model CERCLA De Minimis Landowner Consent Decree and Revised Model CERCLA De Minimis Landowner Administrative Order on Consent (May 13, 2004).

82. 54 Fed. Reg. at 34,237.

agency will actually be able to determine whether a landowner qualifies for the innocent landowner defense without such information. Rather than wait for EPA, it would appear to be incumbent upon the landowner named as a PRP to step forward and make the innocent landowner case as soon as possible. EPA's guidance states that "the Agency will place on the landowner the burden of coming forward with information establishing his eligibility for a de minimis settlement."[83] Thus, in addition to information pertinent to the issue of "all appropriate inquiry," the PRP landowner must submit information relevant to the issues of whether due care was exercised and whether the landowners contributed to the release or threat of release through any act or omission.[84]

In 2002, Congress passed the Small Business Liability Relief and Brownfields Revitalization Act,[85] which, among other things, clarified the requirements for demonstrating all appropriate inquiries. Following passage of the 2002 act, EPA published Interim Guidance Regarding Criteria Landowners Must Meet in Order to Qualify for Bona Fide Prospective Purchaser, Contiguous Property Owner, or Innocent Landowner Limitations on CERCLA Liability (Common Elements).[86] PRPs seeking to qualify for such status and settle on that basis with EPA should consult this guidance to evaluate the merits of their position.

Settlement terms with a landowner meeting the criteria of section 122(g)(1)(B) of CERCLA will depend on the strength of the landowner's case. Where there is a "very high probability" that the landowner would prevail in establishing a third-party defense at trial, the agency will typically require site access and due care assurances; less convincing cases may also require cash payments for response activities at the site. In exchange, the landowner can expect contribution protection and a covenant not to sue with a reopener designed, in part, to protect EPA from the submission of false or inaccurate information.

With the passage of the Small Business Liability Relief and Brownfields Revitalization Act, parties to commercial transactions have become

83. *Id.* at 34,239.
84. *Id.* at 34,238–39.
85. Pub. L. No. 107-118, 115 Stat. 2356 (2002).
86. This March 6, 2003, guidance is published on EPA's website.

increasingly cognizant of the need to exercise all appropriate inquiry at the time of acquisition. Given the higher standard of investigation placed on parties to commercial transactions, contamination will usually be detected by all but the most cursory of commercially based investigations. Practically speaking, section 122(g)(1)(B) of CERCLA is far more likely to be applicable (1) to unsophisticated pre- and post-CERCLA purchasers of property outside of the commercial context and (2) to more commercially sophisticated parties facing possible liability under section 107(a)(2) of CERCLA for periods of former ownership that substantially predate the enactment and amendment of CERCLA. In the latter instance, it may be arguable that the level of inquiry expected of a purchaser of real property in, for example, the 1950s or 1960s was not as great at that time because of the lower public awareness of the hazards associated with contaminated property.[87]

### 2. *Mixed Funding Settlements*

In SARA, Congress gave EPA a means to avoid a recurring stumbling block to settlement: reluctance among PRPs to fund more than their fair share of cleanup costs, which often arises where some PRPs either refuse to participate, are impecunious or cannot be located. Section 122(b)(1) of CERCLA specifically authorizes EPA to enter into "mixed funding" arrangements.[88] Under a mixed funding arrangement, EPA uses Superfund money to cover the nonparticipating PRPs' share of the cleanup costs. In other words, mixed funding can relieve the participating PRPs of the burden of funding 100 percent of the cleanup costs, and having to incur the transactional cost of a contribution suit against nonparticipating PRPs to recoup their overpayment.

EPA addressed the use of mixed funding in a guidance document titled Evaluating Mixed Funding Settlements under CERCLA.[89] The Mixed Funding Guidance notes that factors of primary importance in assessing a mixed funding settlement offer will include (1) the individual strengths

87. *See* United States v. Serafini, 706 F. Supp. 346 (M.D. Pa. 1988).

88. EPA's decision to use this provision, however, is completely discretionary and beyond judicial review. *See* CERCLA § 122(a).

89. 53 Fed. Reg. 8279 (Mar. 14, 1988).

of the liability cases against the settling PRPs and the nonsettling PRPs; (2) the "Government's options in the event that settlement negotiations fail"; (3) the "Size of the portion or operable unit for which the Fund will be responsible"; and (4) "Good-faith negotiations and cooperation of [settling PRPs] and other mitigating and equitable factors."[90] Optimal candidates for mixed funding are cases in which "[t]he potential portion or operable unit to be covered by the Fund is small, or the settling PRPs offer a substantial portion of the total cost of cleanup" and "[t]he Government has a strong case against financially viable non-settling PRPs. . . ."[91] Not surprisingly, a "poor" candidate for mixed funding is one in which the case against settling parties is "strong" and the "potential Fund portion is large."[92]

### a. Forms of Mixed Funding

Mixed funding may take one of three forms: (1) "preauthorization"; (2) "cash-out"; and (3) "mixed work."[93] In a preauthorization arrangement, the form specifically authorized by section 122(b)(1) of CERCLA, the PRPs agree to perform a response activity, and EPA agrees to reimburse the PRPs for a portion of their response costs. One advantage of a preauthorization agreement is that the Fund will be responsible for costs of remedy failure up to the same proportion as the Fund contributed in the original agreement.[94] EPA's Mixed Funding Guidance makes it clear, however, that remedy failure attributable to PRP negligence will not be covered by the Fund.[95]

Under the "cash-out" form of mixed funding, EPA agrees to conduct the response action, and the PRPs pay EPA up front for a portion of the response costs to be incurred. EPA's Mixed Funding Guidance suggests that cash-out settlements will be most appropriate if they (1) occur before the commencement of a Fund-led action and (2) "involve PRPs

90. *Id.* at 8281.
91. *Id.*
92. *Id.* at 8282.
93. *Id.*
94. CERCLA § 122(b)(4).
95. 53 Fed. Reg. at 8282.

which have contributed a low percentage of the waste to a site, and are not technically or financially capable of conducting the entire response action."[96] Unlike the preauthorization form of agreement, PRPs remain liable for all costs resulting from remedy failure.

If the response action involved at a given site can be conducted in discrete portions, the Mixed Funding Guidance sanctions the use of a "mixed work" arrangement under which EPA and the PRPs agree to separately conduct and pay for such discrete portions.[97]

**b. "Orphan" and "Recalcitrant" PRPs**

Since the enactment of section 122(b)(1) of CERCLA, considerable debate has arisen over the extent to which response costs attributable to "orphan" PRPs (e.g., unknown or insolvent PRPs) should be borne by the Fund. The conference report for section 122(b)(2) appears to indicate that mixed funding should be used to pay for the shares of orphan PRPs, recalcitrant PRPs, or both:

> Section 122(b)(1), which addresses mixed funding for site response actions, provides that [EPA], where appropriate and in the public interest, may reimburse parties for certain costs of actions under the agreement by using monies from the Fund on behalf of parties who are unknown, insolvent, similarly unavailable, or refuse to settle.[98]

EPA's policy, however, is quite different. Pursuant to the Mixed Funding Guidance, "the Agency will not approve mixed funding simply on the basis that a share of wastes at a site may be attributable to an unknown or financially non-viable party."[99] Notably, however, pursuant to a 1996 Policy EPA may compromise up to 25 percent of its claim against settling liable parties where an orphan share exists at the site to be remediated.[100]

96. *Id.* at 8283.
97. *Id.* at 8283–84.
98. H.R. CONF. REP. NO. 99-962 at 252 (1986).
99. 53 Fed. Reg. at 8281.
100. *See* Memorandum from Steven A. Herman, Assistant Adm'r, EPA Office of Enforcement & Compliance Assurance, to Reg'l Adm'rs, Interim Guidance on Orphan Share Compensation

## C. Procedural Considerations

Section 122 of CERCLA contains a number of provisions addressing procedures to be followed in settlements under sections 106 and 107. Following internal approval by the Department of Justice, a settlement agreement with a PRP with respect to remedial action under Section 106 of CERCLA must be entered in the appropriate federal district court as a consent decree.[101] The proposed consent decree must be filed with the court at least 30 days before it can become final.[102] Before the proposed consent decree becomes final, the Department of Justice must provide an opportunity to persons who are not parties to the consent decree to submit comments on its contents.[103] The Department of Justice is required to consider, and file with the court, any written comments, views, or allegations pertaining to the proposed consent decree.[104] If such comments "disclose facts or considerations which indicate that the proposed judgment is inappropriate, improper or inadequate," the Department of Justice may withdraw or withhold its consent to the proposed decree.[105] As a consequence, citizen groups that oppose the remedy often submit comments in opposition and lobby others to do the same. PRPs that are not parties to the proposed consent decree also have a right to submit comments and frequently do, albeit for different reasons. Their focus usually concerns the allocation of response costs paid by the settling PRPs. As noted *supra*, one effect of the settlement is to reduce the government's claim against nonsettlors by the dollar amount of the settlement. This, combined with the fact that settling PRPs receive contribution protection, often raises a concern among nonsettlors that the government may have accepted a dollar amount from the settling PRPs that is less than their equitable share of liability, thereby exposing the nonsettlors to a disproportionately large balance of the government's response costs.

for Settlors of Remedial Design/Remedial Action and Non-Time Critical Removals, at I-X (June 3, 1996).

101. CERCLA § 122(d)(1)(A).
102. CERCLA § 122(d)(2)(A).
103. CERCLA § 122(d)(2)(B).
104. *Id.*
105. *Id.*

As discussed in chapter 6, courts tasked with approving a proposed consent decree must independently scrutinize the terms, and cannot approve it unless it concludes that the settlement is fair, reasonable, and consistent with both the Constitution and the mandates of Congress.[106] That said, the settlement of litigation is highly desirable from the perspective of a court reviewing a consent decree. Indeed, a court can approve a settlement if provided with almost any logical rationale for apportioning liability.[107]

If settlement was reached during the prelitigation stage, the entry of the consent decree in federal district court will have to be contemporaneously filed with a complaint stating a section 107 cause of action against the PRP. Since, under section 122(d)(1) of CERCLA, EPA is *not* required to include a finding of "imminent and substantial endangerment" in the consent decree, PRPs should always argue against the inclusion of such a finding. To further hedge against possible toxic tort exposure, PRPs should take advantage of section 122(d)(1)(C) of CERCLA, which allows EPA to fashion a decree so that its entry and subsequent compliance do not constitute an admission of liability.

A section 107 claim for total response costs exceeding $500,000 (excluding interest) may be compromised and settled only with the prior written approval of the Department of Justice.[108] Any federal or state agency with authority to undertake a response action may independently settle a section 107 claim for total response costs not exceeding $500,000 (excluding interest) without prior Department of Justice approval if the claim has not been referred to the Department of Justice for further action.[109]

Section 122(h)(2) of CERCLA also authorizes the use of arbitration procedures to resolve such a claim. Special arbitration procedures were published by EPA.[110]

---

106. *See, e.g.*, United States v. City of Fort Lauderdale, 81 F. Supp. 2d 1348, 1350 (S.D. Fla. 1999). For further case law and discussion, see chapter 6, section VI.

107. See chapter 6 for a comprehensive discussion of the relevant case law.

108. CERCLA § 122(h)(1).

109. *Id.*

110. *See* 54 Fed. Reg. 23,174 (May 30, 1989).

# CHAPTER 6

# Judicial Review

## I. INTRODUCTION

As noted in the Overview, Congress enacted CERCLA to provide the federal government with the authority and funding to promptly clean up abandoned contaminated sites and, where possible, to shift the expense of such remediation from the taxpayers to the responsible parties. In the early years of the Superfund program, the goal of prompt cleanup was frustrated by legal challenges to every aspect of the scheme—from the constitutionality of Congress imposing liability retroactively, to the premise that liability was intended to be retroactive, strict, joint, and several.[1] These legal challenges mired the program in delays and dramatically increased the cost of cleanup.

Recognizing that the goals of prompt and efficient cleanup were being jeopardized by a litigation tsunami, Congress amended CERCLA in 1986, adding section 113(h), which severely limits the jurisdiction of federal courts to review removal and remedial actions before they are completed. This near-ban on pre-enforcement judicial review reflects congressional sentiment for "shooting first and asking questions later."[2] Notwithstanding the addition of section 113(h), questions regarding the

1. See discussion of these legal challenges in chapter 4, *supra*.

2. *See* 132 Cong. Rec. 28,409 (1986) (statement of Sen. Stafford); *see also* B.R. MacKay & Sons, Inc. v. United States, 633 F. Supp. 1290, 1294 (D. Utah 1986) ("It appears that with the dangers or potential dangers caused by hazardous substances, shooting first and asking questions later was the intent of Congress, making it clear that under CERCLA the EPA should have and has full reign to conduct or mandate uninterrupted cleanups for the benefit of the environment and the populous." (sic)).

availability of pre-enforcement judicial review persist. In this chapter, we examine the availability, scope, and standard for judicial review, as well as the court's reasoning where judicial review has been denied.

## II. PRE-ENFORCEMENT REVIEW OF SELECTED REMEDY

### A. Pre-enforcement Review of EPA Action Barred Under Section 113(h)

EPA's response authority under section 104 of CERCLA permits EPA to take short-term removal actions to respond quickly to a release of a hazardous substance and long-term remedial action at sites listed on the National Priorities List (NPL).[3] As originally drafted, CERCLA did not explicitly bar pre-enforcement judicial review of removal and remedial actions. In a series of decisions prior to the 1986 amendments, however, courts repeatedly held that Congress intended to bar such pre-enforcement review.[4]

Section 113(h), which was added to CERCLA in 1986, codifies the case law prohibiting pre-enforcement review of response actions selected by EPA pursuant to section 104. Specifically, section 113(h) of CERCLA denies the federal courts subject-matter jurisdiction "to review any challenges to removal or remedial action selected under section 9604 of this title, or to review any order issued under section 9606(a) of this title, in any action," except for the following five specified situations:

3. *See* chapter 1.

4. *See, e.g.*, J.V. Peters & Co. v. Adm'r, EPA, 767 F.2d 263, 265 (6th Cir. 1985) ("[p]re-enforcement review would lead to considerable delay in providing cleanups, would increase response costs, and would discourage settlements and voluntary cleanups"); Wheaton Indus. v. EPA, 781 F.2d 354, 356–57 (3d Cir. 1986) ("the district court correctly ruled that judicial review was not available under section 104 of CERCLA at this time"); Lone Pine Steering Comm. v. EPA, 777 F.2d 882 (3d Cir. 1985), *cert. denied*, 476 U.S. 1115 (1986); Barnes v. U.S. Dist. Ct., 800 F.2d 822 (9th Cir. 1986); Wagner Seed Co. v. Daggett, 800 F.2d 310 (2d Cir. 1986); Pac. Resins & Chems., Inc. v. United States, 654 F. Supp. 249, 253 (W.D. Wash. 1986); *B.R. MacKay & Sons, Inc.*, 633 F. Supp. 1290; Wagner Elec. Corp. v. Thomas, 612 F. Supp. 736 (D. Kan. 1985); and Aminoil, Inc. v. U.S. EPA, 599 F. Supp. 69 (C.D. Cal. 1984).

(1) An action under [section 107] to recover response costs or damages for contribution;
(2) An action to enforce an order issued under [section 106(a)] or to recover a penalty for violation of such order;
(3) An action for reimbursement under [section 106(b)(2)];
(4) An action under [section 310] (relating to citizens suits) alleging that the removal or remedial action taken under [section 104] or secured under [section 106] was in violation of any requirement of this chapter. Such an action may not be brought with regard to a removal where a remedial action is to be undertaken at the site;
(5) An action under [section 106] in which the United States has moved to compel a remedial action.[5]

The legislative history of section 113(h) indicates that the congressional intent behind the provision was to prevent time-consuming litigation from delaying cleanup action.[6]

Adding section 113(h) to CERCLA, however, did not end attempts to challenge the remedy selected prior to enforcement. Emblematic of these attempts is *Boarhead Corp. v. Erickson.*[7] The plaintiffs sought to challenge an EPA cleanup plan for a contaminated parcel of land on which the plaintiffs alleged there were American Indian remains and artifacts. The plaintiffs sought to challenge EPA's proposed cleanup plan under the provisions of the National Historic Preservation Act. The plaintiffs argued that if they were not permitted to pursue their

5. CERCLA § 113(h)(1)–(5).

6. *See, e.g.*, 132 Cong. Rec. H9583 (daily ed. Oct. 8, 1986) ("Clearly the conferees did not intend to allow any plaintiff, whether the neighbor who is unhappy about the construction of a toxic waste incinerator in the neighborhood, or the potentially responsible party who will have to pay for its construction, to stop a cleanup by what would undoubtedly be a prolonged legal battle. It was for this very reason that the conferees included section 113(h)."). *See also Wagner Seed Co.*, 800 F.2d at 315 (citing the above legislative history as establishing Congress's intent to prevent time-consuming litigation from delaying prompt cleanup of Superfund sites); Boarhead Corp. v. Erickson, 923 F.2d 1011, 1019 (3d Cir. 1991) (same); Alabama v. EPA, 871 F.2d 1548, 1560 (11th Cir. 1989) (same), *cert. denied*, 493 U.S. 991 (1989); Reynolds v. Lujan, 785 F. Supp. 152, 153 (D.N.M. 1992) (same); Neighborhood Toxic Cleanup Emergency v. Reilly, 716 F. Supp. 828, 832 (D.N.J. 1989) (same).

7. 923 F.2d 1011.

legal challenge, EPA would violate the National Historic Preservation Act, which in turn would result in irreparable harm to irreplaceable artifacts. The government argued that allowing the plaintiffs to pursue pre-enforcement review of the remedy would violate the jurisdictional bar under section 113(h). The Third Circuit agreed with the government's position, albeit reluctantly:

> Although post-study judicial review cannot rectify damage to historical artifacts or remains on this landmark site that occurs in the course of EPA's clean-up, we must presume Congress balanced the problem of irreparable harm to such interests and concluded that the interest in removing the hazard of toxic waste from Superfund sites outweighed it. Boarhead's remedy lies with Congress, not the district court.[8]

Many other circuit courts have rejected pre-enforcement challenges to remedial plans based on similar reasoning.[9]

### B. Citizen Suit Exception to the Pre-enforcement Review Bar

One of the enumerated exceptions to the pre-enforcement review bar has proven to be elusive, specifically, the exception for citizen suits "alleging that the removal or remedial action taken under Section [104] . . . or secured under Section [106] . . . was in violation of" the Act.[10] Courts reviewing this issue, however, including at least one circuit court, have looked at the express language of section 113(h)(4), as well

8. *Id.* at 1023.

9. *See, e.g.*, Schalk v. Reilly, 900 F.2d 1091 (7th Cir. 1990) (rejecting pre-enforcement challenge to remedy that had been premised on EPA's failure to prepare an environmental impact statement prior to adopting the remedy, which they alleged constituted a violation of the National Environmental Policy Act, a position rejected by the court); Ark. Peace Ctr. v. Ark. Dep't of Pollution Control & Ecology, 999 F.2d 1212, 1218 (8th Cir. 1993), *cert. denied*, 511 U.S. 1017 (1994) (reversing lower court's grant of injunction against EPA removal plan, noting use of past tense in section 113(h), specifically, that challenge to removal permissible only where "the removal or remedial action taken . . . was in violation of any requirement of [CERCLA]") (citation omitted); N. Shore Gas Co. v. EPA, 930 F.2d 1239, 1244 (7th Cir. 1991); *Alabama*, 871 F.2d 1548 (rejecting pre-enforcement review, noting use of past tense in section 113(h)); *In re* Combustion Equip. Assocs., Inc., 838 F.2d 35 (2d Cir. 1988).

10. CERCLA § 113(h)(4).

as the legislative history of the provision, and have held that section 113(h)(4) applies only *after* a removal action or some discrete phase of the remedial action is completed.

The first case to reject use of the citizen suit provision to circumvent the general prohibition of section 113(h) against pre-enforcement review was *Cabot Corp. v. EPA*.[11] Section 310(a)(2) of CERCLA grants any person the right to bring a civil action on his own behalf against EPA for any failure to perform any duty that is not discretionary. In this case, a potentially responsible party (PRP) brought suit against EPA under the citizen suit provision, claiming that the agency had failed to perform a nondiscretionary duty, namely, selecting the most cost-effective remedy. The PRPs brought suit under this provision claiming that EPA had violated its duties under CERCLA and the National Contingency Plan (NCP) by failing to take certain steps to limit the cost of the cleanup it chose to undertake at Moyers Landfill in Pennsylvania. The PRPs argued that they, too, were citizens and that this provision permitted immediate review of their claim.

The court reviewed the legislative history of the 1986 amendments to determine how section 310, which allows citizens to challenge EPA's remedial plans should be reconciled with section 113(h)(1), which postpones review of EPA actions under section 104 until EPA files a cost recovery action under section 107. The court first found that Congress intended that citizen suits be exempt from the bar to pre-enforcement review under section 113(h)(1), since section 113(h)(4) expressly exempts from the bar to pre-enforcement review "[a]n action under Section [310] . . . (relating to citizens suits) alleging that the removal or remedial action taken under Section [104] . . . or secured under Section [106] . . . was in violation of any requirement of [CERCLA]." The court then found, however, that the limitation of section 113(h)(1) did apply to the PRP suit at bar. To hold otherwise would allow PRPs "in the guise of citizens" to raise challenges that they, as PRPs, otherwise would not be able to raise until a cost recovery action had been instituted. That

11. 677 F. Supp. 823 (E.D. Pa. 1988).

result would "eviscerate subsection [113(h)(1)]."[12] The court then found that section 113(h)(4) must be interpreted to apply only to those claims that would not otherwise be postponed under the other subsections of section 113(h).[13]

In so holding, the court distinguished between the allegations of "typical" suits brought by PRPs as opposed to typical citizen suits. The court stated that the focus of the typical PRP suit is an allocation of the expense of the cleanup (i.e., a monetary matter), while the typical citizen suit that was discussed during the debate in Congress over the 1986 amendments is a suit alleging that EPA had violated cleanup standards or other requirements of law and that public health or the environment would be threatened if the proposed actions were undertaken. Matters of public health and environmental hazards were recognized as requiring prompt attention rather than review after the completion of an inadequately protective response action.[14]

It should be noted, however, that the court stated in dicta that a PRP, like any other person, is empowered to bring a citizen suit irrespective of its status as a PRP, provided that the gravamen of such suit is that EPA's response action "posed a risk of irreparable harm to health or the environment."[15] At least one other court has accepted use of the citizen suit provision by a PRP.[16] Notably, however, that case was later overruled.[17]

The possibility of "true" citizen suits prior to implementation of a remedy proved to be short-lived. In *Neighborhood Toxic Cleanup Emergency v. Reilly*, the court, after reviewing the legislative history, held that section 113(h)(4) precludes review of EPA's actions until some action is taken and "that a remedial plan must go forward before review

12. *Id.* at 826 n.4.
13. *Id.* at 828.
14. *Id.* at 829.
15. *Id.*
16. *See* United States v. Princeton Gamma-Tech, Inc., 31 F.3d 138, 148 (3d Cir. 1994) (permitting pre-enforcement challenge to EPA remedy posing irreparable health and environmental threat, in part, based on use of citizen suit provision by PRP).
17. Clinton Cnty. Comm'rs v. EPA, 116 F.3d 1018 (3d Cir. 1997).

is available."[18] The court went on to consider how much action was required before review is available. It rejected the view that the mere formulation of the remedial plan is sufficient action to permit judicial review in a citizen suit. In the court's opinion, the most authoritative legislative history (the conference committee report) supports the view that a citizen suit to challenge EPA's choice of remedy would not be available "until after a distinct phase of the cleanup is completed."[19] The court stressed that "in no event is judicial review to delay the *start* of a cleanup."[20] The court thus suggests that once a distinct portion of the remedial action is completed, the entire plan is subject to review.

The trend to deny jurisdiction over citizen suits prior to implementation of the remedy continued in *Alabama v. EPA*.[21] In this case, the State of Alabama sought to challenge EPA's remedy at a site in Texas that called for shipment of wastes to a facility in Alabama. The circuit court held that the plain language of section 113(h)(4) indicates that section 113(h)(4) "applies only after a remedial action is actually completed."[22]

The latest defeat of parties seeking to use the citizen suit provision came in *Pakootas v. Teck Cominco Metals, Ltd.*[23] In this case, the Ninth Circuit upheld the dismissal of a citizen suit under section 310(a)(1) of CERCLA seeking civil penalties for a PRP's failure to comply with a section 106 administrative order.[24] The defendant in this case, Teck Cominco Metals, Ltd. (Teck Cominco), owns a smelter in British Columbia. As part of its smelting operations, Teck Cominco would dump slag into the Columbia River in Canada, which flowed downstream into the State

18. 716 F. Supp. at 833.

19. *Id.* at 834.

20. *Id.* (emphasis added).

21. 871 F.2d at 1557.

22. *See also* Frey v. EPA, 751 F.3d 461, 468 (7th Cir. 2014), *cert. denied*, 135 S. Ct. 494 (2014) (court may review citizens' claim about EPA's initial cleanup action at site, even if EPA has "new, concrete plans to conduct further remediation," provided the new plans do not affect prior, completed work).

23. 646 F.3d 1214 (9th Cir. 2011).

24. Section 310(a)(1) of CERCLA provides in pertinent part that "any person may commence a civil action on his own behalf (1) against any person (including the United States and any other governmental instrumentality or agency, to the extent permitted by the eleventh amendment to the Constitution) who is alleged to be in violation of any standard, regulation, condition, requirement, or order which has become effective pursuant to this Chapter."

of Washington. EPA issued a section 106 administrative order to Teck Cominco compelling the company to conduct a remedial investigation/feasibility study. Teck Cominco, however, did not comply with the order, and EPA did not attempt to enforce it. Two citizen plaintiffs sued Teck Cominco under the CERCLA citizen suit provision to enforce the EPA order. The citizen plaintiffs sought a declaration that Teck Cominco was in violation of EPA's order, an injunction compelling compliance with the order, penalties for failure to comply with the order, and attorneys' fees and costs. Teck Cominco moved to dismiss for lack of subject-matter and personal jurisdiction and for failure to state a claim upon which relief could be granted. One of Teck Cominco's arguments was that the citizen suit was an extraterritorial application of CERCLA. The district court denied the motion to dismiss, but certified its order to the Ninth Circuit for interlocutory appeal. On appeal, the Ninth Circuit determined that the citizen suit was not an extraterritorial application of CERCLA, and upheld the district court's denial of the motion to dismiss.[25]

Before that appeal had been decided, EPA and Teck Cominco entered into a settlement agreement, which required the company to perform remedial activities at the Upper Columbia River site. Under the settlement, EPA agreed not to sue the company for noncompliance with the unilateral administrative order if Teck Cominco fulfilled its obligations under the agreement. In return, Teck Cominco consented to jurisdiction in federal district court in the event EPA brought a suit to enforce the contractual agreement and seek penalties for noncompliance with the unilateral administrative order. Upon executing the settlement agreement, EPA withdrew the section 106 administrative order, declining to collect penalties for the 892 days that Teck Cominco failed to comply with the order.

Thereafter, the citizen plaintiffs amended their complaint to seek only civil penalties for noncompliance with the section 106 administrative order and for attorneys' fees and costs. Teck Cominco moved to dismiss the amended complaint. The district court granted the motion and dismissed the citizen plaintiffs' claims for lack of subject-matter jurisdiction

25. Pakootas v. Teck Cominco Metals, Ltd., 452 F.3d 1066 (9th Cir. 2006).

because the lawsuit was barred by CERCLA's pre-enforcement review provision, section 113(h). Once again, the citizen plaintiffs appealed to the Ninth Circuit, and once against the Ninth Circuit affirmed the district court's dismissal of the citizen's suit.

The Ninth Circuit determined that a citizen suit seeking penalties for noncompliance with a section 106 administrative order was barred by CERCLA's pre-enforcement review provision, section 113(h) for a number of reasons. First, the Ninth Circuit ruled that the lawsuit qualified as a "challenge" under section 113(h). The court noted that Congress's purpose in creating the section 113(h) pre-enforcement review bar was to prevent lawsuits that would interfere with the expeditious cleanup of contaminated sites. Although the citizen plaintiffs did not seek to modify any of the requirements of the cleanup, the lawsuit was nevertheless a challenge because it interfered with Teck Cominco's agreement to perform the cleanup. The court reasoned that one of the bargaining chips in the settlement agreement was the fact the EPA would not pursue 892 days of penalties for not complying with the section 106 administrative order provided that Teck Cominco performed the remediation. If Teck Cominco did not perform the cleanup, EPA retained the right to pursue recovery of 892 days of penalties: "They are the EPA's hammer, held over Teck Cominco's head."[26] Allowing the citizens to independently pursue such penalties would deprive EPA of this hammer.

Second, if Teck Cominco were forced to pay a large penalty for its past noncompliance, its ability to fund the remainder of the cleanup effort could be jeopardized. The court reasoned that Teck Cominco faced up to $24 million in penalties plus attorneys' fees if Pakootas prevailed in the case. Although the court did not consider Teck Cominco's ability to pay in the event it lost the lawsuit for penalties, the court noted that it was certainly possible that a judgment of that magnitude could interfere with a performing party's ability to pay for a cleanup. Thus, a suit of this nature conflicts with Congress's rationale behind the pre-enforcement review prohibition: ensuring the expeditious cleanup of contaminated sites.

26. *Pakootas*, 646 F.3d at 1222.

The Ninth Circuit also rejected the citizen plaintiffs' second basis for arguing that the pre-enforcement bar did not apply, namely, that the suit was an exempt action "to enforce an order issued under [section 106(a)] of this title or to recover a penalty for violation of such order."[27] The citizen plaintiffs stated that they were seeking to recover penalties for violations of the withdrawn administrative order, and that based on the plain language of the exception, the pre-enforcement bar did not apply to their action. The court disagreed, reasoning that although the plaintiffs were indeed seeking penalties in their suit, they were not seeking to *recover* penalties. The court noted that the penalties available under section 106(a) are payable to EPA and the Superfund, not to citizen plaintiffs. "The penalty money is Superfund money, payable to the government, not payable to citizens who bring lawsuits."[28] Because the citizen suit is not one "to recover penalties," the section 113(h)(2) exception to the pre-enforcement review bar does not apply. The court further observed that it was logical for EPA create an exception for EPA to pursue penalties while a cleanup is still underway so that EPA can "swing its hammer without waiting for years until the cleanup is done."[29]

In sum, based on the case law to date, citizen suit actions that precede implementation of a remedial action will rarely survive motions to dismiss. The lone exceptions are instances in which citizens allege that implementation of a remedy poses an irreparable health and environmental threat.

## III. PRE-ENFORCEMENT JUDICIAL REVIEW BASED ON CONSTITUTIONALITY

As noted above, section 113(h) prohibits pre-enforcement review of EPA action under CERCLA, subject to narrowly defined exceptions. Courts are split as to whether section 113(h) bars pre-enforcement review of challenges that are based on the constitutionality of an EPA action or an aspect of CERCLA. Courts that refuse to review challenges based

27. *Id.* at 1222.
28. *Id.* at 1223.
29. *Id.* at 1224.

on constitutionality tend to focus on the language of section 113(h), which does not explicitly exempt constitutional claims from the prohibition on pre-enforcement review, and Congress's intent to prohibit pre-enforcement review of all claims.[30] Courts that have entertained constitutional challenges, notwithstanding section 113(h), have based such decisions on their belief that Congress does not have the power to completely deny judicial review of constitutional claims.[31]

In the discussion below, we examine two areas of such challenges, section 106(a) orders and section 107(l) liens.

## A. Section 106 Administrative Orders

As discussed in chapter 3, where EPA determines that "an imminent and substantial endangerment to the public health or welfare or the environment" exists "because of an actual or threatened release of a hazardous substance from a facility,"[32] Section 106 provides EPA with a range of enforcement alternatives. EPA may seek an injunction from the federal, district court "in the district in which the threat occurs" to compel PRPs to undertake a particular response action.[33] Alternatively, EPA may issue to PRPs a section 106(a) unilateral administrative order (UAO) requiring cleanup of a site and directing them to take specified actions.[34] These authorities exist as alternatives, or additions, to the authority provided in section 107, by which the government may recover funds it expends on cleanup activities, pursuant to section 104.[35]

If a PRP subject to a UAO fails or refuses to comply, EPA can go to district court to seek penalties of up to $25,000 per day against any PRP "who, without sufficient cause, willfully violates, or fails or refuses to comply with [the] order."[36] EPA can also defer going to court to collect

30. *See, e.g.*, Barmet Aluminum Corp. v. Reilly, 927 F.2d 289 (6th Cir. 1991) (court rejected pre-enforcement constitutional challenge to remedy).

31. *See, e.g.*, Reardon v. United States, 922 F.2d 28 (1st Cir. 1990), discussed *infra*.

32. CERCLA § 106(a).

33. *Id.*

34. *Id.*

35. However, section 106 may not be used as an alternative basis for cost recovery. United States v. Stringfellow, No. CV-83-2501-MML, 1984 WL 3206 (C.D. Cal. Apr. 5, 1984).

36. CERCLA § 106(b)(1). See chapter 3 for a detailed discussion of what constitutes "sufficient cause."

penalties and can proceed to perform the necessary response action itself, using Superfund money, and later sue the noncomplying PRP under section 107(c)(3) for punitive damages of up to three times the amount of its incurred response costs. As in a penalty action under section 106(a), the PRP is liable for such punitive damages if its refusal to obey EPA's order was "without sufficient cause." In theory, EPA can seek *both* daily penalties *and* treble damages in the same case.

A recipient of a UAO who believes that the order may be defective (e.g., because their liability has not been clearly established or because the cleanup requirements are not consistent with the NCP) faces a difficult choice. If the recipient complies, the recipient will incur substantial costs and must file a separate action for reimbursement under section 106(b)—a lengthy and somewhat uncertain pathway. In the recipient refuses to comply with the UAO, it receives its opportunity to present defenses rather than incur huge costs and wait to seek reimbursement; but the noncomplying PRP will face substantial penalties and/or treble damages if the PRP fails to convince a court that there was sufficient cause for refusal to comply. Obviously, PRPs could eliminate these risks if they could obtain judicial review of a section 106(a) order at the time it is issued. As discussed below, however, it is well established that PRPs cannot obtain a judicial determination of the sufficiency of their objections until EPA decides to go to court (to obtain a court-issued section 106(a) order, to seek section 106(b) penalties, or to recover costs and/or treble damages under sections 107(a) and (c)).

Prior to the 1986 amendments, the courts had found that the judiciary lacks subject-matter jurisdiction to review the merits of a section 106(a) order prior to an EPA enforcement action.[37] Most courts based this decision on the conclusion that pre-enforcement review of the merits of a section 106(a) order—that is, selection or scope of remedy—would

37. Solid State Cirs., Inc. v. EPA, 812 F.2d 383, 386 n.1 (8th Cir. 1987); *Wheaton Indus.*, 781 F.2d at 356; *Wagner Elec. Corp.*, 612 F. Supp. at 740; *Aminoil, Inc.*, 599 F. Supp. at 71. One court, however, did grant pre-enforcement review of a section 106(a) administrative order. Lone Pine Steering Comm. v. EPA, 600 F. Supp. 1487 (D.N.J. 1985) (holding that EPA had no rational basis for commencing a response action where no preliminary assessment had been made), *aff'd*, 777 F.2d 882 (3d Cir. 1985).

improperly delay expeditious cleanup.[38] A handful based their denial of pre-enforcement review on a finding that a section 106(a) order, absent EPA enforcement action, is not "final agency action" and, therefore, is not ripe for review.[39]

Section 113(h), added to CERCLA by the 1986 amendments, expressly provides that section 106(a) orders are not subject to pre-enforcement review. Section 113(h) postpones judicial review until after an action is instituted seeking enforcement of the order or seeking recovery of penalties or treble damages for its violation.[40]

An open question is whether section 113(h) bars pre-enforcement review of the constitutionality of UAOs issued by EPA under section 106(a). This question was at the heart of *General Electric v. Jackson*.[41] General Electric had received multiple UAOs directing the company to perform remediation at multiple sites. In 2000, GE filed suit in the U.S. District Court for the District of Columbia challenging the constitutionality of UAOs that EPA had issued to GE under CERCLA § 106(a). GE argued that the statutory regime violates the Fifth Amendment to the U.S. Constitution because it "'deprive[s] persons of their fundamental right to liberty and property without . . . constitutionally adequate procedural safeguards.'"[42] According to GE, UAOs impose "'a classic and unconstitutional 'Hobson's choice'" because "refusing to comply 'risk[s] severe punishment [i.e., fines and treble damages]'" leaving them with only one rational option, namely, to "'comply . . . before having any opportunity to be heard on the legality and rationality of the underlying order.'"[43] The district court dismissed GE's complaint for lack of jurisdiction under section 113(h) of CERCLA.

The D.C. Circuit reversed and remanded the case, stating that section 113(h) did not bar GE's lawsuit because GE was not challenging

38. *Aminoil Inc.*, 599 F. Supp. at 71.

39. *See, e.g.*, United States v. United Nuclear Corp., 610 F. Supp. 527, 528 (D.N.M. 1985); Earthline Co. v. Kin-Buc, Inc., Civ. No. 83-4226, 1984 U.S. Dist. LEXIS 17623 (D.N.J. Apr. 13, 1984).

40. *See* CERCLA § 113(h)(2).

41. 610 F.3d 110 (D.C. Cir. 2010), *cert. denied*, 131 S. Ct. 2959 (2011).

42. *Id.* at 116 (alterations in original).

43. *Id.* (alterations in original).

"'any particular action or order by EPA.'"[44] On remand the district court granted EPA's motion for summary judgment on GE's facial due process challenge, stating that the statutory scheme was constitutionally sufficient because a PRP can refuse to comply with a UAO, and then challenge the UAO when EPA seeks to enforce it in federal court. In this regard, the district court also rejected GE's position that the fines and treble damages are so severe that they effectively preclude judicial review. Finally, the district court addressed GE's claim that EPA's policies and procedures for issuing UAOs exacerbated CERCLA's unconstitutional deficiencies (GE's "pattern and practice" challenge). The court held that GE had suffered significant injuries (e.g., loss of stock value) as a result of issuance of multiple UAOs, which are entitled to due process protection. However, the court concluded that such "consequential injuries"—when balanced against the government's interest and risk of errors in UAO procedures—were insufficient to render EPA's practices unconstitutional.[45]

On appeal, the D.C. Circuit affirmed the district court's rejection of GE's constitutional challenge to UAOs. Citing the Supreme Court's decisions in *Reisman v. Caplan*[46] and *Oklahoma Operating Co. v. Love*,[47] the court noted that statutes imposing fines—even enormous fines—for noncompliance may satisfy due process concerns if such fines are subject to good faith or reasonable grounds defenses. CERCLA provides such safeguards because to impose fines and treble damages a court must find that (1) the UAO was proper, (2) the PRP willfully failed to comply without sufficient cause, and (3) in the court's discretion the fines and treble damages are appropriate.[48]

The D.C. Circuit also affirmed the district court's ruling on GE's "pattern and practice" claim. GE relied on the Supreme Court's decision in *Connecticut v. Doehr*,[49] in which the Supreme Court held that a state

---

44. *Id.*
45. *Id.* at 116–17.
46. 375 U.S. 440, 446–50 (1964).
47. 252 U.S. 331, 338 (1920).
48. 610 F.3d at 118–19.
49. 501 U.S. 1 (1991).

statute authorizing ex parte prejudgment attachment of real estate violated due process. GE asserted that the consequential effects of UAOs, such as declining stock price, are deprivations of property interests (similar to the effects of prejudgment attachment of real estate) that cannot be remedied by asserting and prevailing on a sufficient cause defense. The D.C. Circuit, however, disagreed with GE's reading of *Doehr*. The court stated that "although *Doehr* does hold that direct, partial impairments of property rights may well warrant due process safeguards, nothing in that opinion implies that consequential injuries, standing alone, merit due process protection."[50] The court concluded that other Supreme Court precedent foreclosed GE's constitutional challenge relating to consequential losses:

> Stripped of its reliance on *Doehr*, GE's case boils down to this: by declaring that a PRP is responsible for cleaning up a hazardous waste site, a UAO harms a PRP's reputation, and the market, in turn, devalues its stock, brand, and credit rating. Viewed in this way, GE's argument is foreclosed by *Paul v. Davis*, 424 U.S. 693.[51]

In *Paul v. Davis*, the Supreme Court ruled that a sheriff's inclusion of Davis's name and photograph in a flyer captioned "Active Shoplifters" did not violate due process, notwithstanding the consequential damage to reputation.

In sum, case law to date confirms that UAOs are not subject to pre-enforcement review, with the possible exception of where such review is based on constitutional grounds.[52] Constitutional challenges to UAOs, however, have been unsuccessful.

---

50. 610 F.3d at 120.

51. *Id.* at 121.

52. Note that several courts have rejected even a constitutional challenge as a basis for pre-enforcement review of a UAO. *See* City of Rialto v. W. Coast Leasing Corp., 581 F.3d 865, 867 (9th Cir. 2009) (holding that the district court lacked jurisdiction to decide a claim that EPA's pattern and practice of enforcing CERCLA violated the plaintiff's due process rights); United States v. Capital Tax Corp., No. 04-C-4138, 2007 WL 488084, at *6 n.2 (N.D. Ill. Feb. 8, 2007) (stating that plaintiff lacks standing to challenge EPA's pattern and practice of enforcement under CERCLA).

## B. CERCLA Liens

In *Reardon v. United States*,[53] a court not only agreed that it had jurisdiction to review a constitutionality challenge to a provision of CERCLA, notwithstanding the pre-enforcement review bar of section 113(h), but ruled the provision unconstitutional. At issue was EPA's use of its lien authority under section 107(l) of CERCLA. The plaintiff in this case—the Reardons—had acquired a 16-acre parcel of land that had been part of a larger tract where an electrical manufacturing plant had operated. The prior owner separated the 16-acre plot and sold it to the Reardons who intended to develop the land commercially. EPA, however, discovered polychlorinated biphenyl (PCB) contamination on both the Reardon property and the adjoining industrial land where electrical manufacturing had occurred. EPA remediated both properties and thereafter filed a notice of lien on the Reardon's property to recoup some of the cleanup costs. The Reardons sought declaratory relief to remove the lien from their property, stating, inter alia, that the imposition of a lien deprived them of a property right without due process.

The district court dismissed the action, stating it lacked jurisdiction to hear the case under section 113(h). It reasoned that placing a lien on property is an enforcement activity, and thus it was subject to the pre-enforcement review bar under section 113(h). The court also noted that the lien does not constitute a substantial property interest protected by the due process clause, noting that although the lien clouds title, it does not interfere with an owner's ability to use, possess, or transfer the property.

The First Circuit reversed the district court. It first ruled that the court did have jurisdiction to hear the constitutionality challenge because section 113(h) does not preclude judicial review of statutory challenges to a CERCLA lien.[54] In this regard, the court held that a lien was not one of the types of enforcement activity for which pre-enforcement review is barred under section 113(h) in that it is not removal or remedial activity nor is it related to such activity. On petition for rehearing, the First

53. 947 F.2d 1509 (1st Cir. 1991).
54. *Reardon*, 922 F.2d 28.

Circuit, sitting en banc vacated the panel's decision.[55] It first held that a lien was in fact the type of enforcement as to which section 113(h) bars pre-enforcement review. The First Circuit, however, ruled that it could hear the constitutionality challenge, notwithstanding section 113(h). It thereafter held that the provision authorizing liens, section 107(l), violated due process. It noted that under Supreme Court precedent even a temporary or partial impairment of a property right merits due process protection and that before the government may impose a lien, which impairs one's ability to sell property, the owner is entitled to due process. Thus, imposing a lien without such due process is unconstitutional. To pass constitutional muster, the CERCLA lien must provide, at a minimum, at least notice and a preattachment hearing.[56]

## IV. JUDICIAL REVIEW OF NOTICE LETTERS AND INFORMATION REQUESTS

As discussed in chapter 1, if EPA determines that there has been a release of a hazardous substance from a facility and identifies PRPs associated with the facility, EPA will often contact the PRPs before responding to the release by issuing a notice letter. The notice letter informs PRPs of their potential liability for response costs and initiates negotiations between EPA and the PRPs. Prior to or concurrent with a notice letter, EPA may also issue an information request to parties the agency believes may have information relevant to cleanup of a site.

Some PRPs, upon receiving these notice letters and information requests, have attempted to strike preemptively by filing an action seeking declaratory judgment that they are not liable for response costs. To establish "injury in fact"—a prerequisite to declaratory relief—recipients of notice letters claim that the letter has cast a cloud on their financial status, raised potential SEC reporting issues, and so on. These intangible harms, they argue, are sufficient "injury in fact" to create a controversy ripe for judicial review and to create standing on their part to seek redress in the district courts.

55. 947 F.2d 1509.
56. *Id.* at 1523.

The courts that have considered this issue, however, have refused to rule on the merits of the liability issue, based on both ripeness and standing grounds.[57] As to ripeness, the courts have accepted EPA's position that the issuance of a notice of potential liability and request for information that does not require any action on the part of the recipient other than to supply information to the agency is not a final agency determination of liability, but something more preliminary and tentative. Thus, a party's receipt of a notice letter and information request does not make the recipient's potential liability under section 107 ripe for adjudication.[58]

Similarly, courts have found that notice letter recipients lack standing to challenge agency action. For example, the court in *Pacific Resins & Chemicals, Inc. v. United States* held that the possibility of future liability is insufficient to confer standing on the recipient of a general notice letter and information request, that CERCLA does not authorize review of EPA's response activities prior to initiation of an enforcement action under section 106(a) or a cost recovery action under section 107, and that the Declaratory Judgment Act does not provide an independent basis for subject-matter jurisdiction because ripeness and finality of agency action are requirements for relief thereunder.[59]

In sum, where the only action taken by EPA is to issue a general notice letter and information request, attempts to seek judicial review of a PRP's liability have been uniformly unsuccessful.

57. *See, e.g.*, United States v. Cohen, 914 F. Supp. 252 (N.D. Ill. 1996) (court refused to order EPA to rescind information requests and refused to grant defendants a declaratory judgment as to whether the salvage operation in question resulted in the disposal of hazardous substances, ruling that such actions would constitute pre-enforcement review, which is barred under CERCLA § 113(h)); *see also In re Combustion Equip. Assoc., Inc.*, 73 B.R. 85 (S.D.N.Y. 1987), *aff'd*, 838 F.2d 35 (2d Cir. 1988); *Pac. Resins & Chems., Inc.*, 654 F. Supp. 249; D'Imperio v. United States, 575 F. Supp. 248 (D.N.J. 1983).

58. *In re Combustion Equip. Assocs., Inc.*, 73 B.R. at 87–88; *see also Pac. Resins & Chems. Inc.*, 654 F. Supp. at 252–53.

59. 654 F. Supp. at 253–54. *Accord D'Imperio*, 575 F. Supp. at 252 (receipt of a general notice letter not a determination of liability of sufficient immediacy to warrant declaratory relief).

## V. JUDICIAL REVIEW OF NPL LISTING

PRPs also have sought judicial review of EPA's decision to list a site on the National Priorities List (NPL). A challenge to the inclusion of a site on the NPL must be filed in the U.S. Court of Appeals for the District of Columbia Circuit, which has exclusive jurisdiction under section 113(a) of CERCLA.[60]

Most challenges to NPL listings have failed. Emblematic of these failed NPL challenges is *Eagle-Picher Industries, Inc. v. EPA*.[61] In this case, the petitioners contended that the NPL listing was improper because the mining wastes and fly ash contaminating the sites at issue are not "hazardous substances" as defined under section 101(14) of CERCLA. Accordingly, the petitioners contended that EPA was without authority to list the sites on the NPL. After reviewing the merits of the challenge, the court rejected the petitioners' contention that mining wastes and fly ash are not hazardous substances, and, therefore, affirmed the NPL listing.[62]

60. Section 113(a) of CERCLA provides that "[r]eview of any regulation promulgated under this chapter may be had upon application by any interested person only in the Circuit Court of Appeals of the United States for the District of Columbia. Any such application shall be made within ninety days from the date of promulgation of such regulations. Any matter with respect to which review could have been obtained under this subsection shall not be subject to judicial review in any civil or criminal proceeding for enforcement or to obtain damages or recovery of response costs." Attempts to challenge NPL listings in courts other than the D.C. Circuit have been rejected on venue grounds. *See, e.g.*, U.S. Ecology, Inc. v. Carlson, 638 F. Supp. 513, 518 (C.D. Ill. 1986) (district court held that it lacks subject-matter jurisdiction to consider an NPL listing because it was a challenge to a regulation and, therefore, the Circuit Court of Appeals for the District of Columbia had exclusive jurisdiction); United States v. Ottati & Goss, 694 F. Supp. 977 (D.N.H. 1988), *aff'd in part*, 900 F.2d 429 (1st Cir. 1990) (holding that district court was without jurisdiction to hear a challenge to EPA's decision to list arguably discrete contiguous sites as a single unified site on the NPL); *but see* SCA Servs. of Ind., Inc. v. Thomas, 634 F. Supp. 1355 (N.D. Ind. 1986) (limited constitutional review of NPL available in district court; NPL found to satisfy requirements of due process).

61. 759 F.2d 922 (D.C. Cir. 1985).

62. *See also* CTS Corp. v. EPA, 759 F.3d 52 (D.C. Cir. 2014) (petitioner impermissibly sought to rely on information that was not contained in the administrative record and failed to demonstrate that EPA's listing decision was arbitrary and capricious); U.S. Magnesium, LLC v. EPA, 630 F.3d 188 (D.C. Cir. 2011) (challenge to NPL listing based on argument that HRS scoring artificially inflated the site's score rejected because argument was in essence a challenge to underlying regulation which could no longer be asserted); RSR Corp. v. EPA, 102 F.3d 1266 (D.C. Cir. 1997) (challenge to NPL listing ruled untimely because it was raised three years after listing).

There have, however, been a handful of successful NPL challenges.[63] In *Tex Tin v. EPA*, the petitioner successfully challenged EPA's NPL listing of the Tex Tin Superfund site by focusing on EPA's use of the Hazard Ranking System (HRS). The HRS is a set of criteria that measure the potential for harm to humans or the environment from migration of hazardous substances away from the site through either groundwater, surface water, or air. It is a composite of separate scores for each of these routes.[64] A site is placed on the NPL on the basis of the site's numerical score under the HRS; where the composite HRS score exceeds 28.50, the site is listed on the NPL. In *Tex Tin v. EPA*, the petitioner focused on the air route score, which is composed of two elements—the observed release score and the toxicity score. The D.C. Circuit ruled that Tex Tin had waived its right to challenge the observed release score by failing to raise the issue in comments on the proposed listing. The court, however, agreed with the challenge to the toxicity score, ruling that EPA had failed to adequately respond to petitioner's contention that arsenic would not migrate because it is bound tightly to rock. The listing was reversed and remanded to EPA for the agency to supplement the administrative record. Upon remand, EPA supplemented the administrative record with an explanation of its position on arsenic. Tex Tin submitted comments enumerating flaws in the agency's position, and when EPA did not change its position, Tex Tin once again petitioned for review of the NPL listing. This time, the D.C. Circuit rejected EPA's justification for the NPL listing, ruling that the agency's "conclusory statements" were insufficient to satisfy the remand order, and directed EPA to delete the site from the NPL.[65]

---

63. *See, e.g.*, Mead Corp. v. Browner, 100 F.3d 152 (D.C. Cir. 1996) (vacating the NPL listing of a site, which included three noncontiguous areas, one of which was a mile away from the others); Nat'l Gypsum Co. v. EPA, 968 F.2d 40 (D.C. Cir. 1992) (vacating and remanding NPL listing to EPA in light of the agency's failure to adequately explain the scientific basis for the listing as well as its failure to offer substantial evidence in support of its decision); Tex Tin v. EPA, 935 F.2d 1321 (D.C. Cir. 1991) (discussed above). Additional cases are discussed in chapter 6.

64. *See* 40 C.F.R. pt. 300, App. A, Sec. 1.0.

65. 992 F.2d 353 (D.C. Cir. 1993).

# VI. STANDARD AND SCOPE OF JUDICIAL REVIEW

## A. Limitations Imposed by Section 113 on the Review of EPA's Selection of Remedy

Section 113(j) of CERCLA, added by the 1986 amendments, codified the standards of review that had developed in the pre-Amendments case law with regard to the selection of any response actions taken by EPA. As discussed below, these standards are extremely deferential to EPA; challenges to EPA's choice of response action will be subject to several significant limitations.

First, judicial review of any issues concerning the selection of any response action taken or ordered by EPA will be limited to the administrative record.[66] Section 113(k) of CERCLA requires EPA to establish an administrative record upon which it will base its selection of a response action, and to promulgate regulations establishing procedures for the participation of interested parties, including PRPs, in the development of the administrative record. EPA intends that the following types of documents be included in the administrative record file: (1) documents containing factual information, data, and analysis of such information and data (e.g., sampling data, site investigation reports, Agency for Toxic Substances Disease Registry health assessments, and the RI/FS); (2) guidance documents, technical literature and site-specific policy memoranda; (3) documents received, published, or made available to the public (e.g., the proposed remedial action plan and any public comments thereon); (4) decision documents such as the record of decision; (5) enforcement orders; and (6) an index of documents that make up the administrative record.[67]

Second, the court will uphold EPA's decision in selecting the response action "unless the objecting party can demonstrate, on the administrative record, that the decision was arbitrary and capricious or otherwise not in accordance with law."[68]

66. CERCLA § 113(j)(1).

67. 53 Fed. Reg. 51,515 (Dec. 21, 1988).

68. CERCLA § 113(j)(2); *see also* United States v. P.H. Glatfelter Co. & NCR Corp., 768 F.3d 662, 670 (7th Cir. 2014) (court rejected challenge to remedy because the "decision to

Finally, a finding that EPA's choice of response action is arbitrary and capricious will not prevent EPA from recovering those portions of its incurred costs not inconsistent with the NCP, even if EPA's remedy selection violated the NCP. The court must still award to EPA the response costs or damages that are not inconsistent with the NCP, and other relief as is consistent with the NCP.[69]

### B. De Novo Review of Section 106 Actions for Injunctions

Although section 106(a) administrative orders will be accorded only postenforcement review on the administrative record, there is substantial precedent that a civil action by EPA under section 106(a) for equitable relief will be given a full hearing on the merits, including complete discovery and a full trial on both liability and remedy.[70] This is one of the reasons that EPA rarely pursues equitable relief through civil actions, as opposed to using its administrative order authority.

In *United States v. Hardage*, the court found CERCLA's record review provision inapplicable to actions for injunctive relief, and held that the defendants were entitled to *de novo* review of the remedy selected by EPA.[71] The court reviewed the language of section 113(j)(1), which limits "judicial review of any issues concerning the adequacy of any response action taken or ordered by the President" to the administrative record. It distinguished between actions taken or ordered by the president and actions ordered by the court, and held that review of only the former is limited to the administrative record. An action requesting court-ordered

maintain a preference for dredging in the amended remedy was rationally related to the facts" and thus not "arbitrary and capricious"); United States v. Broderick Inv. Co., 955 F. Supp. 1268, 1272 (D. Colo. 1997) (court applied "arbitrary and capricious" standard of review to determine whether EPA's response action was "not inconsistent with the NCP"), *aff'd in part and rev'd in part*, 200 F.3d 679 (10th Cir. 1999); United States v. Vertac Chem. Corp., 33 F. Supp. 2d 769, 784 (E.D. Ark. 1998) (defendants held liable for full $100 million in response costs where they could not establish that any of the government's cleanup methods were "arbitrary, capricious, or otherwise not in accordance with the law").

69. CERCLA § 113(j)(3).

70. *See* United States v. Conservation Chem. Co., 661 F. Supp. 1416, 1424–31 (W.D. Mo. 1987); *see also* United States v. Hardage, 663 F. Supp. 1280, 1284 (W.D. Okla. 1987). *But see* United States v. Bell Petrol. Servs., Inc., 718 F. Supp. 588 (W.D. Tex. 1989); United States v. Seymour Recycling Corp., 679 F. Supp. 859, 861–63 (S.D. Ind. 1987).

71. 663 F. Supp. at 1283–84.

relief is not covered by section 113(j)(1), and review of the response action should be de novo.[72]

On the other hand, in *United States v. Bell Petroleum*, the court held that the limitations on review set forth in section 113(j) were intended to govern judicial review of EPA's choice of remedy in all cases, including those in which EPA seeks injunctive relief under section 106.[73] In so holding, the court noted that this case originated under section 107 and was, therefore, distinguishable from *Hardage*, but that EPA intended to issue a section 106(a) order to the defendants upon completion of remedial design.[74] The court went on to dismiss the distinction drawn in *Hardage* as "hypertechnical and inconsistent with the plain meaning of the statutory language."[75]

## VII. JUDICIAL REVIEW OF CONSENT DECREES

A settlement agreement under CERCLA that takes the form of a consent decree must be approved by the federal district court in which it is lodged before it can take effect.[76] Although most consent decrees are approved with little or no fanfare, a PRP should never assume that a court will simply "rubber stamp" the decree. From time to time, courts *do* reject proposed consent decrees.[77] This section discusses the standard of review applicable to a proposed consent decree and the finality and modification of the decree once it is approved.

72. *Id.* at 1284.

73. 718 F. Supp. at 591.

74. *Id.*

75. *Id.* EPA specifically rejects the argument that section 113(j)(1) does not apply to injunctive actions under section 106(a) and states that judicial review of *any* response action is limited to the administrative record. 53 Fed. Reg. 51,465 (Dec. 21, 1988).

76. An example of a settlement that must take the form of a consent decree is an agreement by a PRP to perform remedial work pursuant to section 106. *See* CERCLA § 122(d)(1)(A).

77. *See In re* Acushnet River & New Bedford Harbor: Proceedings re Alleged PCB Pollution, 712 F. Supp. 1019 (D. Mass 1989) (Proposed partial consent decree could not be approved until the federal natural resources trustee agrees to the United States' covenant not to sue. Additionally, the court rejected the proposed consent decree because the covenant not to sue did not contain a reopener clause, as required under section 122(f)(6) of CERCLA.).

### A. Intervention

When the federal government settles with a PRP, it must lodge the proposed settlement, in the form of a consent decree, with the appropriate district court, which then reviews the settlement and, if the settlement is fair, reasonable, and consistent with CERCLA, enters the consent decree.[78] Often a PRP that is not a party to a settlement between another PRP and the federal government may object to the terms of the settlement. In these situations, the nonsettling PRP may seek to intervene in the litigation filed by the government in which the government and the settling PRPs are seeking entry of a consent decree. There is a substantial body of case law examining when and under what circumstances a nonsettling PRP may intervene in such an action. The results have not been consistent. A discussion of this case law is set forth below.

Both CERCLA and the Federal Rules of Civil Procedure grant nonparties to litigation a right to intervene in that litigation under certain circumstances. Rule 24(a)(2) of the Federal Rules of Civil Procedure (Rule 24(a)(2)) provides:

> (a) Intervention of Right. On timely motion, the court must permit anyone to intervene who: . . . (2) claims an interest relating to the property or transaction that is the subject of the action, and is so situated that disposing of the action may as a practical matter impair or impede the movant's ability to protect its interest, unless existing parties adequately represent that interest.[79]

Section 113(i) of CERCLA provides:

> In any action commenced under this chapter or under the Solid Waste Disposal Act [42 U.S.C. 6901 et seq.] in a court of the United States, any person may intervene as a matter of right when such person claims an interest relating to the subject of the action and is so situated that the disposition of the action may, as a practical

78. *See* CERCLA § 122(d)(1)(A) and § 122(g)(1).
79. Fed. R. Civ. P. 24(a)(2).

> matter, impair or impede the person's ability to protect that interest, unless the President or the State shows that the person's interest is adequately represented by existing parties.[80]

The similarities in the language are obvious. The only material differences between the two statutes are the burden of proof as to the adequacy of representation and the lack of an explicit timeliness requirement in CERCLA section 113(i). This similarity has led courts to apply the same standards in interpreting both Rule 24(a)(2) and CERCLA section 113(i) as to all but the adequacy of representation element.[81]

Both CERCLA section 113(i) and Fed. R. Civ. P 24(a)(2) require that the applicant for intervention have "an interest" relating to the subject of the litigation. The interest required under both statutes has been characterized by courts as a "legally protectable interest."[82] The U.S. Supreme Court used the term "significantly protectable interest" and, in a later case, qualified that the interest must be "legally protectable,"[83] but has given little guidance to the lower courts as to how these broad terms should be interpreted. The majority rule that has emerged in the lower courts is that the interest required by Rule 24(a)(2) is one that is direct, substantial or significant, and legally protectable (DSL Rule).[84]

In *New Orleans Public Service, Inc. v. United Gas Pipe Line Co.*,[85] the court defined the interest as being "direct," "substantial," and "legally protectable," not just a mere economic interest, but rather, "one which

---

80. CERCLA §113(i) (alteration in original).

81. *See, e.g.*, United States v. Union Elec. Co., 64 F.3d 1152, 1157–58 (8th Cir. 1995); United States v. Acton Corp., 131 F.R.D. 431, 433 (D.N.J. 1990).

82. *See* United States v. Alcan Aluminum, Inc., 25 F.3d 1174, 1181 (3d Cir. 1994).

83. Donaldson v. United States, 400 U.S. 517, 531 (1971); *see also* Tiffany Fine Arts, Inc. v. United States, 469 U.S. 310, 315 (1985).

84. *See, e.g.*, Edwards v. City of Houston, 78 F.3d 983, 1004 (5th Cir. 1996) ("To demonstrate an interest relating to the property or subject matter of the litigation sufficient to support intervention of right, the applicant must have a 'direct, substantial, legally protectable interest in the proceedings'") (citation omitted); Reich v. ABC/York-Estes Corp., 64 F.3d 316, 322 (7th Cir. 1995) ("We require that the potential intervenor's interest be a direct, significant legally protectable one") (citations and quotations omitted); Med. Liab. Mut. Ins. Co. v. Alan Curtis, L.L.C., 485 F.3d 1006, 1008 (8th Cir. 2007) ("An interest is cognizable under Rule 24(a)(2) only where it is 'direct, substantial, and legally protectable'") (citation omitted).

85. 732 F.2d 452 (5th Cir. 1984).

the substantive law recognizes as belonging to or being owned by the applicant."[86] The interest that most nonsettling plaintiffs seek to protect through intervention is their potential liability, because once a settlement between the government and the PRP is approved, the government can collect the unpaid balance of its claim from a nonsettlor. The nonsettlor's right to seek contribution from the settlor for any overpayment is extinguished by the settlement under the contribution protection provision, hence the legally protectable interest.

Courts have reached different conclusions where nonsettling PRPs moved to intervene in CERCLA cases between the government and settling PRPs. Early cases were hostile to the nonsettling PRP seeking intervention, with the lone outlier being *United States v. Acton Corp.*, where the court ruled that the PRPs had a right to intervene.[87] The *Acton* decision was heavily criticized by other district courts that had determined that nonsettling PRPs do not have a right to intervene.[88] But since these early victories for the opponents of intervention, the Eighth,[89] Tenth[90] and Ninth[91] Circuits began bucking what had become known as the majority rule and instead held that nonsettling PRPs *should* be allowed to intervene.

## B. Standard of Review

Section 122(m) of CERCLA specifically provides that "general principles of law regarding the setting aside . . . of consent decrees" entered under CERCLA shall not be affected or precluded in any way by the provisions of CERCLA. Operating both before and subsequent to the enactment of section 122(m) in 1986, courts reviewing proposed CERCLA consent decrees have fashioned a number of rules. Generally, a reviewing court

86. *Id.* at 464.
87. 131 F.R.D. at 436.
88. *See, e.g.*, Ariz. v. Motorola, Inc., 139 F.R.D. 141, 145 (D. Ariz. 1991); United States v. ABC Indus., 153 F.R.D. 603, 608 n.3 (W.D. Mich. 1993) ("Having carefully reviewed [the *Acton* decision], in light of other contrary decisions and the CERCLA statutory scheme, this Court respectfully declines to follow the holding of *Acton Corp.*").
89. *Union Elec. Co.*, 64 F.3d at 1170–71.
90. United States v. Albert Inv. Co., 585 F.3d 1386, 1390, 1399 (10th Cir. 2009).
91. United States v. Aerojet Gen. Corp., 606 F.3d 1142, 1153 (9th Cir. 2010).

will not substitute its judgment for that of the parties and will make only a limited inquiry into the technical terms and the factual disputes underlying the proposed settlement.[92] In this regard, a reviewing court does not have the right or the duty to reach any ultimate conclusions on the issues of fact and law involved in the case.[93] When a government agency committed to the protection of the public interest has participated in and endorsed the settlement agreement, there is a presumption in favor of settlement.[94] "The presumption is particularly strong where a consent decree has been negotiated by the Department of Justice on behalf of a federal agency, like the Environmental Protection Agency . . . , [which has] substantial expertise in the environmental field."[95] Notably, however, courts will accord less deference where the consent decree was negotiated by a state agency.[96]

Against these declarations of deference to the parties must be weighed the right of the reviewing court to assure itself that the terms of the proposed decree are fair and adequate and not unlawful, unreasonable, or against public policy.[97] This level of scrutiny is warranted for CERCLA consent decrees because of the general need to protect the "public interest" inherent in such decrees and to guard against "sweetheart" deals.[98]

92. United States v. City of Fort Lauderdale, 81 F. Supp. 2d 1348, 1350 (S.D. Fla. 1999); United States v. Hooker Chems. & Plastics Corp., 540 F. Supp. 1067, 1072 (W.D.N.Y. 1982), *aff'd*, 749 F.2d 968 (2d Cir. 1984).

93. City of New York v. Exxon Corp., 697 F. Supp. 677, 692 (S.D.N.Y. 1988).

94. United States v. BP Exploration & Oil Co., 167 F. Supp. 2d 1045, 1050 (N.D. Ind. 2001) (citing Metro. Hous. Dev. Corp. v. Vill. of Arlington Heights, 616 F.2d 1006, 1014 (7th Cir. 1980)); United States v. Cannons Eng'g Corp., 899 F.2d 79, 84 (1st Cir. 1990); United States v. Hooker Chem. & Plastics Corp., 776 F.2d 410, 411 (2d Cir. 1985).

95. *BP Exploration & Oil Co.*, 167 F. Supp. 2d at 1050 (citing United States v. Akzo Coatings of Am., Inc., 949 F.2d 1409, 1426 (6th Cir. 1991)); *see also Hooker*, 776 F.2d at 411; SEC v. Randolf, 736 F.2d 525, 529 (9th Cir. 1984); Kelley v. Thomas Solvent Co., 717 F. Supp. 507, 515 (W.D Mich. 1989).

96. *See* City of Bangor v. Citizens Commc'ns Co., 532 F.3d 70 (1st Cir. 2008) (noting that because the consent decree involved the State of Maine, rather than the federal government, the court would accord lesser deference to the state's expertise); Opinion, Arizona v. Ashton Co. Contractors & Eng'rs, No. 12-15691 (9th Cir. Aug. 1, 2014) (reversing district court approval of consent decree between Arizona and a group of de minimis PRPs, stating that the district court improperly accorded deference to the state environmental agency's assurance that the settlement was fair and reasonable).

97. *Hooker Chems. & Plastics Corp.*, 540 F. Supp. at 1072.

98. *See* Amoco Oil Co. v. Dingwell, 690 F. Supp. 78, 85 (D. Me. 1988), *aff'd sub nom.* Travelers Indem. Co. v. Dingwell, 884 F.2d 629 (1st Cir. 1989).

Accordingly, a court reviewing a proposed consent decree under CERCLA will rely upon what has come to be known as the "legality, fairness and reasonableness" test for determining whether the proposed decree should be approved.[99]

Looking to the element of legality first, the proposed consent decree must be "consistent both with the Constitution and with the mandate of Congress."[100] In the context of CERCLA, the legality of a proposed consent decree would, for example, depend upon whether the remedial action to be implemented is consistent with section 121 of CERCLA and the NCP, and whether the proposed consent decree itself has been lodged with the court in accordance with the procedures of section 122 of CERCLA.[101] In addition, insofar as a proposed consent decree includes settlement provisions codified in section 122 (e.g., covenant not to sue, contribution protection), the terms of the proposed consent decree must be consistent with the substantive requirements of section 122.[102]

In evaluating the fairness of a proposed consent decree, the Ninth Circuit has said that a district court must independently evaluate the apportionment scheme and must find that it is "based upon, and roughly correlated with, some acceptable measure of comparative fault, apportioning liability among the settling parties according to rational (if necessarily imprecise) estimates of how much harm each PRP has done."[103]

An inquiry into the fairness of the proposed consent decree may also involve consideration of the impact of settlement on the apportionment

---

99. One court has suggested, without ruling on the question, that EPA's determination that parties are eligible for a de minimis settlement under section 122(g) might be subject to a different (i.e., lower) standard of review than normal agency action since the de minimis determination is "a matter of litigation strategy within the prosecutorial discretion of EPA and the Department of Justice." United States v. Rohm & Haas Co., 721 F. Supp. 666, 684 (D.N.J. 1989).

100. United States v. Seymour Recycling Corp., 554 F. Supp. 1334, 1337 (S.D. Ind. 1982); United States v. Conservation Chem. Co., 681 F. Supp. 1394, 1415 (W.D. Mo. 1988).

101. *Conservation Chem. Co.*, 681 F. Supp. at 1416–17.

102. *See, e.g., In re Acushnet River & New Bedford Harbor*, 712 F. Supp. 1019.

103. Arizona v. Raytheon Co., 761 F.3d 1005, 1012 (9th Cir. 2014) (citation omitted) (vacating district court order approving and entering the consent decree, finding that the district court had failed to compare the total projected costs of a cleanup with the liability attributed to each party and had given undue deference to the Arizona Department of Environmental Quality).

of liability upon nonsettling parties, as well as settling parties, for a court "will not tolerate either a 'windfall' or a 'wipe out' which results in an apportionment of responsibility which arbitrarily or unreasonably ignores the comparative fault of the parties, where there is a reasonable basis for allowing that comparison to be made."[104] The effect on nonsettling parties is, however, only one factor to be considered.[105] Furthermore, where the amount of a settlement bears some reasonable relation to a plausible estimate of the settling parties' contribution, any unfairness to the nonsettling parties will not be cause for the court to disapprove of the decree.[106] As a corollary to the concept of fairness, the proposed settlement must be entered into in good faith, a situation that can be evidenced by arms-length negotiations and a data base sufficient to make an informed decision.[107]

Somewhat overlapping the concept of fairness is the requirement that a consent decree be reasonable. The court in *Rohm & Haas* articulated several factors "potentially relevant to determining whether a CERCLA settlement is fair and reasonable," including "the strength of the Government's case, the good faith efforts of the negotiators, the possible risks of and transaction costs involved in litigation under CERCLA, and the effect of the proposed settlement on non-settling parties."[108]

### C. PRP Participation in the Consent Decree Review Process

In determining whether a proposed consent decree meets the "legality, fairness and reasonableness" test, the reviewing court will consider, at a minimum, the terms of the proposed decree itself and any comments filed with the court by the Department of Justice (DOJ).[109] Some district

104. United States v. Conservation Chem. Co., 628 F. Supp. 391, 401–02 (W.D. Mo. 1985), *modified by* 681 F. Supp. 1394 (W.D. Mo. 1988).

105. *In re Acushnet River & New Bedford Harbor*, 712 F. Supp. at 1029.

106. *Rohm & Haas Co.*, 721 F. Supp. at 685–86; *see also* United States v. Se. Pa. Transp. Auth., 235 F.3d 817, 824 (3d Cir. 2000) ("As long as the measure of comparative fault on which the settlement terms are based is not 'arbitrary, capricious and devoid of a rational basis,' the district court should uphold it") (citation omitted).

107. *City of New York*, 697 F. Supp. at 693.

108. 721 F. Supp. at 680.

109. Under section 122(d)(2)(B), when DOJ lodges a consent decree with a district court, it must provide interested persons with an opportunity to comment on the proposed decree.

courts have been willing to conduct a hearing to review the adequacy of the proposed consent decree. Both before and after CERCLA was amended in 1986, persons who were not parties to the proposed consent decree were allowed to appear at such hearings and present arguments against entry of the proposed consent decree.[110] At one such hearing, the court's ultimate decision to approve a contested decree was affected, in part, by the failure of persons objecting to the decree to present more than written statements and the arguments of counsel (e.g., evidence such as sworn testimony).[111]

Settling defendants should not necessarily rely on DOJ to defend the proposed consent decree against its opponents, for example, nonsettlors, public interest groups, and so on. Rather, they should consider submitting comments to DOJ regarding the legality, fairness, and reasonableness of the decree. As noted earlier, DOJ must file such comments with the district court charged with reviewing the consent decree. For a consent decree involving remedial action, it may be appropriate to submit comments that address the technical adequacy of such action.

Nonsettling defendants may wish to challenge a proposed consent decree that they believe fails the "legality, fairness and reasonableness" standard of review. The first step would be to file comments with DOJ opposing the proposed decree. It is possible that DOJ may withdraw the proposed decree based on such comments, as it did in *United States v. Pepper Steel and Alloys, Inc.*[112] Assuming DOJ does not withdraw the proposed decree, the next step would be to oppose entry of the proposed decree when the United States moves its entry. Where a nonsettlor wishing to oppose entry of the decree is not a party to the action to enter the decree—that is, where the United States has sued only those

DOJ must then consider, and file with the court, such written comments. Typically, DOJ will also file its responses to such comments.

110. *See, e.g.*, *Seymour Recycling Corp.*, 554 F. Supp. at 1337; *Conservation Chem. Co.*, 681 F. Supp. 1394.

111. *Seymour Recycling Corp.*, 554 F. Supp. at 1338, 1339.

112. 720 F. Supp. 164 (S.D. Fla. 1989).

companies that signed the consent decree—the nonsettlor must intervene in the action.[113]

It should be noted that courts usually reject arguments of nonsettling PRPs to set aside proposed consent decrees.[114] Nevertheless, given the right facts, courts may rule—and have ruled—in favor of challenges by nonsettling PRPs.[115]

113. Section 113(i) of CERCLA provides that "[i]n any action commenced under" CERCLA, "any person may intervene as a matter of right when such person claims an interest relating to the subject of the action and is so situated that disposition of the action may, as a practical matter, impair or impede the person's ability to protect that interest, unless the President or the State shows that the person's interest is adequately represented by existing parties." The CERCLA intervention provision is similar to Fed. R. Civ. P. 24(a)(2).

114. *See, e.g.*, Opinion, United States v. Fed. Res. Corp., No. 12-36065 (9th Cir. Sept. 16, 2014) (intervenor failed to establish that the lower court abused its discretion by forgoing a comparative fault analysis, which the lower court deemed irrelevant in the case of a PRP with limited ability to pay); United States v. Coeur d'Alenes Co., 767 F.3d 873 (9th Cir. 2014) (settlement with defendant for less than its proportionate share of liability was "substantially fair" because of defendant's limited financial ability to pay); Kelley v. Thomas Solvent Co., 717 F. Supp. 507 (W.D. Mich. 1989); United States v. McGraw-Edison Co., 718 F. Supp. 154 (W.D.N.Y. 1989); and United States v. Cannons Eng'g Corp., 720 F. Supp. 1027 (D. Mass. 1989), *aff'd*, 899 F.2d 79 (1st Cir. 1990).

115. *See Arizona v. Raytheon Co., supra.*

# CHAPTER 7

# Organizing PRPs at Multiparty Sites

## I. INTRODUCTION

As noted elsewhere in this book, settlement with EPA is often the best option available to potentially responsible parties (PRPs) confronted with strict, joint, and several liability for cleanup of a Superfund site. The cost of litigating a Superfund case can often exceed a PRP's likely share of the cleanup costs, particularly when the waste attributable to a PRP is minimal (de minimis) when compared to waste sent by other PRPs to the site. It should be noted that EPA also has an incentive to settle rather than litigate Superfund cleanup cases. Litigation is a resource suck, which diverts EPA personnel from the core objective of cleaning up the largest number of sites in the shortest time at the lowest cost—a diversion that the historically cash-strapped agency can ill afford.

Despite the mutual interest of EPA and PRPs in settlement, negotiating such a settlement could become untenable were the agency to attempt multilateral negotiations with all the PRPs individually. While a few National Priorities List (NPL) sites involve only one PRP, more typically EPA finds scores—if not hundreds—of companies that have either sent waste to the site, transported waste to the site, hold title to the site, or held title at the time of waste disposal, each of whom may be held jointly and severally liable for the response costs. Given the multiplicity of parties that may be held liable at a Superfund site, and the complications that multilateral negotiations would engender, EPA's long-standing practice has been to negotiate with the PRPs as a whole, rather than individually.[1] Almost invariably, the PRPs develop an organizational structure—sometimes informal but, more often formalized by a written agreement—and undertake their activities, such as negotiating with EPA, by committee.

In this chapter, the functions, structure, and operations of PRP organizations are analyzed.

## II. FORMATION AND STRUCTURE OF PRP ORGANIZATIONS

The need for PRP organization becomes apparent early in the process, and EPA itself provides the catalyst for the PRPs to so organize. As described in chapter 1, EPA begins the claims process when it issues a notice letter informing the recipient that the agency regards it as a PRP at the NPL site and, as such, potentially liable for response costs incurred or to be incurred. In addition to the provisions identified in chapter 1, the notice letter often provides the names and current addresses of PRPs identified by EPA, exhorts the PRPs to organize themselves, and states EPA's insistence that the PRPs collectively negotiate with the agency. In many cases, the agency will go one step further and call a meeting of all PRPs in order to facilitate their organization. In the absence of such an EPA initiative, representatives of one or more of the larger-volume generators will often call such an all-PRP meeting. By either route, an organization of all PRPs identified at the site is created, commonly referred to as the steering committee.

The actual organization of the PRPs is subject to innumerable variations, reflecting the difference in the makeup and profile of the PRPs at each site. These variables include, but are by no means limited to, the number of PRPs, the available data regarding the waste contributions of each PRP to the site, whether there are a relatively few PRPs that clearly sent larger volumes of waste to the site (who often will assume a leadership role), and, more subtly, how experienced the PRPs are in responding to Superfund claims.

1. 50 Fed. Reg. 5036 (Feb. 5, 1985).

Notwithstanding the variability in the makeup of the PRP organization, all organized PRPs should share certain basic, common goals and should perform certain common functions in pursuit of those goals. These are the most common goals:

(1) to be organized, and, thereby, to be able to function cohesively and efficiently;
(2) by acting in community with one another, to define their common interests, from which they can define their site-specific long-term goals and the strategies necessary to achieve such goals;
(3) to allow each PRP to reduce its individual transaction costs, particularly for those PRPs that sent smaller volumes of waste to the site; and
(4) ultimately, to obtain the best possible resolution of their liability, if any, to the government, either through a negotiated settlement or, when necessary, by litigation.

The functions necessary to achieve these goals are numerous and, again, can vary from site to site. These necessary functions include the development of an overall strategy for confronting EPA's claims, and the development of mechanisms necessary to resolve the somewhat inevitable points of contention among the organized PRPs. More specifically, PRP organizations must grapple with the issue of cost allocation (both administrative and remedial); will often develop a single set of technical comments on EPA's studies and/or remedial proposals; and will conduct negotiations with EPA and the Department of Justice, and sometimes with other entities, such as nonparticipating PRPs.

There are no written rules regarding who may and who may not be a member of, or represented by, the PRP organization, although typically, generators cluster with other generators, and sometimes with transporters. In those instances in which the owners and operators of the hazardous waste disposal site are available and viable, they pose clear conflict problems—the other PRPs' natural inclination is to portray owners and operators as the real culprits at the site, or at least the parties with the greatest culpability. As a result, these PRPs may regard their conflict of interest with the owner-operators as too severe to be accommodated within a single organization.

The degree of formality observed by PRP groups in conducting their business varies greatly. When the PRPs decide to formalize their relationships with one another, they do so in the form of a PRP organization agreement. The agreement will establish the procedural rules to be observed (e.g., voting power, notice of meetings, financial contributions), as well as the substantive duties and functions of the steering committee and its subcommittees. PRP agreements minimize misunderstanding among PRPs and, more importantly, help them reach agreement on common strategy for minimizing liability. It allows PRPs to focus on areas of agreement—such as the most cost-effective way to remediate a site—rather than become consumed entirely by areas of disagreement. Of course, entering an agreement does not guarantee smooth sailing,[2] but it does improve the odds of a desirable outcome.

The PRP agreement almost always includes a common interest or joint defense provision. Typical of such provisions is the following:

> Each Member agrees that all shared information received from any other Member or its counsel, or technical consultant retained for the Group pursuant to this Agreement shall be held in strict confidence by the receiving Member and by all persons to whom such confidential information is revealed by the Receiving Member, pursuant to this Agreement, and that such information shall be used only in connection with asserting any common claims or defenses in connection with the Site and conducting such other activities that are necessary and proper to carry out the purposes of this Agreement.[3]

2. *See, e.g.*, Goodyear Tire Rubber Co. v. W.R. Grace Co., Case No. 1:00-CV-249 (W.D. Mich. June 22, 2000) (party to PRP agreement filed suit alleging breach of PRP agreement by allegedly repudiating an agreed-upon position regarding a proposed cleanup level).

3. *See* THE INFORMATION NETWORK FOR SUPERFUND SETTLEMENTS, PRP ORGANIZATION HANDBOOK 1–14 (1989).

Such a provision is required to ensure that information exchanged by members of the PRP group are treated as confidential communications covered by the attorney-client privilege. As background, the attorney-client privilege protects communications between a lawyer and a client seeking legal advice. However, the privilege protects only confidential communications, that is, those intended to remain private between or among the lawyer, the client, and their respective representatives. If other parties are present, the general rule is that the communication is not privileged.[4] There is an exception, however, where the communication takes place among persons with a common interest.[5] Under the common interest privilege (or joint defense privilege), the attorney-client privilege extends to communications that are "part of an ongoing and joint effort to set up a common defense strategy."[6] In order to establish the existence of a joint defense privilege, the party asserting the privilege must show that (1) the communications were made in the course of a joint defense effort, (2) the statements were designed to further the joint effort, and (3) the privilege has not been waived.[7]

Joint defense agreements can be verbal,[8] but it is best to reduce the agreement to writing to formalize the privilege and ensure that the privilege is not waived, despite disclosure. Disclosure of PRP communications is not merely a theoretical concern. Indeed, parties have gone to court to compel disclosure of not only PRP communications, but the PRP agreement itself.[9]

As the list of NPL sites has grown, PRP agreements have become more a matter of routine and their basic provisions more standardized.

4. *See, e.g.*, Matter of Fischel, 557 F.2d 209, 211 (9th Cir. 1977).

5. *See, e.g.*, *In re* Teleglobe Commc'n Corp., 493 F.3d 345, 363–66 (3d Cir. 2007).

6. Eisenberg v. Gagnon, 766 F.2d 770, 787 (2d Cir. 1985).

7. Matter of Bevill, Bresler & Schulman Asset Mgmt. Corp., 805 F.2d 120, 126 (3d Cir. 1986) (citations omitted).

8. United States v. Schwimmer, 892 F.2d 237, 244 (2d Cir. 1989).

9. *See, e.g.*, Port of Portland v. Or. Ctr. for Envtl. Health, 243 P.2d 102 (Or. Ct. App. 2010) (joint defense agreement between PRPs at a Superfund site held exempt from disclosure under Oregon's public records law).

Model PRP agreements now circulate among the community of companies identified as PRPs at large numbers of sites.[10]

At sites with large numbers of PRPs, the PRP organization will elect an executive committee. Depending on the size and complexity of the site and the nature of EPA's claims, the executive committee may perform all functions necessary to be undertaken on behalf of the PRP group as a whole or, alternatively, may form subcommittees for certain discrete tasks. The makeup of the executive committee can sometimes itself be controversial. In order for all of the PRPs to perceive that they are being adequately represented (and as an inducement to the participation of all PRPs in the organization), the executive committee should *not* comprise solely representatives of the large-volume PRPs. Every effort should be made to include representatives of some of the smaller PRPs.[11] However constituted, and once formed, the executive committee becomes the focal point for confronting EPA's Superfund claims.

## III. TASKS OF THE STEERING COMMITTEE

Steering committees—the organization of all PRPs at a site—have both organizational and substantive tasks. Organizational tasks are necessarily the first order of business. While no two sites are quite the same, typical examples of organizational tasks are the election of a chairperson and other committee officers; the formation of subcommittees (e.g., technical, negotiation, legal, allocation); collection of administrative fees; retention and supervision of consultants (e.g., technical consultants to review EPA's proposed remediation or to assess data collected by EPA); and, at some sites, retention of outside legal counsel to represent the PRPs as a group.

Substantive tasks of a steering committee include

- gathering and disseminating information;
- identifying additional PRPs;

10. The Information Network for Superfund Settlements' *PRP Organization Handbook* (1989) contains several model PRP agreements.

11. See chapter 5, "Settlements," regarding eligibility for, and pursuit of, de minimis settlements for "small volume" generators.

- developing consensus positions among PRPs;
- establishing communication with agencies and nonmember PRPs;
- influencing/addressing/attacking the administrative record;
- negotiating tentative settlement with the government;
- allocating cost of settlement among PRP members; and
- implementing any settlement reached.

Many of these substantive tasks are delegated to subcommittees. A brief description of the more important subcommittees and their tasks follows.

### A. Executive Committee

At sites with large populations of PRPs, where the steering committee is too large to effectively administer its many tasks, an executive committee will be formed. The executive committee will coordinate activities of subcommittees, speak for the group in its relations with EPA and the public, and develop policies and proposals for action and consideration by the steering committee as a whole. The executive committee usually consists of chairpersons of any other subcommittees and may also include representatives from companies with the greatest potential exposure at the site.

Without question, the most challenging task handled by the executive committee is negotiating a settlement agreement with the agencies involved (although sometimes a negotiation subcommittee is formed or common counsel is retained to perform this task). The terms and conditions of settlements and negotiating strategies are significant matters that must be dealt with by the steering committee, and the executive committee can play a crucial role in such matters.

### B. Technical Subcommittee

The technical subcommittee provides input to the steering committee regarding relevant technical issues that are the subject of negotiations with EPA, reviews and analyzes technical data, supervises the work of consultants retained by the steering committee, and, at some sites, can also be responsible for overseeing all site work performed by, or on behalf of, the steering committee. As its name suggests, the technical subcommittee typically comprises technical representatives of PRPs. It is sometimes advisable, however, to have a close relationship between the technical subcommittee and the legal subcommittee, given the complex legal/technical issues that arise at Superfund sites, such as the cleanup standards under section 121 of CERCLA.

Technical data and reports produced by consultants often are used to support the steering committee's comments on EPA's RI/FS or proposed remedial action plan.[12] One of the more important tasks of the technical subcommittee is to review draft reports, communicate the points made in the reports, and ensure that the final reports effectively advocate the steering committee's position.

### C. Allocation Subcommittee

Perhaps the toughest job falls to the allocation subcommittee, which must develop a proposed allocation scheme for sharing the costs of the remedial action. The problem, of course, is that every PRP has an argument as to why it should be allocated a smaller share of liability than it ultimately receives. Imagination and diplomacy are prerequisites to success in this area.

At landfill sites, the allocation of the costs often involves a document called a "waste-in" list, which sets forth for each PRP the volume of waste it contributed to the site, expressed in terms of total volume or weight and relative percentage. Frequently, EPA prepares such a waste-in list on the basis of waste manifests and similar shipping records obtained from PRP responses to section 104(e) information requests. The agency then provides the waste-in list to the PRPs to facilitate resolution of allocation issues. When EPA fails to prepare a waste-in list, the allocation subcommittee itself must create the list. Generally, the allocation subcommittee will look to shipping records from the site and/or PRPs, responses to section 104(e) letters, and specific questionnaires to PRPs; many other creative methods might also generate useful data.

Whether the waste-in list is prepared by EPA or the PRPs, many such lists are based on largely incomplete records of the operations

12. See chapter 4.

and disposal activities at the site. Given the number of years or even decades over which disposal occurred and the era in which it occurred, it is hardly surprising that records may have been lost or destroyed or that, as measured by today's standards, the records that survive may be woefully inadequate in describing the nature and extent of disposal. Almost inevitably, reliance on such records produces confusion and controversy between EPA and the PRPs and, more importantly, among and between PRPs attempting to use these records to reach consensus on a reasonable allocation.

The deficiencies in these waste-in lists are boundless, but a few problems are often seen. For example, in trying to decipher and transcribe contemporaneous records, such as log entries of waste shipments, typographical errors and/or omissions are simply inevitable; they can become the focus of "make or break" controversies among the affected parties. So, too, with the improper conversion of units of volume or weight. In addition, where records at the disposal site are unavailable or inadequate, the records of individual generators and transporters become critical. Unfortunately, no two companies appear to have ever kept records in quite the same way. Differences in retention periods, details, units of measure, and so on, create and compound all of the problems identified above.

Finally, the waste-in lists usually do not take into consideration important factors bearing on allocation, such as the differing nature of the material shipped by PRPs (i.e., the relative toxicities of materials, the relative persistence of PRP wastes, whether certain wastes are "driving the remedy," etc.). While not every allocation addresses toxicity or persistence, when such issues do arise (for example, where the presence and effects of polychlorinated biphenyls (PCBs) or dioxins is an issue) they can become the focal point of the allocation controversies. Nevertheless, for better or for worse, volume is always a major, and sometimes the only, basis for allocation.

Given the numerous factors affecting allocation, the allocation subcommittees face a daunting task in developing an allocation scheme that strikes a balance among them. Generally, however, one of three allocation methods (or a variation thereon) is proposed by the allocation subcommittee for ratification by the full steering committee:

1. *Straight Line:* The straight line method is an allocation in which each PRP's share of the cost of cleanup is the same as its share of the total units on the waste-in list, either by volume or by weight.
2. *Tiered:* The tiered method is a modification of the straight line method pursuant to which PRPs with volume contributions falling within a given range are clustered in groups. Each PRP within the same volume group pays the same share of the cost.
3. *Per Capita:* On occasion, parties adopt an equal cost or "per capita" allocation. This method, however, is rarely adopted. Generally, it can only work where the cost of remediation is low and the number of PRPs is small. It might also work where disposal records are nonexistent.

## IV. TO JOIN OR NOT TO JOIN THE STEERING COMMITTEE

A threshold question every PRP receiving an EPA notice letter must address is whether to join the steering committee and attempt to settle with EPA. A number of factors can influence a PRP's decision. For example, PRPs that sent large volumes of waste to the site, have an interest in the management, and in the outcome, of negotiations with EPA that makes their participation a virtual certainty. On the other end of the spectrum, companies that sent only small volumes of waste to the site do not have the same incentive. Particularly where such de minimis volume companies are themselves small companies, the substantial cost and time commitments attendant to joining the steering committee can act as a deterrent.

Sometimes the decision not to join the steering committee by those PRPs whose volumetric contribution of waste to a site is de minimis is a far more calculated one. Such companies may choose not to join in the hope that EPA will regard them as not worth the trouble of litigation and, therefore, they can avoid liability at the site altogether. This strategy, referred to as "lying in the weeds," met with some success in the early years of Superfund. EPA enforcement, understaffed as it was, did not place a high priority on recovering the remaining sums from such "recalcitrant" PRPs through costly litigation. Thus, the only legitimate exposure for response costs for recalcitrant PRPs was the threat of contribution actions by settling PRPs. Like EPA, however, settling PRPs often did not bring contribution actions against de minimis, or even moderate volume, PRPs, given the high costs of litigation.[13]

Over time, however, the risks associated with lying in the weeds have increased. EPA began pursuing an enforcement strategy to aggressively seek from nonsettling PRPs any costs that are not recovered from the settling PRPs. For example, in negotiating a settlement with PRPs to perform a remedial action, the government may carve out a portion of the cleanup from the settlement and unilaterally order the nonsettlors to perform that portion, pursuant to its authority under section 106 of CERCLA. Failure to comply with a section 106 order would subject the nonsettlors to daily penalties and potential treble damages, as discussed in chapter 3. Similarly, EPA may forgive a portion of its past costs as to the settling PRPs, and bring a section 107 action to recover these costs from the nonsettlors. In either case, if the nonsettlors' costs exceed their respective fair share of the total cost, they could be without legal recourse, since the government generally grants contribution protection to settlors under the consent decree. Thus, the stakes of lying in the weeds have gone up. There may be situations in which it is the best strategy (such as having a good defense to liability), but in all cases its risks must be carefully weighed.

One other option may be considered, although its success is by no means certain. A PRP can decide not to join the steering committee, and attempt, instead, to make direct contact with EPA. Such PRPs can offer their own settlement, for example, as de minimis parties. EPA generally disfavors such separate negotiations, and would likely push such PRPs to join the steering committee. Under the right circumstances, however, individual PRPs may achieve a separate settlement or may simply participate in the final, larger settlement, without having borne the expense of steering committee participation. Under less favorable circumstances, such PRPs may be labeled recalcitrant nonsettlors, and face subsequent EPA or PRP litigation.

13. *Contra* United States v. Conservation Chem. Co., 619 F. Supp. 162 (W.D. Mo. 1985) (four defendants named by EPA brought contribution actions against scores of other PRPs).

# CHAPTER 8

# Litigating a Federal Cost Recovery Action

## I. INTRODUCTION

Where EPA and potentially responsible parties (PRPs) are unable to resolve a Superfund dispute through settlement negotiations, the Department of Justice will file a lawsuit on behalf of EPA against some or all of the PRPs. The complaint may include (1) a count under section 107(a) of CERCLA for the recovery of costs the government has expended to date; (2) a count for declaratory judgment of liability for costs the government expects to expend in the future; (3) a count under section 106(a) of CERCLA seeking a court-ordered injunction compelling defendants to implement a remedy at the site; and (4) perhaps also a count under section 7003 of the Resource Conservation and Recovery Act (RCRA) seeking injunctive relief. At a multiparty site, the government may sue only a small number of PRPs, usually those having the largest volume of waste at the site. Those defendants then must consider whether to file third-party complaints against additional PRPs, further complicating the litigation.

This chapter describes how the Department of Justice prosecutes the typical federal cost recovery action, focusing on multiparty cases. Although very few enforcement actions have proceeded through a trial on the merits, many others have been settled during the pretrial stage as a result of effective litigation. Throughout this chapter, we will describe strategies that defendants have considered and used in litigating Superfund cases.

## II. COMPLAINT

As mentioned above, the government may include not only claims under sections 106 and 107 of CERCLA, but also a claim under section 7003 of RCRA for injunctive relief to abate an imminent and substantial endangerment. In the past, the government routinely included a RCRA count in Superfund actions because RCRA provides a basis for liability for disposal not only of hazardous waste, but also of solid waste. However, reliance on section 7003 has presented some strategic problems for the government. Unlike section 107 of CERCLA, section 7003 of RCRA does not specify applicable defenses, and defendants have successfully argued that the presence of a RCRA count must defeat efforts to limit available affirmative defenses to the statutory CERCLA defenses listed in section 107(b), namely, act of God, act of war, and acts of an unrelated third party.[1] In addition, RCRA, unlike CERCLA, contains no provision limiting judicial review to an administrative record. PRPs have also argued with success that the inclusion of a RCRA count entitles them to full discovery and a de novo trial in which the trial judge is free to exercise his or her full panoply of equitable powers.[2] In light of these risks, EPA has on at least one occasion voluntarily dismissed its RCRA count.[3]

## III. ANSWER, AFFIRMATIVE DEFENSES, AND COUNTERCLAIMS

The government's complaint in a Superfund case typically withstands any motion to dismiss brought under Rule 12 of the Federal Rules of Civil Procedure, unless service of process is defective.[4] Prior to the passage of the Superfund Amendments and Reauthorization Act of 1986 (SARA), the defense of lack of personal jurisdiction was raised

1. *See* United States v. Hardage, 116 F.R.D. 460, 465 (W.D. Okla. 1987).
2. *See id.* at 467.
3. *See* United States v. Hardage, 28 Env't Rep. Cas. (BNA) 1485 (W.D. Okla. 1987).
4. *See, e.g.*, United States v. Witco Corp., 865 F. Supp. 245 (E.D. Pa. 1994) (denying defendant's motion to dismiss for failure to state a claim under Fed. R. Civ. P. 12(b)(6), despite the court's view that the government's oversight costs were not recoverable)

successfully by defendants in several cases.[5] However, CERCLA now provides that when the United States brings a cost recovery action under section 107, it has explicit authority to serve process on defendants in every jurisdiction, regardless of any nexus between the jurisdiction and the disposal site at issue.[6]

Defendants' motions to dismiss for failure to join an indispensable party likewise have met with little success. Courts have reasoned that CERCLA provides for joint and several liability and that the plaintiff may always choose among joint tortfeasors when bringing an action.[7] Further, even if joint and several liability were not imposed, the courts have reasoned, there would of necessity be a finding that harm at the site is divisible, and each defendant could be held liable only for its apportioned share and would not run the risk of liability for all response costs.[8]

Other possible motions to dismiss include lack of subject-matter jurisdiction, which may have some success in the context of an environmental group's suit under the private right of action added by SARA, but is unavailing against a government-filed enforcement action and failure to comply with statutory prerequisites to recovery.[9] Lack of proximate cause has also been rejected by the courts as grounds for failure of a claim.[10] In general, defendants choose to answer the complaint and to preserve any available defenses by pleading them as affirmative defenses.

CERCLA sets out three basic defenses to liability in section 107(b): act of God, act of war, and act of a third party. As noted in chapter 4, additional statutory defenses to liability have been added to CERCLA since enactment in 1980, such as the de micromis defense under section 107(o), the exemption from liability for disposal of municipal solid waste

5. *See, e.g.*, Wehner v. Syntex Agribusiness, Inc., 616 F. Supp. 27, 29 (E.D. Mo. 1985) (motion to dismiss for lack of personal jurisdiction over out-of-state parent corporation granted).

6. CERCLA § 113(e).

7. *See, e.g.*, United States v. A & F Materials Co., 578 F. Supp. 1249, 1256 (S.D. Ill. 1984).

8. *See also* Idaho v. Bunker Hill Co., 635 F. Supp. 665, 676–77 (D. Idaho 1986); United States v. Chem-Dyne Corp., 572 F. Supp. 802, 810 (S.D. Ohio 1983).

9. *See* Thompson v. Thomas, 680 F. Supp. 1, 3 (D.D.C. 1987); Regan v. Cherry Corp., 706 F. Supp. 145, 147–49 (D.R.I. 1989).

10. *See, e.g.*, United States v. Marisol, Inc., 725 F. Supp. 833, 840–41 (M.D. Pa. 1989).

under section 107(p), the contiguous property defense under section 107(q), the prospective purchaser defense under section 107(r), and the recyclable materials defense under section 127. Although some courts have held that these statutory defenses are exclusive in all CERCLA actions, many courts have rejected this notion and defendants will generally plead a long list of additional affirmative defenses, many of which technically are not affirmative defenses.[11] These include

- the absence of a cooperative agreement between the United States and the state;
- statute of limitations;
- response costs not consistent with the National Contingency Plan (NCP);
- laches, unclean hands, waiver, and other equitable defenses; and
- constitutional defenses.

Although some courts have held that, in addition to procedural and express statutory defenses, sections 106 and 107 of CERCLA permit the assertions of equitable defenses, the government continues to take the position that only the statutory defenses enumerated in CERCLA are proper in a Superfund case and frequently files a motion to strike all other affirmative defenses. Defendants have defeated such motions in both section 106[12] and section 107[13] actions.

In some cases, defendants will have counterclaims against the government, which should be included in their answer. If federal agencies themselves sent waste to the site, then claims for contribution against

11. *See* chapter 4.

12. Section 106 on its face provides for equitable defenses. It specifically states that a court may "grant such relief as the public interest and the equities of the case may require." CERCLA § 106(a). Numerous courts have ruled that equitable defenses are applicable in section 106 actions. *See, e.g.*, United States v. Conservation Chem. Co., 619 F. Supp. 162, 205 (W.D. Mo. 1985).

13. Equitable defenses have also been held applicable to section 107 actions. *See, e.g.*, Mardan Corp. v. C.G.C. Music, Ltd., 600 F. Supp. 1049, 1057–58 (D. Ariz. 1984); United States v. Hardage, 116 F.R.D. at 465. Other courts have held that equitable defenses are merely one of the factors to be considered in apportioning damages. *See, e.g.*, Southland Corp. v. Ashland Oil Inc., 696 F. Supp. 994, 1003 (D.N.J. 1988).

such federal PRPs for contribution to site costs must be brought as counterclaims against the unitary plaintiff, the United States.[14] Likewise, if defendants have themselves spent money on investigation or remediation of the site, they will want to sue any federal PRPs via counterclaim to pay their fair share of such response costs under section 107. Generally, claims that are in the nature of recoupment have been allowed to proceed against the United States without running afoul of the principle of sovereign immunity.[15] Thus, claims that seek to reduce or eliminate the liability or relief sought by the United States in its complaint—either by reducing the amount of costs recoverable or by shaping the injunctive relief granted by the court—have been held to be proper. Some examples of facts giving rise to such counterclaims include the government's failure to exercise due care, its delay in undertaking mitigative measures at the site, and any failure by the government to afford defendants notice and an opportunity to participate in remedy selection decisions at the site.[16]

## IV. THIRD-PARTY PRACTICE

In its complaint, the government frequently names only a small number of parties that lead the volume list of PRPs at the site. (In earlier

14. The time for a federal defendant to assert such counterclaims has, itself, been the source of litigation. In *Raytheon Aircraft Co. v. United States*, 532 F. Supp. 2d 1316 (D. Kan. 2008), the plaintiff Raytheon had incurred response costs performing removal activities ordered by EPA at a contaminated airport which at one time had been an army airfield. Raytheon sought cost recovery from the U.S. Army Corps of Engineers, a co-PRP at the site. After filing its answer, the United States filed a counterclaim for cost recovery on behalf of EPA. Raytheon filed a motion to dismiss contending that the counterclaim was compulsory under Fed. R. Civ. P. 13(a). The court denied the motion ruling that even if it were compulsory the United States has broad discretion under CERCLA to decide when to pursue cost recovery.

15. The United States, under the principle of sovereign immunity, is immune from suit, including suit by counterclaims, unless permission to sue has been granted by Congress. However, it is well established that a counterclaim based on the equitable doctrine of recoupment is excepted from this rule. The principle of recoupment is that by initiating an action the federal government consents to the court's jurisdiction over a defendant's counterclaim where that claim arises out of the same transaction or occurrence as the original claim and does not seek independent, affirmative relief from the government, but seeks only to defeat or diminish the government's recovery. *See* United States v. U.S. Fid. & Guar. Co., 309 U.S. 506, 511 (1940); Frederick v. United States, 386 F.2d 481, 488 (5th Cir. 1967). *See also* United States v. Nicolet, Inc., Civil Action No. 85-3060, 1986 U.S. Dist. LEXIS 15795, at *3–4 (E.D. Pa. Dec. 31, 1986); United States v. Moore, 703 F. Supp. 455, 458 (E.D. Va. 1988).

16. United States v. Hardage, 116 F.R.D. at 466.

enforcement actions, the government was criticized for having selected only "deep pocket" companies as its targets.) These few must then determine whether to bring other PRPs into the suit as third-party defendants. Although suits for contribution against third parties may always be initiated after resolution of the government's claims, most defendants prefer to bring all significant PRPs into the case and to resolve all related claims in a single action. Impleader under Rule 14 of the Federal Rules of Civil Procedure provides the procedural mechanism for defendants in Superfund cases to seek contribution from third parties in the same action. Whether or not impleader will be permitted is a matter of judicial discretion.[17]

There are both advantages and disadvantages to impleading third-party defendants, which must be carefully weighed. The primary advantage is avoiding the necessity of filing a subsequent, separate contribution action if one faces joint and several liability for response costs, which should be shared by the third parties. This advantage, however, may have been diminished somewhat following the Supreme Court's ruling in *Burlington Northern & Santa Fe Railway Co. v. United States*,[18] wherein the Court stated that liability should be apportioned if there is a "reasonable basis for determining the contribution of each cause to a single harm."[19] The primary disadvantage of joining third parties is the additional expense and complication their presence will add to the suit.

If some of the parties sued by EPA are members of a steering committee or PRP group, and not all members were sued, then sued defendants will have to decide whether to bring other PRP group members into the case. In many cases, it may be possible to reach a private agreement with such parties that as long as they continue to participate in the steering committee they will not be sued. This allows all involved to save unnecessary transactional costs, unless nonmembers of the PRP group decide independently to bring them in as third parties. The objective of the defendants should be to maximize the resources available

17. *See* FED. R. CIV. P. 14, Committee Note of 1963.
18. 556 U.S. 599 (2009).
19. *Id.* at 614.

to defend against EPA's claims in a coordinated manner and to minimize the splintering, divisive effect of co-party warfare. To this end, defendants may decide to name third parties in the case, but ask the court to postpone discovery on such claims and the prosecution of the claims themselves until a later phase in the case. Postponement or stay of third-party claims can be accomplished by means of a case management order, discussed below.

## V. CASE MANAGEMENT ORDER

Served on defendants together with the complaint, or soon thereafter, is the government's proposed case management order. The government often will ask the court to bifurcate the trial, with liability issues tried first and remedy issues determined in a second phase. The government will argue that an early liability determination will promote settlement, whereas a remedy-first trial will not result in defendants' willingness to settle so long as the issue of their liability at the site remains open. Also, the government argues, if defendants were to prevail on liability issues, a first phase remedy trial would have been a wasted and expensive exercise. The government often asks in its case management order for a stay of discovery on issues relating to the remedy phase until after resolution of liability. Other features of the typical proposed case management order include deferral of contribution or allocation issues until a third phase of the case, requirements for defendant organization (including a committee with whom the United States can negotiate), appointment of liaison counsel, and schedules for filing dispositive motions.

Defendants, however, may prefer a remedy-first bifurcation of the issues. Not until the appropriate site remedy has been selected, and its cost determined, can defendants begin to consider settlement. Consequently, only a remedy-first trial has any reasonable prospect of expediting the resolution of the litigation. For most generator defendants, liability is a relatively straightforward matter in the typical case in which signed hazardous waste manifests signal defendants' disposal of hazardous substances at the Superfund site. Thus, because few defendants have any serious expectation of being found not liable, an early liability determination would provide little additional impetus for settlement. For these reasons, defendants may oppose the government's proposed case management order and offer an alternative order providing for an early remedy trial.

The courts have generally sided with the government on this issue and ordered liability issues tried first.[20] However, in at least two CERCLA cases, the courts have bifurcated cases with remedy being tried before liability.[21] The court in *Seymour* based its decision on a belief that there was a reasonable prospect of settlement of the case only if a remedy was first selected.[22] Notably, however, the value of trying the remedy first was largely neutralized after Congress amended CERCLA in 1986 to specify that the basis for a remedial decision shall be set forth in the administrative record and shall not be subject to pre-enforcement review.

Perhaps as important as seeking to try remedy issues as the first phase of the case is to obtain an order that allows the defendants to pursue discovery on remedy issues at the same time the government is pursuing its liability discovery against defendants. Defendants may have the opportunity to hit remedy issues hard in early discovery while the government's resources are taxed by the effort of proving each defendant's liability at the site. Also, concurrent discovery on both phases of the case can be urged as a means of assuring that trial of the second phase can follow closely on the heels of the first.

One final point on bifurcation deserves mention. If faced with a liability-first trial that will entail expensive and detailed discovery into a defendant's waste shipments, manufacturing processes, and in some cases even confidential business information and trade secrets, defense counsel should consider the radical strategy of stipulating to liability. Such a stipulation leaves open the issues of appropriate site remedy and each party's share of site costs, but would preclude expensive and time-consuming discovery into defendants' manufacturing process information and would prevent the government from building a record regarding "imminent and substantial endangerment" presented at the site by each

---

20. *See, e.g.*, United States v. Wade, 653 F. Supp. 11, 14 (E.D. Pa. 1984); Kelley v. United States, 25 Env't Rep. Cas. (BNA) 1500 (W.D. Mich. Sept. 19, 1985).

21. *See* United States v. Seymour Recycling Corp., 21 Env't Rep. Cas. (BNA) 1999 (S.D. Ind. 1984); United States v. Price, 20 Env't Rep. Cas. (BNA) 2229 (D.N.J. 1984).

22. *Seymour*, 21 Env't Rep. Cas. at 2001.

defendant's waste that could fuel tort claims by neighboring owners. The advantages of such a strategy must, of course, be weighed against its obvious tactical disadvantages and against the defendant's likelihood of winning the liability issue.

## VI. DISCOVERY

### A. Introduction

The scope of discovery in Superfund litigation may vary from case to case, depending on the scope of the issues to be determined by the court. For example, in a case under section 107 for recovery of costs incurred by the government in cleaning up a site, the case may be decided on a review of the administrative record, under the arbitrary and capricious standard of review.[23] In such a case, the government may seek to limit discovery to a review of the sufficiency of the documents comprising the administrative record. In another case, the court may decide the issues de novo after full discovery.[24] In still a third scenario, the court may rule that its decision will be based on a review of the administrative record, yet permit the parties to undertake full discovery, going beyond the record itself.[25]

The scope of review to be employed by the court is a critical issue in Superfund litigation, and may at times be outcome-determinative. Defendants may want to prompt an early ruling on this issue by seeking to take discovery on EPA's remedy selection process and deliberations. The government will generally respond by seeking a protective order limiting the scope of discovery to the review of the administrative record, based on its theory that the correct standard to be employed by the court is record review. In raising the issue by simply noticing the depositions of key government personnel, defense counsel can put the government, in a position of trying to prevent disclosure of its actions

23. *See, e.g.*, United States v. Nicolet, Inc., 17 Envtl. L. Rep. (Envtl. L. Inst.) 21,091 (E.D. Pa. May 12, 1987). *See generally* chapter 6 on judicial review.

24. *See* United States v. Hardage, 28 Env't Rep. Cas. (BNA) 1485, 1488 (W.D. Okla. 1988).

25. *See, e.g.*, United States v. Bell Petrol. Servs., Inc., 718 F. Supp. 588, 591 (W.D. Tex. 1989).

and decisions, which the court may view as an indication that EPA "has something to hide."

## B. Discovery Scheduling Order

Regardless of the scope of judicial review applicable to a particular Superfund case, extensive discovery will be necessary. It may be helpful or necessary to propose a scheduling order to govern the orderly and efficient conduct of discovery in the litigation. As with a case management order, a discovery scheduling order should be negotiated with the Department of Justice lawyers and jointly presented to the court whenever possible. Such an order may govern:

- discovery, limiting it to issues to be litigated in particular phases of the case;
- procedures for informal discovery, such as the exchange of documents informally requested or the interview of possible witnesses for the purpose of determining whether formal depositions are warranted;
- procedures for discovery dispute resolution;
- limitations on the number and timing of service of interrogatories, requests for admission, and document requests;
- deposition scheduling, including time frames for nonexpert and expert witnesses' depositions; and
- procedures requiring disclosure of expert witnesses' opinions prior to their depositions, as well as the production of all documents relied on by such witnesses in forming their opinions.

Each phase of discovery governed by the discovery scheduling order should have a termination date so that, for example, all document requests and interrogatories have been served and answered before any depositions begin.

## C. Discovery on Liability Issues

The principal objective of the government's discovery on liability issues will be to establish a link between each generator and the Superfund site. It is not necessary for the government to prove that a generator's hazardous waste is currently present at the site or to find the "fingerprint" of a particular defendant's waste.[26] The government will focus on evidence that the generator's waste was hazardous and that it was shipped to the Superfund site. To do this, the government will request documents regarding the generator's manufacturing processes, test results and waste analyses conducted by the generator, general information on the waste stream at the generator's place of business, bills of lading, and other information reflecting waste shipments from the plant, evidence of contractual arrangements with transporters or haulers, and information on raw materials utilized in the generator's manufacturing process. If evidence of the hazardous nature of a defendant's waste is not otherwise available, the government will attempt to collect information that will allow its chemical engineers to reverse engineer the defendant's product to determine what the by-products and the nature of wastes sent for off-site disposal would have been.

After collecting and reviewing available documents, the government will depose persons having knowledge of the defendant's waste disposal practices. Rather than starting with the plant manager or maintenance supervisor, as one might expect, the government often begins by deposing the waste handlers. It may depose any individual who signed a waste manifest, and in the course of such depositions, will authenticate all available documents. In addition, the government will depose employees who loaded trucks, truck drivers, and accounting personnel who contracted for disposal or paid invoices for transportation and disposal of wastes. If further information is still needed, government counsel will then depose management or supervisory personnel familiar with the company's waste disposal practices.

This detailed discovery on liability issues is expensive and time-consuming and is seen by many defendants as a diversion of resources from the central issue in the case, remedy. Of course, a defendant with a valid defense to liability will also prefer to focus its resources on this

26. United States v. Wade, 577 F. Supp. 1326, 1332–33 (E.D. Pa. 1983). *See also* United States v. Ottati & Goss, Inc., 630 F. Supp. 1361, 1402 (D.N.H. 1985) (section 107 "does not require the government to match the waste found to each defendant as if it were matching fingerprints").

liability discovery phase. Note that although the government has taken the position at times that it can prove its liability case with a single, signed manifest from a defendant's company, it still insists on proving the existence of each shipment and its volume and hazardousness.

Short of agreeing to stipulate to liability, as discussed above, defendants have available to them few strategies for limiting the scope of the government's inquiry. They sometimes refuse to answer any questions regarding waste shipments to sites other than the one in question, particularly since such information could be used by EPA to build a database regarding potential liability at other sites.[27] They should refuse to provide information regarding manufacturing processes and waste streams if there is other evidence available to the government, such as waste analyses, to show that the generators' wastes were hazardous. Finally, they should seek a protective order governing the disclosure of any confidential business information to the public or to co-parties, which will ensure that confidential records will be shown only to Department of Justice (DOJ) attorneys and their assistants who are preparing the case.

One issue that frequently arises in the course of depositions of defendants' personnel is "site selection." As discussed in chapter 4, transporters who selected the site for disposal are liable under CERCLA. Records on this issue are often inconclusive, and its resolution will turn on the recollections and testimony of the transporter's and generator's respective employees. In preparing a generator's employees for deposition, it is important to explore fully any basis for claiming that the transporter selected the disposal site and, therefore, should bear some share of liability for the generator's volume of waste at the site.

### D. Discovery on Remedy Issues

In a Superfund case in which the court will determine de novo the appropriate site remedy, discovery on remedy issues becomes the crux of the case. Even in cases limited to an administrative record, defendants may

27. *But see* United States v. Mottolo, 605 F. Supp. 898, 915 (D.N.H. 1985) (information on defendant's disposal of wastes at other sites is discoverable because it allows the government to corroborate the quantity sent to the site in question).

be entitled to full discovery or at least an opportunity to test the sufficiency of the record and to supplement it where warranted. In any of these instances, much effort will be expended on discovery of remedial issues.

Defendants should seek to explore the basis for the government's actions at the site and any remedy selection made. If defendants have in the past furnished technical information to EPA or provided comments on EPA's reports and studies, they should investigate whether such data had any impact on EPA's processes. Evidence that EPA refused to consider PRPs' submissions or that agency personnel rejected viable remedial alternatives without study or for unacceptable reasons is useful in proving that EPA's actions were "arbitrary and capricious" or that the defendants should be given an opportunity to submit their own remedial alternatives or designs to augment the administrative record.

During the depositions phase, defendants typically depose the government's fact witnesses first. These are the people who collected samples and analyzed site conditions, and assisted in early evaluation of risks posed by the site and in the selection of any necessary interim remedial or emergency measures. They include EPA personnel, EPA contractors, laboratory personnel, and any state agency personnel involved. If a remedial investigation/feasibility study (RI/FS) or record of decision (ROD) has been prepared for the site, defendants must discover as much as possible about these actions. Much of the evidence gathered in this phase is also pertinent to the defense of the government's cost recovery claim, since testimony regarding sloppy work, poor sampling and analytical techniques, and inadequate site investigation and data gathering can be used to defend against the government's efforts to recover its costs.

Expert discovery generally follows. Each side will present the technical experts in support of its remedy. These may include geologists, hydrologists, remedial design experts, chemists, risk assessment experts, and other specialists.

Remedy discovery may include the parties' submission of documents describing their preferred remedy for the site. Such remedy reports can be voluminous, and it is advisable to depose the reports' authors. In preparing their own remedy report, defendants must decide whether to prepare a conceptual design, as EPA often does, with final design details to be worked out or agreed upon after field tests or pilot tests or after a final design phase, or whether to present a more detailed remedial design. The detailed design is considerably more expensive and time-consuming to prepare, but should embody sufficient detail such that it could be attached to a final judgment ordering parties to implement defendants' remedy at the site. Defendants must balance the risk of wasting money to develop a design that may or may not be implemented against the certainty provided by having a design that leaves few questions unanswered, and by the strategic advantage to be gained by offering the court a more professional product.

### E. Discovery on Cost Issues

If the case includes a claim for reimbursement of the government's costs incurred at the site, defendants should also take discovery on the cost claim. Generally, the government seeks to recover not only its technical costs at the site, but also the litigation costs, including EPA and DOJ time spent in connection with the site, as well as indirect costs, such as rent on government buildings and a share of the salary of the EPA administrator. EPA witnesses on these claims typically include the chief of the cost accounting section at EPA, who directed the accounting project for establishing the agency's methodology for calculation of EPA's indirect costs; experts in the development of computer models for estimating costs; cost accountants at EPA responsible for the allocation of indirect costs under the Superfund program; Department of Justice financial managers responsible for financial management and cost accounting at DOJ; and outside accounting consultants.

Discovery in defense of cost claims can be effective. A number of defenses to these claims may be pursued during discovery:

- The government's costs are inconsistent with the NCP;
- Technical work done by the government was unnecessary, duplicative, or improperly performed;
- The government paid more than market value for services. Contracts may not have been put out for bid. Labor rates may be significantly in excess of market rates;

- The government's costs are undocumented; and
- The government's indirect costs are not recoverable response costs, but represent fixed costs of government that would have been incurred by DOJ and EPA in any event.

One court disallowed all or a portion of the government's costs on some of the above theories.[28] Two courts, however, held that the government is entitled to recover *all* its response costs and awarded the government summary judgment, foreclosing PRPs from challenging the appropriateness or reasonableness of the individual cost items.[29]

### F. Discovery Disputes

Besides disputes over the scope of discovery in a Superfund case, discussed above, other discovery battles are likely. The government may claim that evidence of its internal deliberations on remedy issues is privileged.[30] In addition, a dispute may arise if defendants withhold documents under a claim of "joint defense privilege" or withhold drafts of reports, notes, and correspondence pursuant to the attorney-client privilege and work-product doctrine. The government may also seek to compel discovery of documents prepared by insurance investigators hired by defendants' insurers to investigate the circumstances of the disposal that gave rise to the litigation.

In order to safeguard the attorney-client privilege or the work-product doctrine in the context of multiparty litigation, it is critical that specific procedures for identification and handling of privileged documents be established and maintained. Separate files of such materials should be maintained; no testifying experts should be allowed to review or rely upon privileged documents; all technical consultants and experts should

28. United States v. Ottati & Goss, 694 F. Supp. 977, 991–92, 995, 997 (D.N.H. 1988), *aff'd in part, vacated in part on other grounds*, 900 F.2d 429 (1st Cir. 1990).

29. *See* United States v. Northernaire Plating Co., 670 F. Supp. 742 (W.D. Mich. 1987), *aff'd sub nom.* United States v. R.W. Meyer, Inc., 889 F.2d 1497 (6th Cir. 1989); United States v. Hardage, 733 F. Supp. 1424 (W.D. Okla. 1989).

30. *See, e.g.*, Coastal States Gas Corp. v. Dep't of Energy, 617 F.2d 854 (D.C. Cir. 1980); Jordan v. Dep't of Justice, 591 F.2d 753 (D.C. Cir. 1978).

be retained and supervised by counsel. Consistent handling of confidential documents according to established procedures can assist a defendant in winning a battle for disclosure.

## VII. SPECIAL MASTERS

The Federal Rules of Civil Procedure provide for the appointment of a "special master" in cases in which the issues are extremely difficult, the computation of damages is complicated, or when there are exceptional circumstances associated with a case.[31] Because CERCLA litigation involves a number of complex legal, technical, and scientific determinations that must be made by the court, some judges have appointed special masters to assist and advise them. Some have been lawyers, relied on by district court judges to hear discovery and liability issues. Two examples are the special masters that have served in the *United States v. Stringfellow* and the *United States v. Conservation Chemical Co.* matters. In *Stringfellow*, the master was assigned to supervise discovery and make recommendations concerning summary judgment.[32] In *Conservation Chemical Co.*, the court appointed a law professor to supervise discovery and a range of other issues.[33] In other cases, the court has selected a technical advisor to serve as special master and make recommendations to the court regarding remedy issues. In the *Hardage* case, an engineer was selected by the court to assist on a wide variety of issues.[34]

The Department of Justice's policy regarding the use of special masters[35] provides that the United States may agree to the appointment of a special master only in limited situations and under tightly controlled circumstances. In short, DOJ will agree to the use of a special master only in areas related to discovery and settlement. Any use of a special master to provide technical advice to the court will be vigorously opposed by DOJ. A broad-ranging order of reference to a special

31. *See* FED. R. CIV. P. 53.
32. United States v. Stringfellow, No. CV-83-2501-MML (C.D. Cal. 1984).
33. *See* 106 F.R.D. 210 (W.D. Mo. May 1, 1985).
34. United States v. Hardage, 26 Env't Rep. Cas. (BNA) 1042 (W.D. Okla. 1987).
35. United States Att'y Gen., Department of Justice Policy Regarding Special Masters (Apr. 15, 1986).

master will be met by a motion from the government to limit the special master's tasks to those consistent with the DOJ policy. For example, in *United States v. Washington*,[36] a special master was appointed to hear the case. The United States petitioned the Ninth Circuit to revoke the authority of the special master to preside at trial and recommend liability. The court determined that appointment of a special master to determine the merits of the case was inappropriate. However, that court held that the master had broad authority to determine pretrial matters, such as discovery and stipulations of fact and, in the case of liability, the authority to determine the extent of the damages.

## VIII. DISPOSITIVE MOTIONS

The government frequently seeks to resolve the issue of a defendant's liability in a Superfund case by summary judgment motion. Following extensive discovery on the nature of a defendant's waste stream and shipments to the site, government counsel can usually demonstrate the necessary elements of liability: that there has been a release or threatened release of a hazardous substance from a facility into the environment and that the defendant is a current owner or operator of the site, a former owner or operator at the time that hazardous substances were deposited there, a person who arranged for the disposal or treatment of hazardous substances, or a person who transported such substances for disposal or treatment to the site. (The elements of liability are fully discussed in chapter 4.) If the discovery record is complete, the government will move for summary judgment on the issue of liability. Although there are cases in which a defendant with a meritorious defense to liability will vigorously contest liability and prevail,[37] in the majority of cases defendants are simply prolonging the inevitable and creating a record about their wastes at the site that may greatly simplify the work of a toxic tort claimant's counsel. At a particularly sensitive site, defendants the government can readily demonstrate are liable may wish to consider negotiating with government counsel for some reasonable and limited

36. 135 F.3d 618 (9th Cir. 1998).
37. *See* United States v. Pac. Hide & Fur Depot, Inc., 716 F. Supp. 1341 (D. Idaho 1989).

liability showing on the record in exchange for defendants' partial or total nonopposition to the government's motion.

If the case includes a claim for reimbursement of the government's past costs at a site, defendants can also anticipate a motion for summary judgment of such costs. Sometimes the motion may be coupled with a motion in liminc (at the outset) to prevent defendants from offering evidence at trial to show that the government is not entitled to recover its costs. Although it might seem relatively easy to point out disputed issues of material facts that would defeat the government's motion, (e.g., that specific costs were not consistent with the NCP because they were payments for poorly performed or unusable work), some courts have construed section 107(a)(4)(A)[38] quite broadly in granting government motions for summary judgment of past costs. In the *Hardage* case, the court seemed to rule that the costs would be found "consistent with the NCP" if the *response actions* they related to were consistent with the NCP. In that case, even though the defendants had challenged particular costs as excessive, unnecessary, or poorly performed, the court ruled that they had not adduced evidence to show that the government's choice of response actions was arbitrary and capricious, and read section 107(a)(4)(A) literally to allow the government to recover *all* its costs under these circumstances.[39]

The final category of significant pretrial motions that deserves mention is motions to clarify or establish the applicable standard of judicial review. As noted earlier, the standard of review is often a critical issue in Superfund litigation, and the government may raise it at an early juncture in an attempt to curtail defendants' discovery efforts or to otherwise streamline the proceedings into a review of the administrative record compiled and offered by EPA. See chapter 6 for a discussion of judicial review.

38. Section 107(a)(4)(A) of CERCLA allows the government to recover "*all* costs of removal or remedial action incurred by the United States Government . . . not inconsistent with the national contingency plan" (emphasis added).

39. 733 F. Supp. 1424, 1432–33 (W.D. Okla. Dec. 8, 1989) (citing United States v. R.W. Meyer, Inc., 889 F.2d 1497 (6th Cir. 1989).

## IX. TRIAL

The trial of the case will follow the order established in the court's case management order, with either liability or remedy issues tried first. Courts have unanimously held that there is no right to a jury trial under CERCLA.[40]

Increasingly, federal district court judges are establishing special courtroom procedures to expedite complex litigation (such as requiring the submission of direct testimony by written affidavit), and are borrowing other practices from administrative proceedings to shorten actual courtroom trial time. Even with such devices, the trial of a government enforcement action can be prolonged and costly.

40. *See, e.g.*, Hatco Corp. v. W.R. Grace & Co., 59 F.3d 400 (3d Cir. 1995) (court held there is no right to a jury trial under either section 107 or section 113 of CERCLA); United States v. Alsol Corp., Civil Action No. 13-CV-0380 (KSH)(CLW), 2014 WL 3447501, at *2 (D.N.J. July 10, 2014) ("CERCLA actions brought to recover [cleanup] costs . . . sound[] in equity and a jury trial is therefore unavailable"); CPI Plastics, Inc. v. USX Corp., 22 F. Supp. 2d 1373 (N.D. Ga. 1995) (court granted motion to strike jury trial demand, noting that sections 107 and 113 share equitable characteristics and that legislative history confirms that such actions are equitable); Thaler v. PRB Metal Prods., Inc., 810 F. Supp. 49 (E.D.N.Y. 1993) (court granted the defendant's motion to strike the parties' jury demand, holding that there is no right to a jury trial under CERCLA); Am. Cyanamid Co. v. King Indus., Inc., 814 F. Supp. 209, 214 (D.R.I. 1993) ("CERCLA action to recover response costs is equitable in nature, and therefore no jury right attaches to it"); GNB Inc. v. Gould, Inc., No. 90 C 2413, 1993 WL 62376, at *2 (N.D. Ill. Mar. 5, 1993) (district court granted the defendant's motion to strike the plaintiffs jury demand in a declaratory judgment action, holding that the claim requested an equitable remedy which "does not give rise to the right to a trial by jury"); *see also* FMC Corp. v. Aero Indus., Inc., 998 F.2d 842 (10th Cir. 1993) (defendant in a private cost recovery case waived any right to a jury trial by signing a pretrial order and participating in a bench trial; the court did not directly address the issue of whether there was a right to a jury trial).

# CHAPTER 9

# Private Party Actions

## I. INTRODUCTION

As noted in the Overview, Congress's prime motivation for enacting CERCLA was to give EPA explicit authority to respond to abandoned hazardous waste sites and other, historic environmental contamination. Congress, however, recognized that EPA would not be able to address the problems at all such sites—there were simply too many contaminated sites and too little federal funding.[1] To fully address the problem, private parties would have to perform cleanup work, and have the ability to recover a share of their expenses from other parties whose wastes or activities caused or contributed to the contamination.

To that end, Congress included language in section 107(a) of CERCLA that enables private parties to recover the costs of cleaning up a contaminated site from liable parties. Whereas section 107(a)(4)(A) provides that PRPs are liable for response costs incurred by the "United States Government or a State or an Indian tribe not inconsistent with the national contingency plan,"[2] Section 107(a)(4)(B) provides that potentially responsible parties (PRPs) are liable for "any other necessary costs of response incurred by any other person consistent with the national contingency plan." Because the phrase "any other person" is placed at the end of section 107(a), it is unclear whether it means anyone other

1. At the time Congress passed CERCLA, EPA estimated that there were between 30,000 and 50,000 hazardous waste sites. H.R. Rep. No. 1016, 96th Cong., 2d Sess., pt. 1, at 18 (1980), reprinted in 1980 U.S.C.C.A.N. 6119.

2. CERCLA § 107(a)(4)(A).

than the four categories of PRPs enumerated in section 107(a)(1)–(4), or alternatively anyone other than the government entities enumerated in section 107(a)(4)(A). Limiting private cost recovery claims to persons who are not PRPs, however, would be peculiar, to say the least, as non-PRPs will rarely if ever have an incentive to perform a cleanup. Allowing a PRP who performs or pays for a response action to recover all or a portion of its response costs from other PRPs makes more sense, but the legislative history of this language is scant.

During the early years of Superfund, courts wrestled with the legal basis for a PRP to recover a portion of its cleanup costs from other PRPs. They characterized an action by one PRP against another PRP as a contribution action. Noting the absence of any express right of contribution under CERCLA, most courts concluded that a right of contribution was implied under section 107(a)(4)(B), as well as federal common law.[3]

Congress sought to correct this omission in 1986 when it amended CERCLA to establish an express right of contribution in two paragraphs of section 113(f). First, under section 113(f)(1), "Any person may seek contribution from any other person who is potentially liable under [section 107(a)], during or following any civil action" under section 106 or section 107(a). Second, under section 113(f)(3)(B), "A person who has resolved its liability to the United States or a State for some or all of a response action or for some or all of the costs of such action in an administrative or judicially approved settlement may seek contribution from any person who is not a party to" that settlement.

These new contributions provisions, however, created as many problems as they solved. For example, they do not address the contribution rights of PRPs who voluntarily perform a cleanup. Given the change in the law, what remedy is available for such PRPs? Can they pursue contribution under section 113(f), notwithstanding that the text of section 113(f) does not expressly cover them, or can they still assert an implied right of contribution under section 107(a)(4)(B), as was the case before

3. *See, e.g.*, Wickland Oil Terminals v. Asarco, Inc., 792 F.2d 887, 890–92 (9th Cir. 1986); Walls v. Waste Res. Corp., 761 F.2d 311, 317–18 (6th Cir. 1985); Philadelphia v. Stepan Chem. Co., 544 F. Supp. 1135, 1140–43 (E.D. Pa. 1982).

the 1986 amendments? And if those PRPs can use section 107(a)(4)(B), can PRPs who are sued by or settle with the United States or a state also elect to use it?

Pursuing cost recovery under section 107 has advantages over a contribution action under section 113. The courts have interpreted liability under section 107(a) as joint and several, and thus a PRP seeking cost recovery in theory may seek to recover all of its response costs.[4] By contrast, liability in a contribution action under section 113(f)(1) is several, meaning that the plaintiff may only recover each defendant's proportionate share of the response costs associated with the site, and bears the burden of proving the proportionate share of liability for each defendant.[5] This can have significant implications when it comes to assigning responsibility for the so-called "orphan share" of liability, which is the share that is attributable to PRPs that are either defunct, dead, insolvent, or unknown. As discussed *infra*, in a cost recovery action the orphan share may be divided up between all of the PRPs through allocation, whereas in a contribution action a court may in theory saddle the plaintiff alone with the orphan share. Thus, section 107 cost recovery provides significant advantages over section 113 contribution.

Defining the boundaries of the private cost recovery and contribution provisions under section 107 and section 113, respectively, has taken the better part of two decades, including two trips to the U.S. Supreme Court. In *Cooper Industries, Inc. v. Aviall Services, Inc.*,[6] the Supreme Court addressed contribution rights under section 113, and in *United States v. Atlantic Research Corp.*,[7] the Supreme Court addressed private cost recovery under section 107(a). Based on these two decisions, the framework for private party actions under CERCLA may now be summarized as follows. PRPs that are being or have been sued in a civil action under section 106 or section 107 can *only* pursue contribution against other PRPs under section 113(f)(1).[8] PRPs that voluntarily

4. See discussion in chapter 4.
5. See discussion in section III of this chapter.
6. 543 U.S. 157 (2004).
7. 551 U.S. 128 (2007).
8. Cooper Indus., Inc. v. Aviall Servs., Inc., 543 U.S. at 158.

perform a response action—separate and apart from a government suit or settlement—may seek to recover response costs from other PRPs in a cost recovery action under section 107(a)(4)(B).[9] The Supreme Court did not rule on whether a PRP that incurs response costs pursuant to obligations under a consent decree with the government is limited to contribution under section 113(f)(3)(B) or could also pursue cost recovery under section 107(a)(4)(B).[10]

Needless to say, the foregoing summary of the private cost recovery/contribution dividing line barely scratches the surface of issues that surround private party actions under CERCLA. We delve deeper into them in this chapter. In section II of this chapter, we examine the elements of private cost recovery under section 107(a)(4)(B) of CERCLA. In section III of this chapter, we explore the evolution of contribution rights under CERCLA, and the nuances of such claims. Parts IV through VIII of this chapter address issues that are common to both private cost recovery and contribution actions, namely, the availability of declaratory judgment, suits against government entities and settlement issues.

## II. PRIVATE COST RECOVERY UNDER SECTION 107

Section 107(a)(4)(B) of CERCLA provides that responsible parties are liable for "any other necessary costs of response incurred by any other person consistent with the national contingency plan." As noted above, the Supreme Court has confirmed that this language creates a private right to recover response costs—a right that extends to PRPs who themselves may bear liability for the contamination that is the subject of their response action.[11]

To assert a claim under section 107(a)(4)(B), a private plaintiff must demonstrate that it has incurred "response costs" that are both "necessary" and "consistent with the national contingency plan." The discussion below addresses each of these elements.

9. *Atl. Res. Corp.*, 551 U.S. at 133.
10. *Id.* at 139 n.6.
11. *Atl. Res. Corp.*, 551 U.S. at 133.

### A. Incurrence of Response Costs

Only response costs may be recovered. Generally, response costs are the costs of investigating and remediating the site, and comprise the costs of removal actions and remedial actions, as defined in section 101 of CERCLA. For example, neither lost profits nor other compensatory damages are recoverable.[12] Similarly, lost income, diminution of property value, and personal injury are not recoverable under CERCLA.[13] (A detailed description of the types of costs that qualify as response costs is set forth in chapter 4.)

A cost recovery action may be commenced any time after response costs have been incurred.[14] Courts are generally in agreement that the claimant must have incurred *some* response costs prior to commencing suit, although it is not necessary that the entire response action be complete.[15] Investigation and monitoring costs are sufficient to establish that response costs have been incurred.[16]

To be recoverable, costs need not be incurred pursuant to a government order so long as they are consistent with the National Contingency Plan (NCP).[17]

### B. Necessary Costs

Response costs must be "necessary" to be recoverable under section 107(a)(4)(B).[18] The NCP does not define or otherwise elucidate the

12. *See, e.g.*, Artesian Water Co. v. Gov't of New Castle Cnty., 659 F. Supp. 1269, 1285–86 (D. Del. 1987), *aff'd in part and remanded in part*, 851 F.2d 643 (3d Cir. 1988) (Congress "clearly manifested an intent not to provide compensation for economic losses or for personal injury resulting from the release of hazardous substances"); Jones v. Inmont, 584 F. Supp. 1425 (S.D. Ohio 1984).

13. Wehner v. Syntex Corp., 681 F. Supp. 651, 653 (N.D. Cal. 1987); Piccolini v. Simon's Wrecking, 686 F. Supp. 1063, 1067–68 (M.D. Pa. 1988); Allied Towing Corp. v. Great E. Petrol. Corp., 642 F. Supp. 1339, 1348 (E.D. Va. 1986).

14. CERCLA § 113(g)(2).

15. *See, e.g.*, United States v. A & F Materials Co., 578 F. Supp. 1249, 1259 (S.D. Ill. 1984).

16. *Wickland Oil Terminals*, 792 F.2d at 892.

17. Int'l Clinical Labs., Inc. v. Stevens, 710 F. Supp. 466, 472 (E.D.N.Y. 1989); *see also* Cooper v. Armstrong Rubber Co., Civ. A. No. J88-0464(L), 1989 WL 60338 (S.D. Miss. Feb. 1, 1989) (no prior government approval of incurrence of costs is required for recovery from a private party).

18. CERCLA § 107(a)(4)(B).

characteristics that qualify response costs as necessary. Courts that have examined the issue have offered a wide variety of determinants for judging the necessity of response costs, but a common thread is evidence that the costs were incurred directly in response to a threat to human health or the environment. For example, in *City of Colton v. American Promotional Events, Inc.-West*,[19] the Ninth Circuit rejected a private cost recovery claim because the plaintiff failed to show a direct connection between the response costs and an actual threat. Similarly, other courts have required a showing that the response costs are closely tied to the actual cleanup to prove they were necessary.[20]

Courts have also evaluated whether the plaintiff's primary reason for undertaking cleanup work was to improve property or to respond to a threat to human health or the environment. As noted by the court in *G.J. Leasing Co. v. Union Electric Co.*, "The statutory limitation to 'necessary' costs of cleaning up is important. Without it there would be no check on the temptation to improve one's property and charge the expense to someone else."[21]

The decision in *Regional Airport Authority of Louisville v. LFG, LLC*[22] illustrates the point. In this case, the Airport Authority filed a cost recovery action under section 107(a)(4)(B) against LFG, the former owner of adjacent property that the Airport Authority had acquired through condemnation for a runway expansion project. At issue was whether the investigative costs incurred by the Airport Authority were necessary response costs. The Airport Authority knew the property had been contaminated by prior industrial activity, as soil sampling data detected the presence of lead. During demolition, the State advised the Airport Authority that it should identify remedial alternatives. The

19. 614 F.3d 998 (9th Cir. 2010).

20. Ellis v. Gallatin Steel Co., 390 F.3d 461, 482 (6th Cir. 2004); *see also* Black Horse Lane Assoc., L.P. v. Dow Chem. Corp., 228 F.3d 275, 297 (3d Cir. 2000) (response costs were not necessary because they did not pertain to a removal or remedial action on the contaminated property); Amoco Oil Co. v. Borden, Inc., 889 F.2d 664, 669–70 (5th Cir. 1989) ("To justifiably incur response costs, one necessarily must have acted to contain a release threatening the public health or the environment.").

21. 54 F.3d 379, 386 (7th Cir. 1995).

22. 460 F.3d 697 (6th Cir. 2006).

Airport Authority, however, elected to build the new runways without first evaluating remedial alternatives. A remedial investigation/feasibility study (RI/FS) was eventually prepared by the Airport Authority's contractor. Although the RI/FS recommended removing contamination, the Airport Authority decided not to do so. Instead, after completing the runway project the Airport Authority proposed a "risk management-based remedy" under which exposure to lead-contaminated soil would be prevented by runway pavement, clean top soil, and vegetation. The State later approved this remedy.[23]

The court rejected the Airport Authority's claim to recover its investigative costs from LFG, finding the Airport Authority's work on the runway project was no greater than what would have been done absent any contamination. As support for this finding, the court noted that the lead contamination in soil was below EPA's level of concern, and indeed was lower than concentrations in the soil of nearby parks.[24] Moreover, the court observed that the Airport Authority did not excavate soil deeper than necessary to complete runway construction:

> Had remediation truly been necessary, the Authority presumably would have (and certainly should have) performed a cleanup of the entire area.
>
> * * *
>
> [A]llowing the Authority to recoup its "response costs" would be tantamount to a reimbursement of its runway construction costs.[25]

Thus, the court held that the Airport Authority's investigative costs were not incurred in response to a threat to human health or the environment and thus were not "necessary" response costs under section

23. *Id.* at 701–05.
24. *Id.* at 705.
25. *Id.*

107(a)(4)(B). Similar rulings have been handed down in a number of cases where remedial activities were found to have been performed primarily for business reasons and thus were not necessary within the meaning of section 107(a)(4)(B).[26]

One approach that courts have rejected for determining whether response costs of a remedial action qualify as necessary is the fact that less expensive remedial alternatives could have been selected. For example, in *Basic Management Inc. v. United States*,[27] the United States, as defendant, argued that the plaintiffs' choice of a $74 million remedy was not necessary because there was a $10 million option for capping the contaminated site. The court found no authority for the proposition that the term "necessary" requires that the least expensive cleanup option be used, and noted that the site's proximity to residential development supported a higher cleanup standard.[28]

## C. Consistency with the NCP

The next requirement contained in section 107(a)(4)(B) of CERCLA is that the costs incurred be "consistent with the National Contingency Plan." In private cost recovery actions, the plaintiff has the burden to show that its costs are consistent with the NCP.[29] The NCP, in addition to setting out the requirements that EPA must follow in site investigation and remediation, also contains provisions that a private party must meet in order to establish that the costs it incurred are consistent with

26. *See, e.g.*, Foster v. United States, 922 F. Supp. 642 (D.D.C. 1996) (cost of an environmental assessment was not a "necessary" cost because it was performed to identify barriers to development); Yellow Freight Sys., Inc. v. ACF Indus., Inc. 909 F. Supp. 1290, 1299 (E.D. Mo. 1995) (costs of investigation and asbestos abatement incurred for business reasons are not "necessary," and thus unrecoverable).

27. 569 F. Supp. 2d 1106 (D. Nev. 2008).

28. *Id.* at 1121.

29. *See, e.g.*, New York v. Almy Bros., Inc., 90-CV-818, 1996 U.S. Dist. LEXIS 341 (N.D.N.Y. Jan 8, 1996); Pneumo Abex Corp. v. Bessemer & Lake Erie R.R. Co., 936 F. Supp. 1250 (E.D. Va. 1996); City of New York v. Exxon Corp., 633 F. Supp. 609, 616 (S.D.N.Y. 1986). By contrast, in an action by the United States to recover costs, the defendant has the burden of proving that the government's costs are inconsistent with the NCP. (See discussion *supra* in chapter 4).

the NCP.[30] Included in these requirements for response actions are selection of a cost-effective response,[31] opportunity for public comment,[32] and compliance with all federal, state, and local requirements, including permitting requirements.[33]

The threshold question in determining whether costs are consistent with the NCP is determining whether the cleanup activity is a removal or remedial action. Several factors must be examined, and even then a clear-cut answer is not always available. As noted in chapter 1, conceptually a removal action is thought of as a short-term, interim response to prevent imminent harm. By contrast, a remedial action is a long-term, permanent solution or remedy for addressing contamination. The statutory definitions of removal action and remedial action, however, are overlapping.[34] Thus it is possible for a response action to be characterized as both a removal action and a remedial action. For example, the excavation of a buried drum may be deemed a removal if it is "necessary to prevent, minimize or mitigate" contamination that would occur if allowed to remain buried. On the other hand, excavating drums may be a remedial action where it constitutes the permanent remedy for a release.

The distinction between removal and remedial actions under the NCP is significant for at least two reasons. First, different statutes of limitation apply to removal and remedial actions. For a removal action, the initial cost recovery suit must be brought within three years of the completion of the removal action.[35] For a remedial action, the limitation period is six years, which begins to run upon "initiation of physical on-site construction of the remedial action."[36] As noted in chapter 1, these differing limitations periods are at the heart of a broad body of

30. *See* 40 C.F.R. § 300.71. The proposed revised NCP retains many of the same requirements for private party actions. *See* 53 Fed. Reg. 51,394, 51,513 (Dec. 21, 1988) (subpt. H).

31. 40 C.F.R. § 300.71(a)(2)(ii)(C).

32. *Id.* § 300.71(a)(2)(ii)(D).

33. *Id.* § 300.71(a)(4).

34. Compare CERCLA § 101(23) (definition of "removal action") with CERCLA § 101(24) (definition of "remedy" or "remedial action").

35. *See* CERCLA § 113(g)(2)(A).

36. *See* CERCLA § 113(g)(2)(B).

CERCLA case law in which courts had to decide whether a response action was a removal action subject to a three-year limitations period or a remedial action subject to a six-year limitations period.[37]

The second reason that the distinction between removal actions and remedial actions is important is because a more extensive set of requirements applies to remedial actions than to removals. In general, remedial actions are more costly to implement than removal actions.

After appropriately characterizing a response action as either a removal or remedial activity, a private plaintiff must demonstrate that the response action is "consistent with the NCP." The NCP sets forth the substantive and procedural requirements for performance of removal and remedial actions. Prior to 1990, there was little guidance on the meaning of "consistent with the NCP," leaving the courts to struggle with the issue. Some courts held that private party plaintiffs must strictly conform with the NCP. For example, in *Versatile Metals, Inc. v. Union Corp.*,[38] after characterizing the response activity as remedial, the court addressed the issue of the consistency of costs with the NCP. The court ruled that a limited assessment of site conditions did not meet the standards for an RI/FS for remedial purposes under the NCP.[39] Because no

37. *See, e.g.*, Schaefer v. Town of Victor, 457 F.3d 188 (2d Cir. 2006) (plaintiff's claim barred because plaintiff initiated on-site construction of remedial action more than six years before initiating its recovery action); Syms v. Olin Corp., 408 F.3d 95, 102 (2d Cir. 2005) (same); GenCorp, Inc. v. Olin Corp., 390 F.3d 433 (6th Cir. 2004) (six-year limitations period for recovery of remedial action costs begins to run when party takes action consistent with the permanent remedy for the site), *aff'd*, 477 F.3d 368 (6th Cir. 2007); California v. Neville Chem. Co., 358 F.3d 661 (9th Cir. 2004) (six-year limitation period for remedial action claims cannot accrue until at least the final approval of the remedial plan has occurred); Douglas Autotech Corp. v. Scott Fetzer Co., Case No. 1:07-CV-1062, 2008 U.S. Dist., LEXIS 4894 (W.D. Mich. Jan. 23, 2008) (late discovery of contamination does not "restart" limitations period for remedial action, unless the new contamination is associated with an entirely separate "facility"); City of Moses Lake v. United States, 458 F. Supp. 2d 1198 (E.D. Wash. 2006) (plaintiff's cost recovery claim was barred under the three-year limitations period applicable to a removal action; court characterized plaintiff's actions—construction of a new reservoir and alteration of wells to draw water from deeper zones—as interim measures intended to minimize imminent harm, rather than the permanent remedy to eliminate the contamination itself); Pneumo Abex Corp. v. Bessemer & Lake Erie R.R. Co., 936 F. Supp. 1250 (E.D. Va. 1996) (surface removal and the Remedial Investigation and Feasibility Study comprise a single removal action for statute of limitation purposes).

38. 693 F. Supp. 1563 (E.D. Pa. 1988).

39. *Id.* at 1581.

alternatives to the selected remedial action were formally investigated, and no comparison of the relative cost-effectiveness of the remedy and any alternative was made, the court held that response costs were not consistent with the NCP and disallowed them completely.[40]

Similarly, in *Amland Properties Corp. v. Aluminum Co. of America*,[41] the court focused on the distinction between removal actions and remedial actions, indicating that the remedial actions undertaken at the site must comply with the more detailed procedural and substantive provisions of the NCP, whereas removals need only comply with the relatively simple NCP requirements.[42] Addressing the consistency of the remedial costs with the NCP, the court had to decide whether strict compliance with the NCP, or merely substantial compliance, is mandated. The court found that the NCP's requirements must be strictly adhered to in order to permit a private party to recover its response costs "unless the party seeking recovery explains why a specific requirement is not appropriate to the specific site and problem."[43] The court found that plaintiff failed to conduct an RI/FS that considered all of the scoping factors set forth in the NCP; failed to conduct an analysis of alternatives according to the criteria of cost, acceptable engineering practices, and effectiveness; failed to document the rationale for screening out technologies not selected; and failed to provide a public comment period on the selected remedy. For all these reasons, plaintiff's remedial costs were ruled not recoverable under CERCLA.[44]

Other courts, however, adopted a less rigid approach. In contrast to the decisions in *Versatile Metals* and *Amland*, the court in *General Electric Co. v. Litton Business Systems, Inc.*[45] ruled that consistency with the NCP does not necessitate strict compliance with its provisions, and emphasized that this view serves a major purpose of CERCLA—"*to*

40. *Id.* at 1582.
41. 711 F. Supp. 784 (D.N.J. 1989).
42. *Id.* at 795.
43. *Id.* at 796.
44. *Id.* at 801.
45. 715 F. Supp. 949 (W.D. Mo. 1989).

*induce* [private parties] *voluntarily* to pursue appropriate environmental response actions with respect to inactive hazardous waste sites."[46]

In 1990, EPA amended the NCP to add procedures to be followed by non-governmental parties when performing a removal or remedial action. Specifically, subpart H (Participation by Other Parties) includes the following key provisions. First, it repeats the statutory mandate that responsible parties shall be liable for necessary costs of response actions incurred by any other person consistent with the NCP.[47] Second, it specifies that for purposes of a private party seeking cost recovery under section 107(a)(4)(B), a private party action will be considered "consistent with the NCP" if the action, when evaluated as a whole, is in "substantial compliance" with the applicable requirements enumerated in 40 C.F.R. § 300.700(c)(5) and (6) and results in a "CERCLA quality cleanup."[48] Notably, EPA states that "immaterial" or "insubstantial" deviations from the NCP will not disqualify a cleanup from being considered "consistent with the NCP." Third, response actions carried out in compliance with the terms of a section 106 order issued by EPA or a consent decree entered pursuant to section 122 of CERCLA will be considered "consistent with the NCP."[49]

The key concepts of "substantial compliance with the NCP" and "CERCLA quality cleanup" have been the subject of controversy and litigation.

### 1. *Substantial Compliance with Potentially Applicable Requirements*

Section 300.700(c)(5) and (6) establish a list of potentially applicable NCP requirements for private response actions that differ from those applicable to government-led response actions. Among the potentially applicable requirements listed in paragraph (c)(5) are worker health and safety requirements (section 300.150), documentation for cost recovery (section 310.160), permit requirements and ARAR identification (section 300.400(e) and (g)). In 40 C.F.R. 300.700(c)(6), EPA states that private

46. *Id.* at 962 (citation omitted).
47. 40 C.F.R. § 300.700(c)(2).
48. 40 C.F.R. § 300.700(c)(3)(i).
49. 40 C.F.R. § 300.700(c)(3)(ii).

parties undertaking response actions should provide an opportunity for public comment concerning the selection of the response action, and identifies sections of the NCP that are potentially applicable in satisfying the public comment requirements.

The question of whether failure to perform one of these numerous "potentially applicable" requirements is material enough to defeat a private cost recovery action is the subject of considerable case law. Courts do not require private parties to adhere strictly to the technical minutia of the remedy selection procedures in the NCP in order to have met the substantial compliance standard.[50] One thing most courts agree on is that failure to provide the public an opportunity to comment on the proposed remedy is per se material and will defeat cost recovery.[51] Courts are divided as to whether state government participation alone may satisfy the public participation request.[52]

---

50. *See, e.g.*, Franklin Cnty. Convention Facility Auth. v. Am. Premier Underwriters, 240 F.3d 534 (6th Cir. 2001) (private plaintiff met the substantial compliance standard, despite shortcomings the RI/FS, the remedial design and record of decision).

51. *See, e.g.*, *Reg. Airport Auth. of Louisville*, 460 F.3d 697 (plaintiff did not demonstrate substantial compliance where the preferred alternative was not presented to the public for comment prior to completion of the project); Cnty. Line Inv. Co. v. Tinney, 933 F.2d 1508, 1514–15 (10th Cir. 1991) (failure to provide opportunity for public comment barred cost recovery); Waste Mgmt. of Alameda Cnty., Inc. v. E. Bay Reg'l Park Dist., 135 F. Supp. 2d 1071 (N.D. Cal. 2001) (plaintiff's failure to provide meaningful opportunity for public participation rendered cleanup not in substantial compliance with the NCP); Union Pac. R.R. Co. v. Reilly Indus., Inc., No. Civ. 4-96-660 DSD/JMM, 1997 WL 684875 (D. Minn. Nov. 3, 1997) (PRP motion for summary judgment granted because plaintiff did not solicit meaningful public input, having begun excavation before end of comment period); Pierson Sand & Gravel, Inc. v. Pierson Twp., 89 F.3d 835 (6th Cir. 1996) (unpublished) (response costs not "consistent with NCP" because plaintiffs failed to provide citizens with an opportunity to participate in the decision-making process that led to selection of a response action for the site); C&C Millwright Maint. Co., v. Town of Greenville, 946 F. Supp. 555 (E.D. Tenn. 1996) (summary judgment against plaintiff granted because of failure to satisfy community relations requirement of NCP); G.J. Leasing Co. v. Union Elec. Co., 854 F. Supp. 539, 566 (S.D. Ill. 1994) (failure to provide opportunity for public comment barred cost recovery), *aff'd*, 54 F.3d 379 (7th Cir. 1995); Sherwin-Williams Co. v. City of Hamtramck, 840 F. Supp. 470, 477 (E.D. Mich. 1993) (same).

52. *See, e.g.*, Am. Color & Chem. Corp. v. Tenneco Polymers, Inc., 918 F. Supp. 945, 957 (D.S.C. 1995); *see also* Bedford Affiliates v. Sills, 156 F.3d 416, 428 (2d Cir. 1998) (concluding public participation requirement is satisfied if "a state environmental agency is substantially involved in the formulation and execution of a preliminary remediation plan"); *but see* Union Pac. R.R. Co. v. Reilly Indus. Inc., 215 F.3d 830 (8th Cir. 2000) (holding a city's involvement in devising a plan to clean up a contaminated site was not an adequate substitute for compliance with NCP requirements where there was insufficient opportunity for public comment);

### 2. *CERCLA Quality Cleanup*

The term "CERCLA quality cleanup" is not defined in the NCP, but is addressed in the preamble to the 1990 changes to the NCP. In the preamble to the 1990 changes to the NCP, EPA stated that to achieve a CERCLA quality cleanup, one must (1) meet the three remedy selection requirements of section 121(b)(1) of CERCLA; (2) attain ARARs according to section 121(d)(4); and (3) provide for meaningful public participation as prescribed in section 117.[53] The three remedy selection provisions require that "the remedial action must be 'protective of human health and the environment,' utilize 'permanent solutions and alternative treatment technologies or resource recovery technologies to the maximum extent practicable,' and be 'cost-effective.'"[54]

EPA stated that the requirements of a CERCLA quality cleanup were not new additions resulting from the proposed rule, but rather that private parties had previously been required to comply strictly with a more detailed list of provisions of the NCP, which included these requirements. However, it is not clear whether cost-effectiveness was previously a requirement of only remedial actions or both remedial and removal actions.

Case law concerning the question of whether a private plaintiff has performed a CERCLA quality cleanup refers to the 1990 preamble for guidance.[55] As with jurisprudence on substantial compliance, however, courts typically regard a cleanup as achieving "CERCLA quality" even with minor deviations from NCP procedure.[56]

---

Bethlehem Iron Works, Inc. v. Lewis Indus., Civ. A. No. 94-0752, 1996 U.S. Dist. LEXIS 14446 (E.D. Pa. Sept. 19, 1996) (communication with government agencies regarding remedy selection does not satisfy the public comment requirements of the NCP).

53. 55 Fed. Reg. 8666, 8793 (1990) (explaining the significance of § 300.700(c) of the NCP). EPA recognized that public participation is not required by CERCLA § 121, but considered it necessary to any proper cleanup. *Id.*

54. *Id.* at 8792.

55. *See, e.g.*, *Franklin Cnty. Convention Facility Auth.*, 240 F.3d 534.

56. *Id.* at 546 ("[T]he NCP requirements are not intended to be a checklist of required actions for private remediation.").

## D. Notice and Government Approval

CERCLA does not require that the private party give defendants advance notice of its intent to file a claim under CERCLA.[57] However, concurrent with the filing of a private cost recovery action, a private party must give notice to the government.[58] Moreover, in addition to providing notice of the private cost recovery action, the private party must provide a copy of the cost recovery complaint to both the United States attorney general and the EPA administrator.[59]

EPA approval of a privately performed response action is not a prerequisite for a section 107 cost recovery action.[60] However, once an RI/FS has been initiated by either EPA, or a PRP performing under a consent order or an administrative order, agency approval of the remedial action is required.[61]

## E. State and Common Law Claims

As noted above, under section 107(a)(4)(B) of CERCLA, plaintiffs in private cost recovery actions can only recover response costs that are "necessary" and "consistent with the National Contingency Plan." Section 107(a)(4)(B) does not authorize recovery of any other types of damages. Private plaintiffs, however, may recover other types of damages under state law. In this regard, section 114(a) provides: "Nothing in this chapter shall be construed or interpreted as preempting any State from imposing any additional liability or requirements with respect to the release of hazardous substances within such State."

An open question is whether a private plaintiff may assert a cause of action for damages other than response costs under federal common law. To date, only one court has ruled on this issue. In *Anderson v.*

57. *See* Cooper v. Armstrong Rubber Co., Civ. A. No. J88-0464, 1989 U.S. Dist. LEXIS 4099, (S.D. Miss. Feb. 3, 1989).

58. 55 Fed. Reg. at 8798 (summarizing § 300.700 of the NCP).

59. *See* CERCLA § 113(l).

60. *See Wickland Oil Terminals*, 792 F.2d at 891; *Cooper*, 1989 U.S. Dist. LEXIS, at *4–5; Mardan Corp. v. C.G.C. Music, Ltd., 600 F. Supp. 1049, 1054 (D. Ariz. 1984), *aff'd*, 804 F.2d 1454 (9th Cir. 1986).

61. CERCLA § 122(e)(6); *see also* Allied Corp. v. Acme Solvents Reclaiming, Inc., 691 F. Supp. 1100, 1105–06 (N.D. Ill. 1988) (EPA approval of response actions under section 122(e)(6) is a prerequisite to private cost recovery).

*Teck Metals Ltd.*,[62] current and former Washington State residents of a town near the Canadian border brought an action to recover damages for personal injury caused by releases at a Canadian smelting facility, which flowed down the Columbia River into the United States. Citing the 2011 Supreme Court ruling in *Connecticut v. American Electric Power Co., Inc.*,[63] the district court ruled that the test for whether Congress excluded federal common law claims "'is simply whether the statute speak[s] directly to [the] question at issue.'"[64] The district court goes on to say "the fact CERCLA does not provide a damages remedy for personal injuries is irrelevant to whether CERCLA displaces and precludes Plaintiffs' federal common law public nuisance claims in the case at bar."[65] Based on the fact that "Congress provided a comprehensive liability and remediation scheme to address releases and threatened releases of hazardous substances" under CERCLA, the court concluded that "Plaintiff's federal common law public nuisance claims have been displaced by CERCLA and therefore, must be dismissed."[66]

## III. CONTRIBUTION ACTIONS UNDER SECTION 113

As noted *supra*, the original text of CERCLA did not contain an express right of contribution. During the early 1980s, PRPs who had been sued by the United States for recovery of response costs, as well as parties who performed voluntary cleanups, argued that a right of contribution was implied by the language of section 107(a)(4)(B) of CERCLA, and was also conferred by federal common law. Most courts agreed.[67] The availability of contribution under CERCLA, however, was far from certain.

62. Order, No. CV-13-420-LRS (E.D. Wash. Jan. 5, 2015), ECF No. 52.
63. 131 S. Ct. 2527, 2537 (2011) (alterations in original) (citations omitted).
64. *Teck Metals*, Order at 18 (alterations in original).
65. *Id.* at 19.
66. *Id.* at 20.
67. *See, e.g.*, *Walls*, 761 F.2d at 318 (noting that district courts have been "virtually unanimous" in holding that CERCLA § 107(a)(4)(B) creates a private right of action for the recovery of necessary response costs); Colorado v. ASARCO, Inc., 608 F. Supp. 1484, 1491 (D. Colo. 1985) (granting right of contribution under section 107, reasoning "Section 107(e)(2) evidences the drafters' view that the common law permits contribution"); *Stepan Chem. Co.*, 544 F. Supp. 1135 (relying on § 107(a)(4)(B) to obtain contribution after conducting voluntary cleanup).

Some district courts had disallowed contribution under CERCLA.[68] Moreover, the Supreme Court had refused to imply contribution rights under other statutes, and it was possible the Court would refuse to imply contribution rights under CERCLA.[69] Congress attempted to remedy this issue when it passed the Superfund Amendments and Reauthorization Act of 1986 (SARA).[70] As discussed below, however, Congress ended up creating as many problems as it solved.

## A. The 1986 Amendments Established an Express Right of Contribution

The 1986 amendments established an express cause of action for contribution for parties that have been sued by or settled with EPA or a state. Under section 113(f)(1) of CERCLA, "Any person may seek contribution from any other person who is liable or potentially liable under [section 107(a)] of this title, during or following any civil action under [section 106] of this title or under [section 107] of this title." Section 113(f)(3)(B) of CERCLA provides a right of contribution for parties who resolve their liability by settlement, rather than civil action: "A person who has resolved its liability to the United States or a State for some or all of a response action or for some or all of the costs of such action in an administrative or judicially approved settlement may seek contribution from any person" who has not itself settled with the United States. Thus, this provision allows parties to settle their liability with the United States or a state, instead of waiting for an enforcement action, and then seek contribution from other potentially liable parties for their portion of the damages under section 113(f)(3)(B).

After Congress added § 113(f) to CERCLA in 1986, private parties who had incurred response costs in efforts to clean up contaminated properties began seeking contribution for those cleanup costs from

68. United States v. Westinghouse Elec. Corp., No. IP 83-9-C, 1983 WL 160587, at *4 (S.D. Ind. June 29, 1983).

69. Tex. Indus., Inc. v. Radcliffe Materials, Inc., 451 U.S. 630, 639–40 (1981) (Sherman Act and Clayton Act); Nw. Airlines, Inc. v. Transp. Workers Union of Am., 451 U.S. 77, 94–95 (1981) (Equal Pay Act and Title VII of the Civil Rights Act).

70. Pub. L. No. 99-499, 100 Stat. 1613 (1986).

other liable parties under § 113(f). Courts uniformly held that section 113(f) provided a sufficient legal basis for such recovery, even if the parties had not been sued for recovery action under section 106 or section 107(a). Indeed, many Courts of Appeals held that section 113(f) offered the *exclusive* remedy for contribution. Each of the 11 federal circuits (excluding the D.C. Circuit, which did not hear a case with this dispute) prohibited private party recovery under section 107(a) between 1994 and 2003.[71] These courts found section 107(a) available only to "innocent" parties, meaning private parties who were not liable or potentially liable themselves. As a result, a liable private party who had voluntarily cleaned up a contaminated site could not sue for cost recovery against other liable parties under § 107(a). The majority of courts, however, allowed PRPs who had performed voluntary remediation to pursue contribution under section 113(f), even though section 113(f) does not expressly authorize such a contribution action. Two Supreme Court decisions, in 2004 and 2007, altered this landscape for private party cost recovery under CERCLA.

## B. The Supreme Court Alters the Landscape

The first of these landmark Supreme Court decisions was *Cooper Industries, Inc. v. Aviall Services, Inc.*[72] In this case, Cooper had sold four aircraft engine maintenance sites in Texas to Aviall Services, who subsequently discovered that the sites were contaminated with hazardous substances. Aviall voluntarily cleaned up the sites, incurring at least $5 million in cleanup costs, and thereafter sought contribution from Cooper Industries under section 113(f). Aviall's original complaint included

71. *See* Dico, Inc. v. Amoco Oil Co., 340 F.3d 525, 530 (8th Cir. 2003); Bedford Affiliates v. Sills, 156 F.3d at 424 (2d Cir. 1998); Centerior Serv. Co. v. Acme Scrap Iron & Metal Corp., 153 F.3d 344, 349 (6th Cir. 1998); Pneumo Abex Corp. v. High Point, Thomasville & Denton R.R. Co., 142 F.3d 769, 776 (4th Cir. 1998); Pinal Creek Grp. v. Newmont Mining Corp., 118 F.3d 1298, 1301–06 (9th Cir. 1997); Redwing Carriers, Inc. v. Saraland Apartments, 94 F.3d 1489, 1496 (11th Cir. 1996); United States v. Colo. & E. R.R., 50 F.3d 1530, 1536 (10th Cir. 1995); United Techs. Corp. v. Browning-Ferris Indus., Inc., 33 F.3d 96, 101 n.9 (1st Cir. 1994); Akzo Coatings, Inc. v. Aigner Corp., 30 F.3d 761, 764 (7th Cir. 1994).

72. 543 U.S. 157.

both a section 107(a) cost recovery count and a section 113(f) contribution count. In an amended complaint, however, Aviall dropped the cost recovery count and retained the contribution count. The district court dismissed Aviall's contribution action, holding that a private party who had not been sued for response costs by the government under section 107(a) cannot bring suit for contribution from other liable parties under section 113(f)(1). This ruling meant that, because Aviall had voluntarily remediated its property without a government suit under § 107(a), Aviall could not subsequently obtain contribution from Cooper under section 113(f)(1). A three-judge panel of Fifth Circuit affirmed the district court ruling. The Fifth Circuit, however, decided to rehear the case en banc, and reversed the panel's decision, finding that a contribution right under section 113(f) did *not* exist where the plaintiff had not been sued by the government. The U.S. Supreme Court reversed.

Justice Thomas, writing for the majority, explains that the Court followed the plain meaning of the text of section 113(f)(1), which states that a party may only seek contribution "during or following any civil action" under section 106 or section 107, because interpreting the statute otherwise would render superfluous the "during or following" language. Likewise, according to the Court, any other interpretation would make section 113(f)(3)(B), which permits contribution actions after settlement, equally redundant. The Court's decision also addressed the so-called "saving clause" of section 113(f)(1), which provides that "[n]othing in this subsection shall diminish the right of any person to bring an action for contribution in the absence of a civil action." The Court rejected the idea that the saving clause expands claims under section 113(f)(1) to authorize actions not brought "during or following" civil action. Instead, Justice Thomas reasoned that the saving clause simply rebuts any presumption that the express right of contribution is the exclusive cause of action for contribution available to a potentially responsible party.

The implications of *Cooper Industries*, combined with the circuit court's blocking of private party recovery under section 107(a), rendered cost recovery difficult or impossible for private parties who undertook voluntary cleanup of hazardous sites. In 2007, the Supreme Court addressed this problem in *United States v. Atlantic Research Corp.*[73] In that case, Atlantic Research had voluntarily cleaned a site it had leased from the United States on which it had retrofitted rocket motors for the United States. The retrofitting operations had contaminated soil and groundwater. The *Cooper Industries* decision eliminated the possibility of Atlantic Research obtaining contribution from the federal government under section 113(f)(1) because there had been no prior litigation or settlement with EPA. Accordingly, Atlantic Research sought to recover costs from the federal government under section 107(a)(4)(B). The district court granted the United States' motion to dismiss, citing pre-*Cooper Industries* case law that limited section 107 private party actions to innocent PRPs. The Eighth Circuit reversed, stating that it "no longer makes sense to view Section 113(f)(1) as the exclusive route by which liable parties may recover costs."

Reversing a decade of circuit court decisions, the Court unanimously affirmed the Eighth Circuit, holding that section 107(a) provides potentially responsible parties with a cause of action to recover from other PRPs for costs incurred in voluntarily cleaning up a contaminated site. Under section 107(a)(4)(B), PRPs may recover from any other person consistent with the national contingency plan. The Court interpreted "any other person" to refer back to the immediately preceding language in section 107(a)(4)(A), which permits suit only by the United States, a State, or an Indian tribes; in other words, "any other person" refers to anyone *other than* those three named parties.[74] Accordingly, PRPs are encompassed by the phrase "any other person." A PRP, therefore, may

73. 551 U.S. 128.

74. The Supreme Court rejected the government's contention that "any other person" refers to any person not identified as a PRP in paragraphs (1)–(4) of section 107(a); in other words section 107(a)(4)(B) only permits suit by a non-PRP: "The Government's interpretation makes little textual sense. In subparagraph (B), the phrase 'any other necessary costs' and the phrase 'any other person' both refer to the antecedents—'costs' and 'persons'—located in some previous statutory provision. Although 'any other necessary costs' clearly references the costs in subparagraph (A), the Government would inexplicably interpret 'any other person' to refer not to the persons listed in subparagraph (A) but to the persons listed as PRPs in paragraphs (1)–(4). Nothing in the text of § 107(a)(4)(B) suggests an intent to refer to antecedents located in two different statutory provisions. Reading the statute in the manner suggested by the Government would destroy the symmetry of §§ 107(a)(4)(A) and (B) and render subparagraph (B) internally confusing." *Id.* at 136.

recover a portion of its voluntarily incurred cleanup costs from other PRPs by asserting a section 107(a) recovery action against them.

### C. Post-*Atlantic Research* Case Law

The decision in *Atlantic Research* left numerous issues unresolved. For example, the Supreme Court did not decide the issue of whether there are circumstances under which a plaintiff could bring both a cost recovery and a contribution claim under CERCLA. Most courts in the post-*Atlantic Research* era have not permitted a plaintiff to pursue a section 107(a) cost recovery claim when a section 113(f)(1) contribution claim is available.[75] These courts have noted the Supreme Court's emphasis on the procedural distinctness of the two CERCLA rights of action and the possibility that, if allowed to pursue a section 107(a) claim, a party that has already resolved its own liability through a settlement could potentially shift full liability onto another PRP.

75. *See, e.g.*, Bernstein v. Bankert, 702 F.3d 964, 976 (7th Cir. 2012) (plaintiff cannot pursue a cost recovery claim when a contribution claim is available), *aff'd in part and rev'd in part*, 733 F.3d 190 (7th Cir. 2013); Solutia, Inc. v. McWane, Inc. 672 F.3d 1230, 1237 (11th Cir. 2012) (response costs that are not voluntary may only be recovered through contribution); Morrison Enters., LLC v. Dravo Corp., 638 F.3d 594, 603 (8th Cir. 2011) (contribution is the "exclusive remedy for a liable party compelled to incur response costs pursuant to an administrative or judicially approved settlement"), *cert. denied*, 132 S. Ct. 244 (2011); Agere Sys., Inc. v. Advanced Envtl. Tech. Corp., 602 F.3d 204, 229 (3d Cir. 2010) (same); Lyondell Chem. Co. v. Occidental Chem. Corp., 608 F.3d 284, 291 n.19 (5th Cir. 2010) (acknowledging district court's implicit decision that plaintiff could not pursue both remedies); Kotrous v. Gos-Jewett Co. of N. Cal., 523 F.3d 924, 934 (9th Cir. 2008) (plaintiffs who perform a voluntary cleanup "must seek cost recovery under § 107, not contribution under § 113, because they have not been subject to an action under § 106 or § 107"); Chitayat v. Vanderbilt Assoc., 2010 U.S. Dist. LEXIS 26745 (E.D.N.Y. Mar. 22, 2010) (PRP that paid state agency's remedial costs, pursuant to an administrative agreement, cannot pursue a section 107(a) cost recovery claim); Appleton Papers Inc. v. George A. Whiting Paper Co., Case No. 08-C-16, 2008 WL 3891304 (E.D. Wis. Aug. 20, 2008) (a plaintiff that has been found liable under a court-approved consent decree may only pursue contribution regardless of whether the costs were incurred pursuant to that consent decree); *see also* Port of Tacoma v. Todd Shipyards Corp., No. C-08-5132, No. C-08-5132, 2009 WL 113852 (W.D. Wash. Jan. 14, 2009) (a PRP who is subject to a third-party contribution claim cannot pursue a section 113 contribution claim because it will pay at most its equitable share of the damage); Niagara Mohawk Power Corp. v. Consol. Rail Corp., 565 F. Supp. 2d (N.D.N.Y. 2008) (plaintiff that incurred costs "involuntarily" under a state-approved consent decree may only seek contribution under section 113), *aff'd in part and rev'd in part on other grounds by* Niagara Mohawk Power Corp. v. Chevron U.S.A., Inc., 596 F.3d 112 (2d Cir. 2010).

Another unresolved issue concerns the contribution bar. Under section 113(f)(2) of CERCLA, a person who has resolved its liability to the United States or a state in an administrative or judicially approved settlement is not liable for contribution regarding "matters addressed in the settlement." In *Atlantic Research*, the United States had argued that allowing a PRP to pursue cost recovery under section 107 would eviscerate the contribution bar because the bar only applies to contribution actions and thus would not preclude a private party from suing a settling party for cost recovery under section 107(a)(4)(B). The Supreme Court rejected that argument, stating that "a district court applying traditional rules of equity would undoubtedly consider any prior settlement as part of the liability calculus."[76] But it remains to be seen whether the Supreme Court's confidence in the liability calculus used by district courts is well placed.

Another unresolved issue is whether a party that has entered a consent decree to perform a cleanup must pursue contribution or cost recovery. Section 113(f)(3)(b) provides that "a person who has resolved its liability to the United States or a State for some or all of a response action or for some or all of the costs of such action in an administrative or judicially approved settlement" may seek contribution from any person not a party to a qualifying settlement. The Supreme Court declined to address the question of whether a PRP that has entered such a settlement is limited to contribution under section 113(f)(3)(B) or may elect instead to pursue cost recovery under section 107(a)(4)(B).[77] Post-*Atlantic Research* case law is split on this question. In *W.R. Grace & Co. v. Zotos International, Inc.*,[78] the Second Circuit held that a party conducting a cleanup under an administrative consent order with a state agency did not resolve its liability, so it could not pursue contribution against other PRPs under section 113(f)(3)(B). Its exclusive remedy was cost recovery under section

76. *Atl. Res. Corp.*, 551 U.S. at 141.

77. *Atl. Res. Corp.*, 551 U.S. at 139 n.6 ("We do not decide whether [consent decree costs] of response are recoverable under § 113(f), § 107(a), or both.").

78. 559 F.3d 85 (2d Cir. 2009).

107(a)(4)(B).[79] By contrast, in *Hobart Corp. v. Waste Management of Ohio, Inc.*,[80] the court held that a party who had entered an EPA administrative order on consent (AOC) was limited to contribution under section 113(f)(3)(B). Hobart had sought cost recovery under section 107 because it had missed the three-year limitations period applicable to a contribution action. Hobart relied on a prior Sixth Circuit decision, *ITT Industries, Inc. v. Borgwarner, Inc.*,[81] which had held that an AOC did not resolve the settlor's liability because EPA reserved its right to nullify the AOC and bring an action against the settlor should it fail to comply with the AOC. But EPA changed the language of the AOC after 2007, and the Sixth Circuit held that under the updated AOC, Hobart had resolved its liability. Accordingly, the court ruled that Hobart's only remedy was contribution under section 113(f)(3)(B).[82]

Yet another area of ambiguity in the wake of *Atlantic Research* is whether a settlement with the government in which the PRP agrees to pay *future* site remediation expenses constitutes a cost incurred that is recoverable under section 107(a). Although these costs will be incurred and paid directly by a private PRP, some courts have recently held that such a settlement forfeits access to cost recovery under section 107(a), essentially reverting to the pre-*Atlantic Research* landscape in which section 113(f)(1) was the preferred mechanism for relief.[83]

---

79. *Id.* at 92–93; *see also* Ford Motor Co. v. Edgewood Props., Inc., Civ. A. Nos. 06-1278, 4266, 2008 WL 4559770 (D.N.J. Oct. 8, 2008) (court held that state administrative consent order did not provide the basis for pursuing contribution under section 113(f)(3)(B)).

80. 758 F.3d 757 (6th Cir. 2014), *cert. denied*, 135 S. Ct. 1161 (2015).

81. 506 F.3d 452, 459–61 (6th Cir. 2007).

82. *See also* Niagara Mohawk Power Corp. v. Chevron U.S.A., Inc., 596 F.3d 112, 124–26 (2d Cir. 2010) (finding that administrative consent order with state agency resolved plaintiff's CERCLA liability, so plaintiff's claim against other responsible parties was governed by CERCLA section 113(f)(3)(B)); *Morrison Enters. LLP*, 638 F.3d at 603 ("§ 113(f) provides the exclusive remedy for a liable party compelled to incur response costs pursuant to an administrative or judicially approved settlement under §§ 106 or 107").

83. For more discussion of this issue, see Steven Ferrey, *Toxic "Plain Meaning" and "Moonshadow": Supreme Court Unanimity and Unexpected Consequences*, 24 VILL. ENVTL. L.J. 1, 55–56 (2013).

## D. Timing of Contribution Action

Unlike a private cost recovery action under section 107(a)(4)(B), a contribution action under section 113 may be brought before any costs are incurred. In this regard, section 113(f)(1) provides that a contribution action may be brought at any time during or following a civil action under section 106 action or section 107(a) cost recovery action. Thus, a PRP who is a defendant in a section 106 or section 107 action need not wait until judgment or settlement of the underlying action before bringing a contribution action against other PRPs. In fact, a contribution action brought by the PRP defendant as a third-party complaint in a cost recovery action may provide several important benefits. A timely contribution action may relieve a defendant that is held liable from paying more than the appropriate share of the response costs. The defendant may also save on transactional costs if all issues with respect to a site are litigated at the same time.

The statute of limitations for a contribution claim under section 113 of CERCLA is three years after (1) the date of judgment in any action under section 107; (2) the date of an administrative order; or (2) entry of a judicially approved settlement concerning costs or damages.[84] Note that the limitations period for a private cost recovery claim is different. As noted *supra*, a section 107 claim for recovering the costs of a removal action is three years from completing the removal, and a section 107 claim for recovering the costs of a remedial action is six years from the initiation of physical on-site construction of the remedy.

## E. Allocation Factors for Contribution

Section 113(f)(1) of CERCLA provides that courts faced with resolving contribution claims "may allocate response costs among liable parties using such equitable factors as the court determines are appropriate." However, courts are by no means in agreement as to how these costs should be allocated. Since there often does not exist an easily divisible harm, common law rules of apportionment are not particularly well suited to CERCLA contribution actions. Although the government and

84. *See* CERCLA § 113(g)(3).

PRPs have agreed to apportion response costs based on the volume of hazardous substances deposited at a site, courts have found that volume is not necessarily a reliable way to divide harm. Courts may impose joint and several liability where the harm is single and indivisible; however, parties may demonstrate that the harm is divisible and the degree to which each party is liable.[85]

Courts often use the so-called "Gore Factors" in apportioning liability among and between PRPs. During debate of the original Superfund legislation, the "Gore Amendment," introduced by then-representative Albert Gore, listed six factors to be considered when allocating site costs among PRPs:

(1) the ability of the parties to demonstrate that their contribution to a discharge, release, or disposal of a hazardous substance can be distinguished from that of other parties;
(2) the amount of the hazardous substance involved;
(3) the degree of toxicity of the hazardous substance;
(4) the degree of involvement by the parties in the generation, transportation, treatment, storage, or disposal of the hazardous substance;
(5) the degree of care exercised by the parties with respect to the hazardous substance concerned, taking into account the characteristics of such hazardous substance; and
(6) the degree of cooperation by the parties with federal, state, or local officials to prevent any harm to the public health or the environment.[86]

Although the Gore Amendment did not pass, the factors were endorsed by Congress when it amended CERCLA in 1986 to add an express right of contribution,[87] and they have been oft-cited and relied

---

85. *See, e.g.*, United States v. Monsanto Co., 858 F.2d 160, 172 (4th Cir. 1988), *cert. denied*, 490 U.S. 1106 (1989).

86. H.R. 7020, 96th Cong., 2d Sess. (1980).

87. H.R. Rep. No. 253, 99th Cong., 1st Sess. 1 (1985), *reprinted in* 1986 U.S.C.C.A.N. 2835, 3042.

upon by courts called upon to allocate CERCLA liability among and between PRPs.[88]

A court may also examine equitable defenses asserted by a party in apportioning liability or damages. While a successfully asserted equitable defense normally will bar liability, courts have been resistant to barring CERCLA liability in this manner in light of Congress's express intentions in enacting CERCLA.[89] At least two courts have held that the equitable defense of unclean hands may completely bar a 107(a) action.[90] Other courts, however, have indicated that equitable defenses may be used in apportioning liability among PRPs.[91]

A second list of factors that are sometimes examined by courts are referred to as the Torres Factors from Judge Torres's opinion in *United States v. Davis*.[92] They include (1) extent to which the costs are related to the waste for which each party is responsible; (2) each party's level of culpability; (3) degree to which the party benefitted from the disposal; and (4) ability to pay.

88. Acushnet Co. v. Coaters, Inc., 948 F. Supp. 128, 138–39 (D. Mass. 1996) (in contribution action by PRP that has entered consent decree against nonsettling PRPs, court allocated zero liability to nonsettlors; citing the Gore Factors, court based its allocation primarily on the finding that the defendants were responsible for very small quantities of wastes which were nonpersistent in the environment); Bethlehem Iron Works, Inc. v. Lewis Indus., Inc., No. Civ. A. 94-0752, 1996 WL 557592 (E.D. Pa. Oct. 1, 1996) (citing Gore Factors); Control Data Corp. v. S.C.S.C. Corp., 53 F.3d 930, 935 (8th Cir. 1995) ("Those parties who can show that their contribution to the harm is relatively small in terms of amount of waste, toxicity of the waste, involvement with the waste, and care, stand in a better position to be allocated a smaller portion of response costs"); United States v. R.W. Meyer, Inc., 932 F.2d 568, 577 (6th Cir. 1991) (affirming district court's allocation of one-third of liability to landowner because of actions he failed to take, fifth Gore Factor); Envtl. Transp. Sys. v. ENSO, Inc., 969 F.2d 503, 510 (7th Cir. 1992) (citing Gore Factors, court allocated 100 percent of liability for PCB spill in single vehicle accident to transporter, which was entirely at fault); Amoco Oil Co. v. Dingwell, 690 F. Supp. 78, 86 (D. Me. 1988).

89. United States v. Conservation Chem. Co., 628 F. Supp. 391, 404–05 (W.D. Mo. 1985).

90. *See Mardan*, 600 F. Supp. at 1057; D'Imperio v. United States, 575 F. Supp. 248, 253 (D.N.J. 1983) (court would not allow an equitable defense to act as a complete bar to a § 107(a) cost recovery action; to do so would require that plaintiffs seeking cost recovery not be liable in any manner for the release of hazardous substances).

91. *See* Chem. Waste Mgmt., Inc. v. Armstrong World Indus., Inc., 669 F. Supp. 1285, 1292 (E.D. Pa. 1987).

92. 31 F. Supp. 2d 45, 63 (D.R.I. 1998), *aff'd*, 261 F.3d 1 (1st Cir. 2001).

## F. Orphan Share Issue

A recurring allocation issue in contribution claims concerns the so-called "orphan share," which is the share of the response costs that equitably should be borne by an insolvent, defunct, dead, or unknown PRP. Disagreement over who should bear the orphan share can arise where a subset of the PRPs either voluntarily agree to fund a cleanup, or are held jointly and severally liable to the government for the cleanup costs in a section 107 action. Those PRPs then adjudicate the question of who should bear the orphan share and/or how to divide the orphan share.

Courts that have directly addressed the issue of orphan share liability in section 113 contribution actions, such as *United States v. Kramer*, typically rule that "notions of fairness and common sense" require that any orphan share amount should be reallocated among all liable PRPs "without limitation to some subset of liable parties."[93] However, the case law on this issue is hardly uniform.

In *Gould Inc. v. A & M Battery & Tire Service*,[94] the court held that defendant-PRPs could not be responsible for *any* part of the orphan shares at the site, and that the section 113 plaintiff would be responsible for the entire amount. The court reasoned that "[s]ince liability under Section 113 contribution action is several, Defendants are only liable for their share of the harm caused."[95] The court rejected plaintiff's argument that it is inequitable to hold plaintiff solely accountable for the orphan shares, and instead found that "it would be inequitable for us to hold Defendants liable for any harm related to the 'orphan shares' when this harm was clearly caused by entities other than Defendants."[96]

However, the court in *United States v. Kramer* found that holdings like these "conflate[] the concept of equitable apportionment of responsibility

93. United States v. Kramer, 953 F. Supp. 592, 598 (D.N.J. 1997) (holding that section 113(f) does not preclude a court from apportioning the orphan share between primary defendant-PRPs and third-party defendant-PRPs); *see also* Town of New Windsor v. Tesa Tuck, Inc., 919 F. Supp. 662, 681 (S.D.N.Y. 1996); Caldwell Trucking PRP Grp. v. Spaulding Composites Co., Civ. No. 94-3531 (WGB), 1996 U.S. Dist. LEXIS 8994 (D.N.J. Apr. 22, 1996); City of Fresno v. NL Indus. Inc., No. CV-F-93-5091 REC/DLB, 1995 WL 641983 (E.D. Cal. July 9, 1995).

94. 901 F. Supp. 906 (M.D. Pa. 1995).

95. *Id.* at 908, 913.

96. *Id.*

among PRP's in establishing each party's share of liability, with the concept that the *judgment* against a party which is only 'severally' liable limits that party's ultimate liability to the amount of its own judgment."[97] Section 113(f)(1) permits a court to consider any appropriate "equitable factors" in arriving at the appropriate liability of a severally liable party, including assignment of the orphan share.[98] To do so does not convert "several" liability into "joint and several" liability because, once the court's "equitable apportionment" is made, "the severally liable party will not be responsible for satisfying the judgment entered against any other party."[99]

## G. Alternative Dispute Resolution

Because of the cost of litigating allocation issues can be very large, some PRPs have turned to alternate dispute resolution (ADR). ADR liberates parties from the strictures of the Federal Rules of Evidence, and allows the parties to devise their own procedures for reaching a fair apportionment of liability. ADR allocations may be binding or nonbinding. Generally, the process calls for the apportionment of a qualified "neutral" or "neutral panel" to serve as both mediator and a decision maker. Following the apportionment, the typical phases of the ADR process include (1) investigation and fact finding; (2) position statements by the PRPs; (3) neutral attempts to achieve agreement through mediation or, if the allocation will be binding, arbitration or a "mini-trial"; (4) the neutral's proposed allocation; (5) comments on that allocation by the PRPs, and (5) acceptance or rejection of the final allocation.

ADR has been used at many sites, including the Bridgeport Rental and Oil Services, Inc. Superfund Site in Logan Township, New Jersey; GE/Housatonic Superfund Site in Pittsfield, Massachusetts; Helen Kramer Landfill Superfund Site in Mantua Township, New Jersey; Auburn Road Landfill Superfund Site in Londonderry, New Hampshire; Landfill & Resource Recovery Site in Coventry, Rhode Island; Iron Mountain

97. 953 F. Supp. at 600 (emphasis added).
98. *Id.* at 601.
99. *Id.*

Mine Site in Redding, California; Milltown Reservoir Sediments/Clark Fork River Superfund Site in Milltown, Montana; Southeast Rockford Groundwater Contamination Superfund Site in Rockford, Illinois; and Rocky Flats Plant (USDOE) Superfund Site in Golden, Colorado.[100]

The Helen Kramer Landfill Site in Mantua Township, New Jersey, illustrates use of ADR. It was a privately owned landfill that received millions of gallons of liquid hazardous wastes, as well as two million cubic yards of municipal solid waste. The responsible parties included approximately 250 private companies and 44 municipalities.[101] Originally, EPA attempted to litigate a cost recovery action, but the judge stayed the litigation for four years to allow the defendants to perform an allocation of liability.[102] When the allocation stalled, the government resumed the litigation, prompting serious settlement negotiations to begin.[103] However, the defendants were still unable to agree upon the allocation.[104] The court ordered mediation, and the mediators became actively involved in the case.[105] They helped the defendants deal with a large orphan share, de minimis parties, defunct companies, insurance companies, and numerous municipalities leading to a settlement among 200 parties that exceeded $100 million.[106]

## IV. AVAILABILITY OF DECLARATORY RELIEF UNDER SECTIONS 107 AND 113

Section 113(g)(2) of CERCLA mandates that in any section 107 cost recovery action, "the court shall enter a declaratory judgment on liability for response costs or damages that will be binding on any subsequent action or actions to recover further response costs or damages." Bear in mind that a declaratory judgment of a defendant's liability under

100. *See EPA Status Report on the Use of Alternative Dispute Resolution in Environmental Protection Agency Enforcement and Site-Related Actions* at 8 (1999).
101. *Id.*
102. *See* EPA summary, *ADR Closes the Deal at the Helen Kramer Landfill*, CLEANUP NEWS at 9, Spring 1999.
103. *Id.*
104. *Id.*
105. *Id.*
106. *Id.*

CERCLA does not preclude such defendant from challenging costs as unrecoverable in a future cost recovery action.[107]

Although section 113(g)(2) mandates declaratory judgment in a section 107 cost recovery action, it is silent as to whether declaratory judgment is authorized in contribution actions. Nevertheless, circuit courts have typically granted declaratory relief in contribution actions because it is "consistent with the broader purposes of CERCLA."[108] Some circuits base declaratory relief on section 113(g)(2).[109] Other circuits, however, find that the Declaratory Judgment Act governs declaratory judgment in CERCLA contribution actions because no specific relief is authorized by CERCLA.[110] Courts also disagree as to whether one must demonstrate recoverability of past costs in order to receive declaratory judgment for future costs. Some courts require such a showing.[111] Other courts do not require such a showing.[112]

## V. INDEMNIFICATION

Nothing in CERCLA prohibits one party from seeking contractual indemnification from another party for CERCLA liability. In fact, section 107(e)(1) preserves the right of private parties to release one another from CERCLA financial responsibility or to transfer such responsibility. Under CERCLA, the indemnity agreement may never be used as a defense

107. *See, e.g.*, United States v. Hardage, 982 F.2d 1436, 1445 (10th Cir. 1992).

108. Boeing Co. v. Cascade Corp., 207 F.3d 1177, 1191 (9th Cir. 2000).

109. *See* United States v. Davis, 261 F.3d 1, 46 (1st Cir. 2001); *Boeing Co.*, 207 F.3d at 1191; *Cnty. Line Inv. Co.*, 933 F.2d 1508.

110. *See, e.g.*, Olin Corp. v. Consol. Aluminum Corp., 5 F.3d 10, 17 (2d Cir. 1993); Reichhold, Inc. v. United States Metals Ref. Co., 522 F. Supp. 2d 724, 729 (D.N.J. 2007); Mercury Mall Assocs., Inc. v. Nick's Mkt., Inc., 368 F. Supp. 2d 513, 520 (E.D. Va. 2005).

111. *City of Coulton*, 614 F.3d 998 (section 113(g)(2), which authorizes declaratory judgment for "further response costs," implies a requirement that plaintiff has already incurred at least some recoverable response costs); Gussack Realty Co. v. Xerox Corp, 224 F.3d 85, 92 (2d Cir. 2000) (ruling declaratory relief unavailable because plaintiffs "have not incurred compensable expenses under CERCLA"); United States v. Occidental Chem. Corp., 200 F.3d 143 (3d Cir. 1999).

112. Trimble v. Asarco, Inc., 232 F.3d 946, 958 (8th Cir. 2000) (plaintiff is not required to actually pay a cost in order to incur it; an incurred cost "may be based on an existing legal obligation"); T&E Indus. v. Safety Light Corp., 680 F. Supp. 696, 709 (D.N.J. 1988); Southland Corp. v. Ashland Oil Co., 696 F. Supp. 994, 1003 (D.N.J. 1988), *rev'd on other grounds*, Civ. No. 88-0700, 1988 WL 125855 (D.N.J. Nov. 23, 1988).

to an action brought by the government or another private party; it may be a defense only between the contracting private parties.[113]

Unlike a contribution action, which is brought under section 113(f) of CERCLA, an indemnification action would be brought as a general common law claim. For example, in *Caldwell v. Gurley Refining Co.*,[114] plaintiff/lessor had leased land to be used as a waste material dumping site. The lessee terminated the lease and, in doing so, represented to the lessor that the waste site had been properly sealed. In reality, the disposal area had been improperly sealed, and hazardous substances had leaked into nearby rivers.[115] The lessor invoked the provisions of the Declaratory Judgment Act[116] and requested the court to determine the rights and obligations of the parties under the lease, particularly the lessor's right of indemnity against the lessee resulting from liability to the EPA for a cleanup operation.[117] The Eighth Circuit affirmed the district court's entry of a declaratory judgment, holding the lessee responsible during the entire lease term for all pollution occurring due to operations at the waste site. Further, the court found that the lessor was entitled to full indemnity from the lessee for any claim or judgment to the federal or state agencies for releases occurring during the full lease term.

In order to maintain an indemnity action, a party must look to state law to determine whether indemnity is available. For example, in *Mardan*,[118] the plaintiff had included an indemnity claim in its complaint. The court disposed of the plaintiff's indemnity claim, holding that, under Arizona law, there is no indemnity among joint tortfeasors. Because the plaintiff in *Mardan* had also deposited waste at the site, it was not entitled to indemnity under Arizona law.

113. *Southland Corp.*, 696 F. Supp. at 1000.
114. 755 F.2d 645 (8th Cir. 1985), *reprinted in* 1986 U.S.C.C.A.N. 2835.
115. *Id.* at 647.
116. 28 U.S.C. § 2201.
117. *Caldwell*, 755 F.2d at 647.
118. 600 F. Supp. at 1052–53 n.5.

# VI. SUITS AGAINST GOVERNMENT ENTITIES

Section 107(a) of CERCLA authorizes a person to recover necessary response costs from any other person. Congress defined "person" in very broad terms and included within its definition the United States government, states, municipalities, commissions, political subdivisions of a state, or any interstate body.[119] Under CERCLA, all political entities are potentially liable under section 107 to any other person who has incurred response costs that are necessary and consistent with the NCP.

## A. Suits against the Federal Government

In addition to including the federal government within the definition of "persons" who may be held liable for response costs, in section 120(a) of CERCLA (formerly section 107(g)), Congress expressly waived the sovereign immunity of the federal government.[120] This is confirmed by the legislative history of the section.[121]

The handful of courts that have directly addressed the issue have confirmed the waiver of sovereign immunity, but only as to causes of action created by CERCLA.[122] Other courts have held that sovereign immunity still precludes the joinder of EPA as an involuntary plaintiff in an action brought against PRPs under CERCLA by a state authority.[123]

Government contractors that incur response costs attributable to acts performed on behalf of or at the direction of the federal government are

119. CERCLA § 101(21).

120. Section 120(a)(1) provides: "Each department, agency, and instrumentality of the United States (including the executive, legislative, and judicial branches of government) shall be subject to, and comply with, this chapter in the same manner and to the same extent, both procedurally and substantively, as any non-governmental entity, including liability under [section 107] of this title. Nothing in this section shall be construed to affect the liability of any person or entity under sections [106] and [107] of this title."

121. *See* H.R. Rep. No. 253, 99th Cong., 1st Sess., pt. 1, at 93 (1985), reprinted in 1986 U.S.C.C.A.N. 2835.

122. United States v. Nicolet, Inc., Civ. A. No. 85-3060, 1986 WL 15017 (E.D. Pa. Dec. 31, 1986); *see also* B.R. MacKay & Sons, Inc. v. United States, 633 F. Supp. 1290, 1296 n.9 (D. Utah 1986) (waiver of sovereign immunity does not make the United States liable for "carrying out a removal action inconsistent with the National Contingency Plan and [for] improperly denominating plaintiffs as responsible parties.").

123. *See, e.g.*, N.J. Dep't of Envtl. Prot. v. Gloucester Envtl. Mgmt. Servs., Inc., 668 F. Supp. 404 (D.N.J. 1987); Missouri v. Indep. Petrochemical Corp., No. 83-2670C, 17 Envtl. L. Rep. 20,241 (E.D. Mo. Oct. 16, 1986).

not limited to CERCLA to recover such costs. They may pursue recovery under contract theories. The leading case on this principle is *Shell Oil Co. v. United States*.[124] During World War II, the United States entered into contracts with several oil companies including Shell Oil, for the production of high-octane aviation gas (called "avgas") to fuel military aircraft as part of the nation's war effort. Production of substantial amounts of avgas resulted in waste products such as spent alkylation acid and acid sludge, and the oil companies contracted with Eli McColl, a former Shell engineer, to dump the waste at a property in Fullerton, California (the McColl site). Significantly, the government and the oil companies understood that production of avgas would result in high volumes of waste products, and as a result, the contracts entered by the government with the oil companies included indemnification provisions. In those provisions, the government agreed to reimburse the contracting oil companies for "any *new or additional taxes, fees*, or *charges* . . . which [the Oil companies] may be required by any municipal, state, or federal law in the United States or any foreign country to collect or pay by reason of the production, manufacture, sale or delivery of [avgas]."[125]

Over 50 years later, the United States and California obtained compensation from the oil companies under CERCLA § 107(a) for the costs incurred in cleaning up the McColl site. The oil companies filed suit in the Court of Federal Claims, arguing that the avgas contracts required the federal government to indemnify them for the CERCLA costs. Although the Court of Federal Claims granted summary judgment for the United States, the Federal Circuit reversed, concluding that the CERCLA costs did, in fact, constitute new or additional charges under the avgas contracts.[126] Situating its analysis within the context of the war effort, the court noted "the stark necessity of increased avgas production" as a driving force in the government's contracts with the oil companies.[127] The court expounded: "The Oil Companies agreed to the avgas contracts' low profits in return for the Government's assumption

124. 751 F.3d 1282 (Fed. Cir. 2014).
125. *Id.* at 1287 (emphasis and alterations in original).
126. *Id.* at 1303.
127. *Id.* at 1296.

of certain risks outside of the Oil Companies' control. . . . The CERCLA charges in this case are one such risk."[128] Therefore, the Federal Circuit held that the contracts must be interpreted as requiring the government to reimburse the oil companies' CERCLA costs arising from avgas production.

In sum, government contractors are not limited to CERCLA § 107(a) and § 113(f)(1) for recovery of cleanup costs where they possess contract rights.

### B. Suits against States

Under the Eleventh Amendment, a private citizen is barred from bringing suit against a state in a federal court. As discussed in chapter 4, in *Pennsylvania v. Union Gas Co.*, the U.S. Supreme Court ruled that Congress may abrogate this immunity and that it had expressed its clear intent to do so under CERCLA.[129] A scant seven years later, however, the Supreme Court overruled this decision. In *Seminole Tribe of Florida v. Florida*,[130] the Supreme Court held that the commerce clause of the Constitution does not authorize Congress to abrogate states' sovereign immunity recognized by the Eleventh Amendment.[131] Thus, the Court held that a private party cannot file a CERCLA action against a state in federal court.

## VII. DEFENSES

### A. Statutory Defenses

A defendant in a private cost recovery action and contribution actions may assert the same statutory defenses that would be available in a federal cost recovery action. First, such defendant may assert that the elements of liability do not exist, that is, that it is not a responsible party under section 107(a) of CERCLA;[132] that response costs have not been

128. *Id.*
129. 491 U.S. 1 (1989).
130. 517 U.S. 44 (1996).
131. *Id.* at 72.
132. *See, e.g.*, New York v. Shore Realty Corp., 759 F.2d 1032 (2d Cir. 1985).

incurred;[133] or that the response action was not "necessary" or "consistent with the NCP."[134] Moreover, the defendant in a private cost recovery action may assert the affirmative defenses of section 107(b) (act of God, act of war, and act of third party), the defenses that were added to section 107 by the Small Business Relief and Brownfield Revitalization Act (i.e., de micromis exemption, municipal solid waste exemption, recycling exemption, adjacent landowner defense, and bona fide purchaser exemption), and the exclusions from liability for certain kinds of releases (e.g., federally permitted releases, the application of pesticides, etc.) and certain kinds of hazardous substances (e.g., those indigenous to petroleum). The elements of these defenses are discussed in chapter 4.

## B. Equitable Defenses

For the most part, courts have denied equitable defenses in cost recovery actions filed by the United States or a state.[135] In the private party context, equitable defenses fare somewhat better. Those courts that permit assertion of equitable defenses adopt the view that, irrespective of whether a private cost recovery action is brought pursuant to section 107 or section 113, the action is essentially one for contribution, which provides for allocation of response costs based on equitable considerations.[136] Because the plaintiff is seeking equitable relief, defendants are allowed to assert equitable defenses.[137] A number of courts, however, have drawn sharp distinctions between private cost recovery actions under section 107 and contribution actions under section 113, and rejected equitable defenses in actions brought under section 107.[138]

133. *See, e.g., Jones*, 584 F. Supp. at 1429.

134. *See, e.g., Amland Props.*, 711 F. Supp. 784.

135. *See* chapter 4.

136. *See, e.g.*, Smith Land & Improvement Corp. v. Celotex Corp., 851 F.2d 86 (3d Cir. 1988), *cert. denied*, 488 U.S. 1029 (1989); PVO Int'l, Inc. v. Drew Chem. Corp., Civ. A. No. 87-3921, 1988 U.S. Dist. LEXIS 18609 (D.N.J. June 27, 1988). *But see* Marden Corp. v. C.G.C. Music, Ltd., 804 F.2d 1454, 1457 (9th Cir. 1986) (doubting that equitable defense of "unclean hands" may be invoked as to a section 107 cost recovery action brought by a PRP).

137. Violet v. Picillo, 648 F. Supp. 1283, 1294–95 (D.R.I. 1986).

138. *See, e.g.*, Thaler v. PRS Metals Prods., Inc., 815 F. Supp. 99, 102 (E.D.N.Y. 1993) (granting summary judgment against defendant's equitable defenses "[b]ecause these defenses do not fall within the very limited exceptions to liability provided in [107(b)]"), *aff'd*, 28 F.3d 102 (2d Cir. 1994); *Gen. Elec. Co.*, 715 F. Supp. 949 (doctrine of unclean hands does not

Examples of equitable defenses raised in private cost recovery or contribution actions are "unclean hands,"[139] laches,[140] and estoppel.[141] It should be noted, however, that there is disagreement as to whether equitable defenses act as a complete bar to liability or are relevant only in apportioning damages.[142] As noted in chapter 4, most courts have held that an equitable defense is merely one of the factors to be considered in apportioning liability for response costs.[143]

## VIII. SETTLEMENT ISSUES IN PRIVATE COST RECOVERY AND CONTRIBUTION ACTIONS

Defendants, seeking to settle in the context of a private cost recovery or contribution action have the same basic motivation as defendants in government cost recovery actions: finality. In government cost recovery actions the government can offer a PRP finality because of the

apply to CERCLA private actions); *Allied Corp.*, 691 F. Supp. 1100 (same); Town of Munster v. Sherwin-Williams Co., 27 F.3d 1268 (7th Cir. 1994) (section 107(a) precludes equitable defense of laches in a private cost recovery action); Velsicol Chem. Corp. v. Enenco, Inc., 9 F.3d 524, 530 (6th Cir. 1993) (equitable defense of laches impermissible in section 107 cost recovery action, but allowed in section 113 contribution action).

139. *Mardan Corp.*, 600 F. Supp. at 1056–58 (9th Cir. 1986) (Ninth Circuit affirmed the result based solely on the contractual release; it did not reach the unclean hands defense). *But see Chem. Waste Mgmt.*, 669 F. Supp. at 1291 n.7 ("the 'unclean hands' doctrine . . . has no place in CERCLA actions"); Gen. Elec. v. Litton Indus. Automation Sys., 920 F.2d 1415 (8th Cir. 1990) (disallowing unclean hands defense).

140. Violet v. Picillo, 648 F. Supp. 1294.

141. *Int'l Clinical Labs., Inc.*, 710 F. Supp. at 471.

142. *See, e.g.*, United States v. Conservation Chem. Co., 619 F. Supp. at 204 (W.D. Mo. 1985) ("Equitable defenses are proper under CERCLA in determining liability, the nature of the remedy and the amount of damage."); *contra Versatile Metals, Inc.*, 693 F. Supp. at 1572 ("Versatile Metals' assertions as to relative fault will be considered in this court's apportionment of necessary response costs, if any"); *Mardan Corp.*, 600 F. Supp. 1049 (plaintiff's claim for response costs barred by, inter alia, the doctrine of unclean hands).

143. *See, e.g.*, *Smith Land & Improvement Corp.*, 851 F.2d at 90 (court rejected the defense of caveat emptor as a complete bar to recovery and held that the doctrine may only be considered in mitigation of the amount due as determined in a contribution action); *Southland Corp.*, 696 F. Supp. at 1003 ("equitable considerations are applicable in allocating the amount of contribution between parties under CERCLA"); Westwood Pharms., Inc. v. Nat'l Fuel Gas Distribution Corp., 737 F. Supp. 1272 (W.D.N.Y. 1990), *aff'd*, 964 F.2d 85 (2d Cir. 1992) (rejecting "caveat emptor" as complete defense to liability, notwithstanding that the "extremely low contract price" may have reflected the parties' understanding that seller would not be liable to purchaser for any condition of the land).

availability of contribution protection, pursuant to section 113(f)(2) of CERCLA.[144]

There is no similar provision in CERCLA applicable to settlements between private parties. Notably, however, there are other legal grounds for a party to obtain finality—to "buy peace"—in a settlement between private parties. Specifically, courts have granted contribution protection to parties that settled private cost recovery claims by reference to two model statutes: the Uniform Comparative Fault Act (UCFA) and the Uniform Contribution among Tortfeasors Act (UCATA).

As background, UCATA and UCFA were both drafted by the National Conference of Commissions on Uniform Laws. Some states have adopted UCATA to govern disputes among joint tortfeasors; others have adopted UCFA. Both model laws specify that a settling tortfeasor is protected from contribution claims by nonsettling tortfeasors. The primary difference is the effect of the settlement on nonsettlers. Under UCATA, the settlement reduces the potential liability of nonsettlers by the dollar amount of the settlement (the pro tanto approach); whereas under UCFA the settlement reduces the potential liability of nonsettling tortfeasors by the equitable share of liability attributable to the settling tortfeasor (pro rata approach).

The court in *American Cyanamid Co. v. King Industries, Inc.*[145] provided a useful illustration of how the UCFA approach works in the context of a CERCLA settlement:

> To clarify how the [proportionate share] rule will operate in this case, consider the following example. A private party sues five defendants for contribution under § 113(f)(1) of CERCLA. Plaintiff seeks recovery for $100,000 for response costs. One defendant settles for $5,000. At trial, it is determined that each of the five original defendants are liable for an equitable share equaling 10% of the $100,000—or $10,000 per defendant. The plaintiff is found

144. For a complete discussion of contribution protection in government cost recovery actions, see chapter 5.

145. 814 F. Supp. 215 (D.R.I. 1993).

> to be liable for 50%—or $50,000. The settling defendant is still responsible for only $5,000. The non-settling defendants, however, may reduce their collective share of liability by the amount of the settling defendant's *equitable* share—in this case $10,000. Thus, the four non-settling defendants are responsible for $10,000 each, for a total of $40,000. The plaintiff bears the "loss" of $5,000 in contribution recovery as a result of compromising for less than the settling party's equitable share. Conversely, the plaintiff may reap any "windfall" from a settlement with one of the five defendants for more than $10,000, or more than the defendant's equitable share. That is the rub of the green.[146]

Which approach—UCATA/pro tanto or UCFA/pro rata—applies in the context of a settlement of a private CERCLA cost recovery claim? The answer appears to be uncertain. Some courts apply the law of the jurisdiction, while others seek to divine the federal common law and apply UCATA given that CERCLA specifies use of the UCATA approach in government cost recovery claims. Yet other courts take into consideration the parties' choice of law.[147] The best that can be said is that federal courts are split on whether to apply UCATA or UCFA in the context of private settlements. Courts adhering to the UCFA approach[148] have reasoned that section 113(f)(1) requires courts to consider "equitable factors" and that this dictates applicable of the pro rata approach in

146. *Id.* at 218.

147. *See* Tyco Thermal Controls LLC v. Redwood Indus., No. C 06-07164 (PVT), 2010 WL 3211926 (N.D. Cal. Aug. 12, 2010).

148. Courts applying UCFA include Tosco Corp. v. Koch Indus., Inc. 216 F.3d 886, 897 (10th Cir. 2000); Adobe Lumber, Inc. v. Hellman, No. Civ-05-1510 WBS EFB, 2009 WL 256553, at *3 (E.D. Cal. Feb. 3, 2009); United States v. GenCorp, Inc., 935 F. Supp. 928 (N.D. Ohio 1996); *Pneumo Abex Corp*., 936 F. Supp. 1274; Am. Cyanamid Co. v. King Indus., Inc., 815 F. Supp. at 218; Barton Solvents, Inc. v. Sw. Petro-Chem, Inc., 834 F. Supp. 342, 348 (D. Kan. 1993); Comerica Bank-Detroit v. Allen Indus., Inc., 769 F. Supp. 1408, 1414–15 (E.D. Mich. 1991); Lyncott Corp. v. Chem. Waste Mgmt., Inc. 690 F. Supp. 1409, 1416–19 (E.D. Pa. 1988); United States v. W. Processing Co., 734 F. Supp. 930 (W.D. Wash. 1990); Edwards Hines Lumber Co. v. Vulcan Materials Co., No. 85 C 1142, 1987 U.S. Dist. LEXIS 11961 (N.D. Ill. 1987), *aff'd on other grounds*, 861 F.2d 155 (7th Cir. 1988).

actions between private parties. Courts selecting the UCATA approach[149] have focused on CERCLA's goal of promoting early settlement, and therefore have applied the pro tanto approach embodied in CERCLA section 113(f)(2) to private settlements.

## IX. DOCUMENTATION AND RECORD-KEEPING PRACTICES TO MAXIMIZE COST RECOVERY

Record-keeping practices (such as record-keeping provisions in agreements with contractors, subcontractors, and consultants) may help a PRP maximize its cost recovery. There will be inevitable logistical difficulties in maintaining and coordinating the production of the mass of data, contracts, cost records, and other evidence generated in a response action. It is important to establish an orderly method, computerized or manual, for retaining that information during the course of a cost recovery action for use during case development, discovery, and trial. The following are common sense practices for maximizing cost recovery:

1. Participants in a CERCLA response activity should maintain documents related to the activity for six years or more after all response activities are completed. The six-year period is necessary because of the possibility that the claim may not accrue upon the first expenditure. Additionally, the litigation may be protracted; documents must be kept until all cost recovery claims have been resolved.
2. Each cost incurred should be directly connected to specific response activities. Courts will award costs only for activities that are related in some way to response activities.

149. Other courts have opted for UCATA, and two circuits have held that the district courts are free to apply UCATA rather than UCFA. *See* Azko Nobel Coatings, Inc. v. Aigner Corp., 197 F.3d 302, 308 (7th Cir. 1999) ("Extending the *pro tanto* approach of § 113(f)(2) . . . enables the district court to avoid what could be a complex and unproductive inquiry into the responsibility of missing parties."); Am. Cyanamid Co. v. Capuano, 381 F.3d 6, 20–21 (1st Cir. 2004); Action Mfg. Co. v. Simon Wrecking Co., 428 F. Supp. 2d 288, 327 (E.D. Pa. 2006), *aff'd*, 287 F. App'x 171 (3d Cir. 2008); Allied Corp. v. Frola, 730 F. Supp. 626, 638 (D.N.J. 1990) (applying pro tanto approach in private party settlement); *see also In re* Acushnet River & New Bedford Harbor, 712 F. Supp. 1019, 1026–27 n.10 (D. Mass. 1989) (noting inconsistency of cases applying UCFA after enactment of CERCLA § 113(f)(2)).

3. For each phase of the response activity (e.g., RI/FS, negotiations, remedial design, implementation), all work assignments, including the people involved and the costs incurred, should be clearly identified. This will assist in connecting the costs to the response activity. If response activity is being taken pursuant to a consent decree, this will help tie all costs to that decree and help maximize recovery.
4. Document all in-house expenditures. These are the costs incurred when a PRP's own employees are involved with the response activities.
   a. Track the time each employee spends on specific response-related activities, using time cards, time sheets, or another technique.
   b. Document any related travel by employees.
   c. Keep records of all PRP employees' salaries related to response activities.
   d. Document all nonpersonnel expenses, such as supplies, overhead, computer time, telephone calls, utilities, and equipment.
5. Retain all correspondence, contractor bids, internal memoranda, etc., relating to selecting and negotiating contractors' agreements. This task will provide documentation that the work contracted for and contract price are reasonable in light of site-specific considerations.
6. Never include expenses for recoverable costs and nonrecoverable costs on the same receipt. At a minimum, require contractors and subcontractors to clearly identify and segregate charges on bills so that a breakdown is possible later.
7. Request contractors to provide detailed descriptions of the job performed, including specific actions taken and time spent, in invoices.
8. Request contractors to generate separate invoices for discrete tasks. This is especially important if a contractor is working on different phases of a response. It is possible that several phases of a response action may be conducted simultaneously, and that cost recovery may proceed on one phase while a second or third phase is still in progress. Separate bills will help minimize confusion and help ensure cost recovery.

9. Request contractors to maintain records of all time and expenses of their employees and subcontractors.
10. Request contractors to maintain all receipts in support of the invoices they generate, including, but not limited to, travel vouchers, airline tickets, meal receipts, receipts for purchasing or renting equipment, and telephone bills.
11. Conduct annual accounting audits of all costs to ensure that they are being properly documented and maintained.

# CHAPTER 10

# Natural Resource Damages

## I. INTRODUCTION

Remedial action undertaken pursuant to CERCLA may not address or mitigate the full extent of environmental injury resulting from the release of hazardous substances. Remediation that satisfies the CERCLA cleanup standards need not fully restore the contaminated site to its original condition. For example, a cleanup that reduces hazardous substance concentrations in environmental media to levels that comply with "applicable or relevant and appropriate requirements" often permits residual levels of contamination to remain on-site.[1] Moreover, some forms of environmental injury simply are not addressed by CERCLA removal and remedial action, such as death of flora and fauna that releases may cause.

In view of the fact that recoverable response costs might not fully compensate the United States or a state for a loss caused by a release of hazardous substances, Congress created a cause of action for "damages for injury to, destruction of, or loss of natural resources, including the reasonable costs of assessing such injury, destruction, or loss resulting from such a release."[2] Natural resource damages pick up where response costs leave off: Response costs are expenses incurred to remediate contamination at a site, whereas natural resource damages provide funds

1. See discussion in chapter 2.
2. CERCLA § 107(a)(4)(C).

either to restore the contaminated site to its original condition, or to provide compensation for the loss in value or use of the resource.[3]

Since the enactment of CERCLA in 1980, the CERCLA litigation landscape has been dominated by cost recovery, and contribution actions focused on response costs. Only a relative handful of cases have specifically addressed natural resource damages. In a sense, recovery of natural resource damages is the last frontier of Superfund.

## II. SCOPE OF LIABILITY FOR NATURAL RESOURCE DAMAGE

The natural resource damages provisions of CERCLA are set forth in three sections: (1) section 101(16) defines "natural resources"; (2) section 107 establishes a cause of action for recovery of natural resource damages; and (3) section 301(c) directs the president to promulgate natural resource damage assessment regulations. Following is an analysis of these provisions, as well as a discussion of some of the substantive and procedural elements of an action for natural resource damages under CERCLA.[4]

### A. Types of Natural Resources

Only certain types of natural resources are within the scope of CERCLA's natural resource damage provisions. Specifically, the term "natural resources" means

> land, fish, wildlife, biota, air, water, ground water, drinking water supplies, and other such resources belonging to, managed by, held in trust by, appertaining to, or otherwise controlled by the United

3. *See* CERCLA § 107(f)(1).

4. There are a number of other federal statutes that provide for recovery of natural resource damages. The Oil Pollution Act of 1990 establishes liability for natural resource damages for discharges of oil. 33 U.S.C. § 2702(b)(2)(A). The Clean Water Act establishes natural resource damage liability for releases of oil or hazardous substances "into or upon the navigable waters of the United States, adjoining shorelines, or upon the waters of the contiguous zone." 33 U.S.C. § 1321(b)(1). The National Marine Sanctuaries Act establishes liability for destroying or causing loss or injury to federal marine sanctuaries. 16 U.S.C. § 1443(a)(1). Last but not least, the Park Resource System Protection Act establishes liability for destroying or causing loss or injury to national park system resources. 16 U.S.C. § 19jj-1(a).

> States . . . any State or local government, any foreign government, any Indian tribe, or, if such resources are subject to a trust restriction on alienation, any member of an Indian tribe.[5]

Critical to this definition of natural resource is the concept that only publicly controlled lands are within the ambit of the natural resource damage provision. Thus, whereas damages resulting from injury to national and state parks and forests are recoverable, damages resulting from injury to private property are not recoverable under section 107(f).[6]

The Department of Interior (DOI), which is responsible for promulgating natural resource damage assessment regulations, has limited the assessment of damages to natural resources with "committed" uses.[7] Committed uses are defined as current or planned public uses "for which there is a documented legal, administrative, budgetary, or financial commitment" that was established before detection of the discharge.[8] The definitional boundaries of the term "natural resources" have not yet been litigated, but are likely to be a controversial issue in the future.

### B. Standing to Sue for Natural Resource Damages

Section 107(f)(1) of CERCLA provides that natural resource damages can be recovered by trustees for the United States or any state, or by any Indian tribe, whose natural resources are damaged by a release of hazardous substances. The statute does not authorize private parties or any other entities to recover natural resource damages.[9]

CERCLA requires the president to designate federal officials to act as trustees of natural resources on behalf of the public.[10] The designated federal trustees are responsible for assessing damages associated with an injury to a natural resource under their trusteeship.[11] In an executive

5. CERCLA § 101(16).
6. *See* United States v. Shell Oil Co., 605 F. Supp. 1064, 1084 (D. Colo. 1985).
7. 43 C.F.R. § 11.84(b)(2).
8. 43 C.F.R. § 11.14(h).
9. United States v. S. Pa. Transp. Auth., Civ. A. Nos. 86-1094, 1986 WL 7565 (E.D. Pa. Jul. 2, 1986).
10. CERCLA § 107(f)(2)(A).
11. *Id.*

order the president designated the Secretaries of Interior, Agriculture, Defense, Commerce, and Energy as federal trustees for natural resources that they manage or control.[12] The NCP further provides (1) the secretary of the Interior shall be trustee for natural resources in cases where the United States acts on behalf of a tribe; (2) the secretary of Commerce shall be trustee for natural resources found "in, under, or using" water bodies; and (3) the head of the relevant department—such as DOI, USDA, DOD, and DOE—shall be trustee for natural resources "located on, over or under land administered by the United States."[13]

State trustees are designated by the governors of each state.[14] Like federal trustees, state trustees are responsible for assessing damages to natural resources under their trusteeship.[15] A state's trusteeship is limited to natural resources "belonging to, managed by, controlled by, or appertaining to such State."[16] Natural resources often are shared by federal and state governments, and in such cases multiple trustees must coordinate and cooperate in order to perform their respective duties.[17]

Courts are split on the question of whether municipalities may properly bring an action for natural resource damages. The minority view is that municipalities may bring natural resource damage claims. For example, in *Mayor & Board of Aldermen v. Drew Chemical Corp.*,[18] the mayor and board of aldermen of the town of Boonton, New Jersey, brought suit against a chemical company for natural resource damages. The court held that the municipality could be considered a state within the statute, or alternatively, was an authorized representative of the state of New Jersey for the purposes of invoking the provisions of CERCLA.[19]

12. Exec. Order No. 12,580, 52 Fed. Reg. 2923 (Jan. 23, 1987).
13. 40 C.F.R. § 300.600(b)(3).
14. CERCLA § 107(f)(2)(B).
15. *Id.*
16. CERCLA § 107(f)(1).
17. *See* 40 C.F.R. § 300.615(a).
18. 621 F. Supp. 663 (D.N.J. 1985).
19. *See also* City of N.Y. v. Exxon Corp., 633 F. Supp. 609, 619 (S.D.N.Y. 1986) (City of New York allowed to seek recovery of natural resource damages on its own behalf); *cf.* Fireman's Fund Ins. Co. v. City of Lodi, 271 F.3d 911 (9th Cir. 2001) (holding that local government could go beyond the cleanup requirements of CERCLA and recover damages to its own natural resources).

The majority of courts, however, have rejected this reasoning and held that only the federal government or a state could bring the action.[20]

Although the United States, a state, Indian tribes, and, arguably, municipalities all have standing to assert a natural resource damages claim against a liable party, double recovery as to the same natural resource is barred under section 107(f)(1).[21] The statute does not clarify, however, whether multiple trustees may assert independent claims for the same natural damages, whether one trustee's interest preempts the interests of other trustees, and/or how trustees may allocate such damages among themselves. The scant case law on the issue is inconsistent. In *Coeur d'Alene Tribe v. ASARCO Inc.*,[22] the court held that a co-trustee must allocate its trusteeship based on the ratio or percentage of actual management and control that it exercises over the injured natural resource as compared to the other co-trustee. The court in *Oklahoma v. Tyson Foods, Inc.*[23] applied the same standard, and then dismissed the state co-trustee's natural resource damages claims because it had failed to join the tribal sovereign and the court could not determine each trustee's percentage of control when one trustee was absent. Then came the decision in *United States v. ASARCO Inc.*, in which the same court that decided *Coeur d'Alene* reversed itself and modified the prior opinion holding that a co-trustee *could* recover 100 percent of natural resource damages (NRD) for natural resources within its trusteeship, less any amount already recovered by another trustee.[24]

20. Consol. City of Indianapolis v. Union Carbide Corp., 02-CV-1340, 2003 U.S. Dist. LEXIS 18182 (S.D. Ind. Oct. 8, 2003) (city did not have standing to bring NRD claim because it had not been designated a trustee by the state and could not be broadly construed to be a state); Werlein v. United States, 746 F. Supp. 887 (D. Minn. 1990); Town of Bedford v. Raytheon Co., 755 F. Supp. 469 (D. Mass. 1991); Artesian Water Co. v. Gov't of New Castle Cnty., 659 F. Supp. 1269 (D. Del. 1987), *aff'd*, 851 F.2d 643 (3d Cir. 1988).

21. Section 107(f)(1) of CERCLA provides: "There shall be no double recovery under this chapter for natural resource damages, including the costs of damage assessment or restoration, rehabilitation, or acquisition for the same release and natural resource."

22. 280 F. Supp. 2d 1094, 1116 (D. Idaho 2003).

23. 258 F.R.D. 472, 480 (N.D. Okla. 2009).

24. 471 F. Supp. 2d 1063, 1067–69 (D. Idaho 2005).

## C. Procedural Requirements for Natural Resource Damage Claims

When a federal or state agency acting as trustee of natural resources is notified by the EPA of a hazardous substance release, as required under section 104(b) of CERCLA, the trustee must determine whether the release is covered under CERCLA. The trustee must then assess the damage to resources in accordance with regulations promulgated under section 301(c) of CERCLA, seek recovery for losses from the persons responsible, and devise and carry out restoration, rehabilitation, and replacement plans.

Section 113(g)(1) of CERCLA provides that the trustee must give the president and the potentially responsible parties (PRPs) notice of intent to file suit 60 days prior to bringing an action for natural resource damages, or before the selection of the remedial action if the president is diligently proceeding with the remedial investigation/feasibility study. This section also requires claims or actions against responsible parties to be filed within three years of the date of discovery of the loss or within three years of the date on which damage assessment regulations are promulgated under section 301(c), whichever is later.[25] As DOI promulgated its first assessment regulations under section 301 in 1986, that three-year period is no longer relevant to determining if a NRD suit was timely filed. The remaining threshold (three years from the date of discovery of the loss) would be determinative of timeliness. Section 113(g)(1) also requires the trustee to commence actions for natural resource damages on a National Priority List site within three years after completion of the remedial action.

## D. Liability for Natural Resources Damages

Section 107(a)(4)(C) of CERCLA establishes liability for "damages for injury to, destruction of, or loss of natural resources, including the reasonable costs of assessing such injury, destruction or loss resulting from such a release" of hazardous substances.

25. CERCLA § 113(g)(1)(A), (B).

In many respects, liability under CERCLA for natural resource damages is the same as liability under CERCLA for response costs, but in several important respects it is different. Section 107(a)(4)(C) imposes liability for natural resource damages on the same class of defendants liable for response costs: the present owner or operator of the facility, the owner or operator at the time of disposal, any person who arranged for disposal or treatment of hazardous substances at the facility, and any person who transported hazardous substances to a facility. Moreover, liability for natural resource damages has been interpreted as being strict, joint, and several.[26]

Further, the same affirmative defenses to liability for response costs are defenses to liability for natural resource damages. Thus, a defendant in a natural resource damage action may assert the affirmative defenses in section 107(b) (act of God, act of war, and act of third party), the defenses that were added to section 107 by the Small Business Relief and Brownfield Revitalization Act (i.e., de micromis exemption, municipal solid waste exemption, adjacent landowner defense, recycling exemption, and bona fide purchaser exemption), and the exclusions from liability for certain kinds of releases (e.g., federally permitted releases, the application of pesticides, etc.) and certain kinds of hazardous substances (e.g., petroleum). The elements of these defenses are discussed in chapter 4. There are several additional defenses that are specific to natural resource damage claims. See discussion *infra* in this chapter.

The most significant difference between liability for response costs and liability for natural resource damages is the requirement to prove causation in a natural resource damage claim. The causation element can be inferred directly from the statutory text: Under section 107(a)(4)(C), responsible parties may be held liable for damages for "injury to, destruction of, or loss of natural resources, including the reasonable costs of assessing such injury, destruction or loss *resulting from* such a release" of hazardous substances.[27] Based on the phrase "resulting

26. *See* Idaho v. Bunker Hill Co., 635 F. Supp. 665, 674–75 (D. Idaho 1986) (the court noted that should damages be capable of apportionment the court could apportion damages).
27. CERCLA § 107(a)(4)(C) (emphasis added).

from," courts have required trustees to establish a causal link between a release for which the defendant is responsible and the injury to natural resources for which damages are being claimed. The causation standard applied by the courts, however, has varied. For example, in *In re Acushnet River & New Bedford Harbor: Proceedings re Alleged PCB Pollution*, the court held that the causation requirement could be met by demonstrating that a defendant's release was a contributing factor to the injury to the natural resource.[28] A more stringent standard was applied *United States v. Montrose Chemical Corp. of California*, wherein the court held that a trustee "must show that a defendant's release of a hazardous substance was the sole or substantially contributing cause of each alleged injury to natural resources."[29]

The other significant differences between liability for response costs and liability for natural resource damages are as follows. First, under section 107(a), a plaintiff may recover only "incurred" response costs. In other words, plaintiff may not recover the cost of performing a remedial action that has not yet been taken. There is no similar limitation on recovery of natural resource damages.[30]

Second, a plaintiff under section 107(a) may recover response costs caused by a release or a *threatened* release of a hazardous substance. By contrast, a plaintiff may recover natural resource damages only from *actual* releases of hazardous substances. A hazardous substance cannot injure a natural resource until it actually contacts the natural resource. Thus, by definition, a threatened release of a hazardous substance cannot cause "injury to, destruction of, or loss of natural resources."

Third, total liability under CERCLA for natural resource damages is limited to $50 million for each release of a hazardous substance.[31] There is no similar limit applicable to liability for response costs.

28. 722 F. Supp. 893 (D. Mass. 1989). *See also Coeur d'Alene Tribe*, 280 F. Supp. 2d at 1102–03 (holding that the contributing factor test is appropriate to determine causality where a release contains commingled hazardous substances).

29. No. CV-90-3122, 1991 U.S. Dist. LEXIS 10128, at *2 (C.D. Cal. Mar. 27, 1991)

30. *See, e.g.*, New York v. Gen. Elec. Co., 592 F. Supp. 291, 298 (N.D.N.Y. 1984) ("[T]here is no requirement that money must be expended by the state before it can seek to recover for damages to natural resources").

31. *See* CERCLA § 107(c)(1)(D).

## E. Statute of Limitations for NRD Claims

CERCLA provides a separate statute of limitations for NRD claims. For sites not listed on the National Priorities List (NPL), NRD claims must be brought within three years of the later of (1) discovery of the injury and its connection to the release in question or (2) the date that damage assessment regulations were promulgated. The assessment regulations became effective in 1987.[32] Thus, for all intents and purposes, the operative limitations period for non-NPL sites typically is three years from discovery of the injury and its cause. For sites included on the NPL, or where remedial action under CERCLA is "otherwise scheduled," trustees have three years after remedial action is completed to bring a claim.[33]

The last case in which a court had to consider the date of promulgation of the NRD assessment regulations in analyzing the applicable limitations period for an NRD claim was *United States v. Montrose Chemical Corp. of California*.[34] The defendants in this case moved for dismissal of the 1990 lawsuit on statute of limitations grounds. There was substantial evidence in the record that the trustee knew of the releases in the early- to mid-1980s, and thus more than three years had passed since the date of discovery of the injury. Thus, the suit would be timely only if had been filed within three years of the date of promulgation of the NRD assessment regulations. As discussed *infra*, there are two types of NRD regulations: the simplified "Type A" assessment regulation, which was promulgated on March 20, 1987, and the more complex "Type B" assessment regulation, which was promulgated on August 1, 1986. Thus, the United States' claim would be timely only if the limitations period was found to commence with the promulgation of the Type A assessment regulation. The district court ruled that the limitations period began to run with the earlier promulgation date of the Type B assessment regulation, but the Ninth Circuit reversed holding that

32. California v. Montrose Chem. Corp., 104 F.3d 1507, 1514 (9th Cir. 1997) (DOI promulgated "Type A" regulations on March 20, 1987).

33. The statute does not specify the circumstances in which a remedial action may be "otherwise scheduled" and thus trigger the three-year limitations period. One possibility, however, would be a site where EPA has issued a section 106 order.

34. 883 F. Supp. 1396 (C.D. Cal. 1995).

CERCLA required both types of NRD assessment regulations to have been promulgated before the limitations period would begin to run.[35]

An issue that has arisen in connection with NPL sites is whether a trustee can extend the statute of limitations by expanding the boundaries of an NPL site. In *United States v. ASARCO Inc.*,[36] the court held that, for NPL sites, the statute of limitations preserved claims only if they related to resources within the designated boundaries of the site. The case concerned the Bunker Hill Mining Site, which EPA designated as an NPL site in 1983. The NPL listing did not set any boundaries for the site, but background documents supporting the listing referred to the Bunker Hill facility as consisting of a 21-square-mile box. In March, 1986, the United States filed suit against several owners of the Bunker Hill facility seeking recovery of natural resource damages relating to the 1500 square mile Coeur d'Alene Basin (which encompassed the Bunker Hill facility). The defendants sought dismissal of the NRD claim under the statute of limitations because areas outside the 21-square-mile area were subject to the shorter 3-year statute of limitations period running from date of discovery, and that inclusion of the larger area was in effect a veiled attempt to expand the NPL listing.

The district court granted the defendant's motion for summary judgment based on the statute of limitations, reasoning that EPA had failed to follow procedural requirements applicable to NPL listing when it attempted to expand the designated boundaries of the site. The Ninth Circuit, however, vacated the decision, holding that only the D.C. Circuit has jurisdiction to review the expansion of an NPL site.[37]

### F. Defenses to NRD Claims

As noted *supra*, the statutory defenses applicable to federal cost recovery claims (e.g., act of God, act of war, and act of independent third party) also apply to NRD claims. (Chapter 4 provides a detailed discussion of

35. 104 F.3d 1507, 1513–18 (9th Cir. 1997).
36. 28 F. Supp. 2d 1170 (D. Idaho 1998).
37. United States v. ASARCO Inc., 214 F.3d 1104 (9th Cir. 2000).

these defenses.) In addition, CERCLA establishes a number of defenses that are specific to NRD claims. Each of these are described below.

### 1. *Exemption from Liability for Authorized Releases*

CERCLA provides an exemption from liability for natural resource damages where:

> [T]he party sought to be charged has demonstrated that the damages to natural resources complained of were specifically identified as an irreversible and irretrievable commitment of natural resources in an environmental impact statement, or other comparable analysis, and the decision to grant a permit or license authorizes such commitment of natural resources, and the facility or project was otherwise operating within the terms of its permit or license.[38]

The court in *Idaho v. Howmet Turbine Component Corp.*[39] addressed the exemption from natural resource damage liability for damages that have been identified in an environmental impact statement. In this case, Idaho brought a CERCLA suit seeking recovery of natural resource damages from the owners of a copper and cobalt mine. Idaho alleged that releases of mining wastes had polluted lakes and streams and caused fish kills. In a motion for summary judgment, the defendants argued that section 107(f) barred the NRD claims because the damage had been identified in an environmental impact statement (EIS) and an environmental assessment (EA). The court held that the EIS exception did not apply because the specific words "irreversible" and "irretrievable" were not used in the EIS.[40]

The Ninth Circuit affirmed the district court's denial of summary judgment but rejected the district court's rationale. It held that the terms "irreversible" and "irretrievable" need not be used for the EIS exception to apply, so long as the EIS contains a clear statement that the natural

38. CERCLA § 107(f)(1).

39. 627 F. Supp. 1274 (D. Idaho 1986), *aff'd in part, rev'd in part*, 814 F.2d 1376 (9th Cir. 1987).

40. *Id.* at 1277.

resource has been specifically irreversibly and irretrievably committed.[41] The court, however, affirmed the mining defendant's liability, noting that damages had been caused by the defendant's pre-EIS activities.[42]

CERCLA also exempts federally permitted releases from liability for natural resource damages:

> Recovery by any person (including the United States or any State or Indian Tribe) for response costs or *damages* resulting from a federally permitted release shall be pursuant to existing law in lieu of this section.[43]

The scope of the federally permitted release exemption was analyzed by the court in *In re Acushnet River & New Bedford Harbor: Proceedings re Alleged PCB Pollution* (Acushnet River).[44] The case arose in the context of government efforts to remediate polychlorinated biphenyl (PCB) contamination of the Acushnet River and New Bedford Harbor. The United States and the Commonwealth of Massachusetts filed suit against several companies seeking, inter alia, recovery of natural resource damages. Defendant Aerovox, Inc., filed a motion for partial summary judgment concerning, inter alia, its federally permitted release defense to liability. Aerovox argued that it had discharged PCBs in accordance with its National Pollutant Discharge Elimination System (NPDES) permit, which EPA had issued pursuant to section 402 of the Clean Water Act. The United States argued, and Aerovox did not dispute, that while some of Aerovox's PCB discharges were federally permitted, the PCBs discharged from rusting capacitors were not federally permitted.

The court stated that a defendant claiming an exemption from, or reduction of, liability based on the federally permitted release exemption bears the burden of proving which releases were federally permitted and the portion of the natural resource damages allocable to such permitted releases. The court noted that case law, including CERCLA

41. Idaho v. Hanna Mining Co., 882 F.2d 392, 396 (9th Cir. 1989).
42. *Id.* at 395.
43. CERCLA § 107(j) (emphasis added).
44. 722 F. Supp. 893.

case law, traditionally places the burden for proving divisibility of harm on the defendant. The court further observed that basic principles of statutory construction place the burden of proof on the party wishing to come within a statutory exemption. To do otherwise would require plaintiffs in CERCLA actions to prove a negative as part of their prima facie case, namely, that no relevant federal permits were issued. Thus, the court denied the defendants' motion for summary judgment ruling because there were genuine issues of material fact sufficient to survive summary judgment.

### 2. *Retroactivity*

Section 107(f) of CERCLA provides "[t]here shall be no recovery . . . where such damages and the release of a hazardous substance from which such damages resulted have occurred wholly before" the enactment of the Act. Therefore, a defendant is not liable for natural resource damages where *both* the release and all of the damages attributable to that release occurred before December 11, 1980.[45] Courts, however, have differed as to when natural resource damages "occur." The court in *United States v. Reilly Tar & Chemical Corp.*[46] held that (1) where a pre-enactment release continues after December 11, 1980, or (2) where any resultant damages from a release occurred after December 11, 1980, the defendant would be liable for all damages, including pre-enactment damages, associated with that release.[47]

The court in *Idaho v. Bunker Hill Co.*[48] held that any postenactment damages were recoverable even though they were caused by pre-enactment releases and that the defendant could not assert section 107(f) as a bar to liability for those postenactment damages.[49] Taken together, the *Reilly Tar* and *Bunker Hill* decisions stand for the proposition that responsible parties are fully liable for injuries to natural resources as

45. *See, e.g.*, Allied Towing Corp. v. Great E. Petrol. Corp., 642 F. Supp. 1339 (E.D. Va. 1986).

46. 546 F. Supp. 1100 (D. Minn. 1982).

47. *Id.* at 1120.

48. 635 F. Supp. at 675.

49. *Id.*

long as either the causal release or the resultant damage continues into the postenactment period.[50]

A different conclusion was reached by the court in *In re Acushnet River & New Bedford Harbor: Proceedings re Alleged PCB Pollution.*[51] The court barred recovery of pre-enactment damages that were divisible from postenactment damages. In this case, the United States and the Commonwealth of Massachusetts asserted a natural resource damages claim against Aerovox, Inc., for contamination of the New Bedford Harbor with PCBs. The harbor had been closed in 1978 as a result of the PCB contamination. The types of natural resource damages sought included aesthetic injury, increased harbor dredging costs, and "lost use" of the harbor by lobstermen. The issue before the court was "the scope of recovery when pre-enactment releases cause injury which results in both pre- and post-enactment damages or when releases continue post-enactment and thus do not occur 'wholly' before December 11, 1980."[52] Stated another way, the court analyzed the question of when damages to the New Bedford Harbor occurred.

First, the court interpreted the term "damages" in section 107(f) to mean the "monetary quantification of the injury done to the natural resources";[53] and the term "occurred" to mean the point at which a party "incurs expenses due to the injury to natural resources."[54] Using these definitions, the court developed the following rules regarding natural resource damages caused by pre-enactment releases. Expenses relating to the injured natural resource that are incurred pre-enactment are not recoverable; but, insofar as expenses can be shown to have been incurred postenactment, they are fully recoverable, even though the injury to the natural resource occurred pre-enactment. The court separated the

50. *See Coeur D'Alene Tribe*, 280 F. Supp. 2d at 1112–14 (damages caused by dumping of mine tailings years before enactment of CERCLA were not barred by the "wholly before" language of CERCLA § 107(f) because continued postenactment passive migration of contaminants constitute a "release" or "re-release," and moreover the "damages associated with such releases occurred post-enactment").

51. 716 F. Supp. 676 (D. Mass. 1989).

52. *Id.* at 679.

53. *Id.* at 683.

54. *Id.*

time of injury from the point at which damages occur and thus set up the possibility of latent damages.[55] Under this interpretation of section 107(f), a release that, for example, injures a natural resource in 1973 is compensable ad infinitum as long as expenses attributable to that release are incurred after December 11, 1980.[56]

Under the court's ruling, the extent to which natural resource damages are recoverable depends on whether the damages are specifically identifiable with either the pre-enactment or postenactment period. In this regard, the court distinguished between divisible and nondivisible damages. Where a pre-enactment release causes both pre-enactment and postenactment damages, if the damages are "divisible," that is, clearly identifiable as either pre-enactment or postenactment, then only the postenactment portion of the natural resource damages are recoverable; if the damages are "nondivisible," then they are recoverable in their entirety.

Applying the divisible/nondivisible principal to the damages sought by the plaintiffs, the court noted that "aesthetic injury to the Harbor may prove" to be a fully recoverable portion of the natural resource damages claim because aesthetic injury is intrinsically nondivisible.[57] On the other hand, the "lost use" of the harbor is divisible. For example, the court noted that lobstermen, forced out of their old lobstering ground when the harbor closed in 1978, had to sail farther out to ply their trade.[58] Thus, the lobstermen incurred increased expenses in the form of fuel, labor, and so on, because of the longer trips.[59]

The court stated that since these losses were measurable in daily increments, they were divisible, and any expenses identified within the pre-enactment period were damages occurring "wholly" before enactment

55. *Id.* at 686.

56. The statute of limitations under section 112(d) is also affected by the concept of latent damages. Section 112(d) precludes presentation of a claim later than three years after discovery of the loss. Since damages occur only upon incurrence of expenses, the loss is not "discovered" and the limitation period does not begin to run, until expenses are incurred by the public trustee. *Id.* at 686 n.16.

57. *Id.* at 686.

58. *Id.* at 685–86.

59. *Id.* at 686.

and, therefore, were barred from recovery by section 107(f). Thus, any daily expenses incurred from the closing of the harbor in 1978 to December 11, 1980 were not recoverable, but any expenses incurred after that date were recoverable.[60]

The court offered a second example of divisible natural resource damages: the "costs of disposing of Harbor dredge spoils."[61] If such costs were incurred pre-enactment, they would not be recoverable.

Another important issue addressed by the court in *Acushnet River* was the burden of proof in determining whether natural resource damages are recoverable under section 107(f). The court held that defendants have the burden of proving which natural resource damages are excluded from recovery. The court contrasted the bar to pre-enactment natural resource damages with the generally retroactive nature of CERCLA and held that any party seeking to come within the section 107(f) statutory exception to retroactive relief bears the burden of proving that the damages fall within that statutory exception.[62]

## III. NATURAL RESOURCE DAMAGES ASSESSMENT REGULATION

CERCLA does not specify how natural resource damages are to be quantified. Instead, section 301(c)(1) of CERCLA directs the president to promulgate regulations for the assessment of natural resource damages. Pursuant to section 301(c)(2) of CERCLA, the regulations shall establish procedures to assess both direct and indirect damage to natural resources, taking into consideration various factors including "replacement value, use value and ability of ecosystem or resource to recover." Any determinations or assessment of damages to natural resources made by a federal or state trustee in accordance with the regulations promulgated under section 301(c) of CERCLA "shall have the force and effect of a rebuttable presumption on behalf of the trustee in any administrative or judicial proceeding."[63] Consequently, the damage assessments

60. *Id.*
61. *Id.*
62. *Id.*
63. CERCLA § 107(f)(2)(C).

made according to the regulations will be presumed by the courts to be accurate, and if the responsible party wishes to challenge an assessment, it must come forward with contrary evidence to persuade a court otherwise.

Section 301(c)(2) specifies that the assessment regulations should take into consideration such factors as "replacement value" and "use value" of the damaged resource, as well as the ability of the damaged resource to recover on its own. The president must promulgate two types of regulations: Type A, simplified regulations for expeditious assessments with minimal field investigation, and Type B, more complex regulations for the individualized assessment of each affected site. DOI, the president's delegatee, promulgated Type A regulations in 1987[64] and Type B regulations in 1986.[65]

Under the DOI regulations, the trustee commences an action under CERCLA by notifying the PRPs, and thereby initiates the first of several phases in the assessment process. First, the trustee must determine whether a discharge of a hazardous substance merits a natural resource damage assessment.[66] The trustee must determine whether the damage to natural resources was of sufficient concern that the potential benefits of an assessment outweigh the costs of performing it. After determining that an assessment is warranted, the trustee must then gather information, determine whether Type A or Type B procedures apply, and prepare an assessment plan to ensure that the assessment is performed in a planned and systematic manner.

Type A regulations apply when the damage to coastal or marine environments results from a relatively minor hazardous substance release. The quantification of injury and estimation of damages in Type A assessments are determined by putting previously gathered data into a computer model, which calculates the path of contamination and its effects, quantifies the effects of the discharge, and provides an estimate of economic damages. The completed result of the modeling process can then

64. 52 Fed. Reg. 9,042 (Mar. 20, 1987) (codified at 43 C.F.R. § 11.40).
65. 53 Fed. Reg. 5,166 (Aug. 1, 1986) (codified at 43 C.F.R. § 11.60).
66. 43 C.F.R. § 11.20–.30.

be used as a measure in recovering a damage award from PRPs under a section 107(a)(4)(C) claim for natural resource damages. The computer model assessment method has been criticized, but has not been reviewed by the courts.

Type B assessment regulations provide alternative methodologies for conducting natural resource damage assessments in individual cases in which Type A procedures are not applicable.[67] Type B assessments require three steps in the assessment process beyond the assessment plan phase. First, the trustee must determine whether an injury to one or more natural resources has occurred and whether the injury resulted from a discharge of oil or release of a hazardous substance. The trustee makes this determination by analyzing samples of the resource pursuant to detailed regulatory criteria that estimate injury to surface water, groundwater, air and geological and biological resources.[68] Second, the trustee must quantify the effects of the releases by determining the extent to which natural resources have been reduced as a result of the injuries identified in the first step.[69] Third, the trustee must estimate the amount of money to be sought as compensation for injury to natural resources resulting from a discharge of oil or release of a hazardous substance.[70]

The hierarchy of valuation methodologies in both the Type A and Type B rules were challenged by a number of state governments, environmental organizations, and industry groups. In two decisions on July 14, 1989, the U.S. Court of Appeals for the D.C. Circuit remanded both sets of rules to DOI, for substantial revisions to be made in accordance with the court's directions.

With respect to the Type A rules, the court in *Colorado v. U.S. Department of the Interior* upheld DOI's decision to apply simplified procedures only to a narrow class of releases or discharges, finding that Congress intended to have Type A rules cover only minor releases, while Type B rules would apply to large or unusually damaging releases.[71] The Type

---

67. 43 C.F.R. § 11.60.
68. 43 C.F.R. § 11.60–.64.
69. 43 C.F.R. § 11.70–.73.
70. 43 C.F.R. § 11.80–.84.
71. 880 F.2d 481, 483 (D.C. Cir. 1989).

A rules were remanded, however, because they improperly called for assessment of damages exclusively on the basis of lost use value, without consideration of replacement or restoration costs.[72] Recognizing that limited data were available to create a standardized model based on restoration costs, the court stated that, nonetheless, such a limitation did not justify ignoring the clear mandate of Congress to consider replacement value in establishing damage awards.[73]

In the companion case addressing the Type B rules, *Ohio v. U.S. Department of the Interior*,[74] the court more fully discussed Congress's intent that restoration or replacement costs be considered in establishing the amount of natural resource damages. The court remanded the Type B rules, in part, because they improperly provided that the measure of damages would be the lesser of restoration or replacement costs as compared with diminution-of-use value.[75] The court found that diminution-of-use value would, in most cases, be the lesser of the two measures and would be insufficient to pay for restoration of the affected natural resource.[76] Finding that CERCLA "unambiguously mandates a distinct preference for using restoration cost as the measure of damages," the court remanded the regulations for establishment of an approach that would, at the very least, make restoration costs the most important factor in determining an appropriate damage award.[77]

The Type B rules were also found to suffer from a flaw in the hierarchy of methodologies by which lost use value was to be measured. As originally promulgated, the regulations designated market value, if available, as the exclusive measure of the value of lost use, which the court found to be an invalid limitation on the factors that Congress intended to have considered in assessing the lost use value of natural resources.[78] The court, however, upheld DOI's use of "contingent value assessments," a hypothetical market value obtained by surveying consumers when

72. *Id.* at 490–91.
73. *Id.*
74. 880 F.2d 432 (D.C. Cir. 1989).
75. *Id.* at 442.
76. *Id.* at 441.
77. *Id.* at 444.
78. *Id.* at 463–64.

no market prices are available.[79] The contingent value assessment can continue to be utilized in assessing natural resource damages, so long as the methodology is used in addition to consideration of restoration or replacement cost estimates.[80]

## IV. PREEMPTION

CERCLA does not expressly preempt state common law claims for injury to public resources. An open question is whether CERCLA preempts such claims under the theory of conflict preemption or field preemption. The question was at the heart of *State of New Mexico v. General Electric Co.*[81] The Tenth Circuit affirmed the lower court's dismissal of New Mexico's $5 billion state law claim for injury to groundwater contamination on preemption grounds. The court reasoned that because the state's claim for money damages was not limited to use for the repair, replacement, or acquisition of equivalent resources, the claim was inconsistent with the CERCLA natural resource damage provision, and therefore was preempted.

The case concerns release of petroleum hydrocarbons and volatile organic compounds from a former nuclear weapons components production and aircraft engine parts manufacturing facility located near the Rio Grande River. These releases contaminated the City of Albuquerque's municipal water supply, which led to the 1981 decommissioning of a well that provided potable water for thousands of residents. In 1983, New Mexico identified the "South Valley Superfund Site" as the highest priority site and placed it on the NPL and began its remedial investigation and feasibility study.

EPA identified GE, ACF, the U.S. Air Force (USAF), and the U.S. Department of Energy (USDOE) as PRPs in 1988. Between 1989 and 1996 GE conducted remediation at the Site under a series of EPA-issued Unilateral Administrative Orders (UAOs). In 1999, the New Mexico

79. *Id.* at 478.
80. *Id.*
81. 467 F.3d 1223 (10th Cir. 2006).

attorney general retained outside counsel and filed companion NRD lawsuits in federal and state court, seeking nearly $5 billion in damages.

The federal lawsuit alleged NRD claims under CERCLA and named as defendants, among others, the PRPs previously identified by EPA. The complaint alleged harm arising from "the presence, migration, and threat of further migration of hazardous chemical wastes and other substances from" the Site.

The state lawsuit alleged various state statutory and common law claims for NRDs, including trespass, public nuisance, and negligence. The named defendants included GE and ACF, but omitted the federal defendants. The complaint alleged exactly the same harm as the complaint in the federal lawsuit. GE and ACF removed the state lawsuit to federal court and the court, on its own motion, consolidated it with the federal lawsuit.

Between April and June, 2004, the district court issued three opinions that defined and narrowed New Mexico's claims, excluded the damages testimony of several of the state's expert witnesses, and ultimately granted the defendants' motion for summary judgment, concluding that New Mexico had failed to prove the existence of groundwater contamination that would not be remediated by EPA's selected remedy.[82]

On appeal, the Tenth Circuit largely affirmed the district court's dismissal of New Mexico's claims. The court concluded that "[a]t most, then, this is a case of conflict pre-emption . . ." and the question presented was whether any portion of New Mexico's NRD claim "stands as an obstacle to the accomplishment of congressional objectives as encompassed in CERCLA."

The court then focused on the *remedy* New Mexico sought, rather than the nature of its causes of action, in rejecting the state's claims. Because New Mexico sought an award of unrestricted money damages that it could use for purposes other than restoration, such as for attorney's fees, the court reasoned that the state's claims "would seriously

82. State of New Mexico v. Gen. Elec. Co., 322 F. Supp. 2d 1237 (D.N.M. 2004), *aff'd in part*, 467 F.3d 1223 (10th Cir. 2006).

disrupt CERCLA's principle aim of cleaning up hazardous waste," and were therefore preempted.[83]

The court concluded that CERCLA's comprehensive NRD scheme preempts *any* state remedy designed to achieve something other than the restoration, replacement or acquisition of the equivalent of an injured resource,[84] notwithstanding CERCLA's saving clauses,[85] which "undoubtedly preserve a quantum of state legislative and common law actions and remedies related to the release and cleanup of hazardous waste."[86] The court raised, but left unanswered, the question whether the potential interference of state law NRD claims like those asserted in *New Mexico* with CERCLA's "obvious preference for claim settlement" was an additional basis for preemption.[87]

The court also rejected New Mexico's argument that it was entitled to recovery of natural resource damages for residual contamination that would remain in the groundwater after implementing EPA's selected remedy. The court regarded this argument as nothing more than a challenge to an EPA-ordered remediation, which is barred from judicial review prior to the completion of the remedy, under section 113(h) of CERCLA.[88] The court, finding that the effectiveness of the remedy and the precise extent of residual injury cannot be determined until the remedy has been completed, dismissed the state's claim for want of jurisdiction, subject to renewal of those claims, if appropriate, following the completion of the remedy.[89]

Finally, the court held that New Mexico was not entitled to interim loss-of-use damages for the loss of the right to appropriate groundwater for beneficial use from the time of release until restoration. The state had argued that, absent the contamination, additional groundwater would be available for appropriation and it was entitled to damages for the deprivation of that right in its capacity as the trustee of the state's

83. 467 F.3d at 1248.
84. *Id.*
85. *Id.*
86. *Id.* at 1246 n.33.
87. *Id.* at 1248 n.37.
88. *Id.* at 1250.
89. *Id.* at 1252.

groundwater. The court found that New Mexico failed to prove that groundwater would be available for appropriation in the absence of contamination, or that any beneficial user of groundwater had actually been denied the use of groundwater as a consequence of the contamination. In the absence of proof of an actual lost use, the court concluded, New Mexico was not entitled to loss-of-use damages.

## V. CONTINGENT FEE PROSECUTION OF NRD CLAIMS

The use of private attorneys to prosecute a state's NRD claims on a contingent fee basis, as was attempted by the state of New Mexico in its $5 billion claim against GE and others, has generated significant controversy. Opponents argue that vesting power in outside counsel with a direct financial interest in maximizing monetary recoveries violates public trust and creates serious conflicts of interest for state attorneys general who may have received large campaign contributions from the same private attorneys. Thus far, legal challenges to such contingent fee arrangements (inspired by the tobacco litigation) largely have focused on state law issues regarding the separation of powers and government attorneys' obligations of neutrality. Discussed below are additional legal issues associated with the use of contingent fee arrangements in NRD litigation—namely, the issues of violation of the state's fiduciary duty and federal preemption.

### A. The Public Trust Doctrine and States' Fiduciary Duties

The theory of the public trust doctrine is that by creating a government, the people delegate specific powers and duties to the government, including the duty to manage the natural resources. The government thus acts as trustee of the land and other natural resources and manages them for the people as beneficiaries. The government, in other words, holds the land and natural resources in trust for the benefit of the public. As trustee, the government owes a fiduciary duty of loyalty to the beneficiary, the citizens of the state, and may not profit at the expense of the beneficiary. Defendants have argued that by appointing outside counsel to litigate NRD claims on a contingency fee basis, the state has breached its duty of loyalty. The argument is that contingency fees by their very nature will allow outside counsel to profit at the expense of the beneficiary because those fees would have otherwise gone to the public.

This argument was tested in the New Jersey State courts in *New Jersey Society for Environmental & Economic Development v. Campbell*,[90] in which six trade associations filed suit against commissioner of the New Jersey Department of Environmental Protection to enjoin the commissioner's use of a special counsel to prosecute NRD claims on a contingent fee basis. The trade associations alleged that the contingency fee arrangement with special counsel violated the Public Trust Doctrine, state law, and the American Bar Association's Code of Professional Responsibility. On June 18, 2004, Superior Court Judge Jack M. Sabatino upheld the contingency fee arrangement, finding the New Jersey Spill Act authorized the commissioner to retain special counsel to seek natural resource damages.

### B. Preemption as Obstacle to Contingent Fee Arrangements

As noted *supra*, section 107(f)(1) of CERCLA requires that NRD recoveries only be used "to restore, replace, or acquire the equivalent of such natural resources." Stated another way, CERCLA prohibits the use of NRD recoveries for anything other than restoration or replacement of the damaged natural resource.[91] A looming issue is whether the state statutory schemes authorizing such retention conflict with CERCLA's mandate to use NRD recoveries only for restoration or replacement of injured resources and therefore are preempted under the theory of conflict preemption. Conflict preemption occurs where it is "impossible to comply with both the federal and state laws or the state law stands as an obstacle to the accomplishment of Congress's objectives."[92]

90. N.J. Sup. Ct., Law Div. (Docket No. MER-L-34-04).

91. *See* Ohio v. U.S. Dep't of Interior, 880 F.2d at 444–45 ("By mandating the use of all damages to restore the injured resources, Congress underscored in § 107(f)(1) its paramount restorative purpose for imposing damages at all"); *see also* 132 Cong. Rec. H9561, H9612-13 (daily ed. Oct. 8, 1986) ("[T]he amendment to section 107(f) clarifies that sums recovered by trustees are to be used only to restore the natural resources without further appropriation. . . . The natural resource regime is not intended to compensate public treasuries. Nor are recoveries to be diverted for general purposes.").

92. *See, e.g.*, United States v. Denver, 100 F.3d 1509, 1512 (10th Cir. 1996).

In *New Mexico v. General Electric*, the defendants challenged the state's authority to recover contingent fees for outside counsel.[93] Defendants argued that the use of outside attorneys on a contingency fee basis violated section 107(f)(1) of CERCLA. In response, New Mexico argued that the payment of attorneys' fees in the context of an NRD claim was consistent with the stated purpose of CERCLA. The district court never directly resolved the issue, but the issue was addressed by the Tenth Circuit when the state appealed the district court's rejection of the state's NRD claim on summary judgment. The Tenth Circuit held that "CERCLA's comprehensive NRD scheme preempts any state remedy designed to achieve something other than the restoration, replacement, or acquisition of the equivalent of a contaminated natural resource."[94] The court specifically rejected the use of monetary damage awards to pay attorneys' fees:

> [I]n a case where an NRD claim is premised upon both CERCLA and state law, a portion of the recovery if earmarked for the state law claims could be used for something other (for example, *attorney fees*) than to restore or replace the injured resource. . . . Clearly, permitting the State to use an NRD recovery, which it would hold in trust, for some purpose other than to "restore, replace, or acquire the equivalent of" the injured groundwater would undercut Congress's policy objectives in enacting 42 U.S.C. § 9607(f)(1).[95]

The court made clear that its holding does not mean that *New Mexico's* nuisance and negligence theories of recovery are completely preempted, but rather that an unrestricted award of money damages is unavailable because it conflicts with CERCLA's comprehensive NRD scheme. In sum, there are strong arguments against the legality of a trustee's using outside counsel to recover natural resource damages. That said, it is far from being settled that such arrangements are improper.

93. 322 F. Supp. 2d 1237.
94. New Mexico v. Gen. Elec., 467 F.3d at 1247.
95. *Id.* at 1248 (emphasis added).

# CHAPTER 11

# CERCLA and SARA Title III Reporting Requirements

## I. INTRODUCTION

The release reporting provisions under CERCLA and the Superfund Amendments and Reauthorization Act of 1986 (SARA) are integral elements of the government's response authority under CERCLA. The reports and notifications required by these provisions inform the government of releases of hazardous substances that may require some form of response. Thus, the reporting provisions trigger EPA's response authority and, ultimately, potentially responsible party (PRP) liability for the cost of responding to such releases.

Section 103 of CERCLA requires that releases of certain hazardous substances be reported to the National Response Center (NRC) so that the federal government can respond to any risks posed by the release. Failure to report a release is punishable with daily fines of up to $25,000[1] and, if such failure to report is knowing, three years' imprisonment.[2]

When Congress enacted the Emergency Planning and Community Right-to-Know Act of 1986 (SARA Title III or EPCRA), the enforcement consequences of nonreporting doubled. Section 304 of this free-standing statute parallels the CERCLA reporting provision. Many of the releases subject to section 103 of CERCLA are also subject to section 304 of SARA.

1. CERCLA §109(a).
2. CERCLA §103(b).

Despite the significant enforcement consequences that can attach to failure to report releases, the circumstances under which releases must be reported pursuant to section 103 of CERCLA and section 304 of SARA remain unsettled. The guidance EPA has issued since the enactment of CERCLA in 1980 has not dispositively answered even the most fundamental questions related to CERCLA reporting (e.g., which chemicals are subject to the reporting requirements? When has a "release" entered the environment? Which releases are federally permitted and thus exempt from CERCLA reporting?). Those issues affect compliance not only with section 103 of CERCLA, but also section 304 of SARA.

Section II of this chapter addresses the release reporting obligations under CERCLA; section III of this chapter addresses the release reporting obligations under SARA Title III.[3]

## II. CERCLA REPORTING REQUIREMENTS

Effective December 11, 1980, CERCLA required that certain releases of "hazardous substances" in "reportable quantities" be reported to the NRC. The statutory language imposing these reporting requirements raised a number of important issues concerning when reporting of releases of hazardous substances was required. The sketchy legislative history of CERCLA sheds little light on these issues.

On May 25, 1983, EPA proposed regulations to, inter alia, "clarify" the requirements of CERCLA concerning the reporting of releases.[4] Final regulations were published in the *Federal Register* on April 4, 1985.[5] Unfortunately, the April 4, 1985 regulations left many important CERCLA reporting issues unresolved. Following is an analysis of some of the critical issues a person subject to the reporting requirements must understand.

3. In addition to the emergency release reporting obligation, SARA Title III contains several "routine" reporting requirements calculated to inform the public of the nature and amount of chemical substances handled by industrial facilities in their communities. Discussion of the routine reporting requirements is beyond the scope of this manual.

4. 48 Fed. Reg. 23,552 (May 25, 1983).

5. 50 Fed. Reg. 13,456 (Apr. 4, 1985).

### A. Reporting Requirements under CERCLA Section 103(a)

Section 103(a) of CERCLA requires:

> Any person in charge of a vessel or an offshore or an onshore facility shall, as soon as he has knowledge of any release (other than a federally permitted release) of a hazardous substance from such vessel or facility in quantities equal to or greater than those determined pursuant to Section 102 of this title, immediately notify the National Response Center established under the Clean Water Act of such release.[6]

Boiled down to its essentials, the CERCLA reporting requirement consists of five elements: (1) A release into the environment (2) of a hazardous substance (3) in a reportable quantity (RQ) (4) from a facility (5) must be reported to the NRC immediately by a person in charge. Note that except for the "reportable quantity" and "person in charge" requirements, the elements of the section 103 reporting requirement are the same as the preconditions to government response authority under section 104 of CERCLA. Thus, the explanation of the reporting requirements that follows must, to some extent, repeat the discussion of the "triggering elements" found in chapter 1.

#### *1. "Releases . . . into the Environment"*

CERCLA defines "release" broadly to include the acts of "spilling, leaking, pumping, pouring, emitting . . . leaching, dumping, or disposing *into the environment*."[7] As defined, a substance must be released into the environment in a reportable quantity before notification of the release is required. The point at which a substance leaves the facility and enters the environment—once thought to be defined by the facility "fence-line"—was clarified by EPA in the preamble to the April 4, 1985, regulations. EPA states that hazardous substances may be released "into the environment" even if they remain on plant grounds:

6. CERCLA § 103(a).
7. CERCLA § 101(22) (emphasis added).

> Examples of such releases are spills from tanks or valves onto concrete pads or into lined ditches open to the outside air, releases from pipes into open lagoons or ponds, or any other discharges that are not wholly contained within buildings or structures. *Such a release, if it occurs in a reportable quantity (e.g., evaporation of an RQ into the air from a dike or concrete pad), must be reported under CERCLA.*[8]

This passage makes clear that reportable quantities of hazardous substances spilled directly onto the natural ground surface or emitted directly into the air within the fenceline of a facility are reportable. A necessary inference of the underscored language, however, is that where a hazardous substances is placed in an impervious engineered structure such as a "dike or concrete pad," a release only enters the environment if the hazardous substance volatilizes into the air or otherwise leaves the confines of the engineered structure.

This inference was at the center of a 1991 legal challenge to a contrary view that EPA had advanced in the preamble to an obscure regulation (adjusting the RQ for radionuclides).[9] In the preamble to that regulation, EPA stated that the mere placement of a hazardous substance into an "unenclosed containment structure" (which EPA defines to include tanks, surface impoundments, and any other "holding device that has an open side with the contained materials directly exposed to the ambient environment") constitutes a release subject to the CERCLA reporting requirements.[10] EPA based this position on its view that mere exposure of a substance to the environment constitutes a release. This interpretation, however, conflicts with the language of the statute since it requires reporting when chemicals are still within the facility, which is defined to include impoundments and lagoons.[11]

8. 50 Fed. Reg. 13,456, 13,462 (Apr. 4, 1985) (emphasis added).
9. 54 Fed. Reg. 22,524, 22,526 (May 24, 1989).
10. *Id.*
11. CERCLA § 101(9).

Several industry groups challenged EPA's "unenclosed containment" interpretation in *The Fertilizer Institute v. EPA*.[12] The D.C. Circuit vacated EPA's position. The court recited the statutory definitions of "release," "facility," and "environment," and opined that the import of these provisions, when read together, is that "CERCLA requires reporting whenever a hazardous substance is released from a facility, such as a tank or a lagoon, into the environment, such as the ground or the air."[13] The court disagreed with EPA, reasoning:

> EPA's interpretation of CERCLA's reporting requirement cannot be reconciled with CERCLA's express terms. Rather than defining a release as the movement of a substance from a facility *into the environment* (as does CERCLA), the EPA defines a release as the placement of a substance into a facility that is *exposed to the environment*. For the EPA's interpretation to come within CERCLA's meaning, the exposure of a substance to the environment must be equivalent to the release of that substance into the environment. But this is not so. A simple example exposes the flaw in the EPA's analysis: a company could place a non-volatile substance into an open-air storage container and the consequences of the open-air storage would be no different from those that would occur if the company had placed the substance to a closed container. Nonetheless, under the EPA's interpretation, the company would have to report the transfer of the substance to the container notwithstanding the non-volatile substance's inability to escape into the air.[14]

Thus, based on the decision, a facility need not report every time it places an RQ or a hazardous substance into an impoundment.

12. 935 F.2d 1303 (D.C. Cir. 1991).
13. *Id.* at 1309.
14. *Id.* at 1309–10.

### 2. *"Hazardous Substance"*

To be reportable under CERCLA, a release must be of a "hazardous substance," as defined by section 101(14) of CERCLA. Section 101(14) of CERCLA defines "hazardous substance" by incorporating lists of hazardous substances developed under a number of other federal environmental statutes, specifically, sections 311 and 307 of the Clean Water Act of 1977, and section 112 of the Clean Air Act. EPA has created a master list of these hazardous substances, which is codified at 40 C.F.R. § 302.4. In addition to these specifically designated hazardous substances, section 101(14)(C) of CERCLA defines hazardous substances to include any waste that EPA has specifically listed as a hazardous waste,[15] and any waste that exhibits one or more of the characteristics of hazardous wastes (ignitability, corrosivity, reactivity, and toxicity) established by EPA under the Resource Conservation and Recovery Act (RCRA).[16]

### 3. *"Reportable Quantity"*

Section 103(a) of CERCLA requires reporting to the NRC only when a "reportable quantity" of a hazardous substance has been released into the environment. The RQ for all substances specifically designated as hazardous under CERCLA is set forth in 40 C.F.R. table 302.4. Hazardous wastes exhibiting the characteristic of ignitability, corrosivity, or reactivity have an RQ of 100 pounds. Hazardous wastes exhibiting the characteristic of toxicity have the reportable quantities listed in table 302.4 for the contaminant on which the characteristic of toxicity is based.[17]

---

15. EPA has promulgated several lists of hazardous wastes. The so-called "F-List" identifies wastes from common manufacturing and industrial processes, such as solvents that have been used in cleaning or degreasing operations. The F-List of hazardous wastes is set forth at 40 C.F.R. § 261.31. The K-List of hazardous wastes is a source-specific list, which identifies the waste by reference to specific industries, such as petroleum refining or pesticide manufacturing. The K-List is set forth at 40 C.F.R. § 261.32. The P-List and U-List are lists of commercial chemical products which are deemed hazardous waste when discarded. *See* 40 C.F.R. § 261.33.

16. The four hazardous waste characteristics are set forth at 40 C.F.R. § 261.21–.24. Exemptions from categories of hazardous substances are discussed in chapter 1.

17. *See* 40 C.F.R. § 302.5(b).

The April 4, 1985 regulations establish 24 hours as the period over which the release of a reportable quantity of a hazardous substance must be aggregated to determine whether reporting is required. Thus, if a reportable quantity of a hazardous substance is released within *any* 24-hour period, the release should be reported to the NRC.[18] The preamble to the April 4, 1985 regulations emphasizes that 24 hours is the period over which a reportable quantity must be released before the release is reportable; 24 hours is *not* the amount of time provided for the reporting of a release of a reportable quantity of a hazardous substance.[19] As soon as a person in charge of a facility has knowledge that a reportable quantity of a hazardous substance has been released during any twenty-four-hour period, the person must *immediately* report the release to the NRC.

#### *4. "Facility"*

To be reportable under section 103(a) of CERCLA, a release of a hazardous substance must be from a "vessel" or a facility. CERCLA defines the term "facility" to include

> (A) any building, structure, installation, equipment, pipe or pipeline (including any pipe into a sewer or publicly owned treatment works), well, pit, pond, lagoon, impoundment, ditch, landfill, storage container, motor vehicle, rolling stock, or aircraft, or
> (B) any site or area where a hazardous substance has been deposited, stored, disposed of, or placed, or otherwise come to be located; but does not include any consumer product in consumer use or any vessel.[20]

Although facility is broadly defined under CERCLA, it is not free from ambiguity in the release reporting context. For example, should releases from different pipes within a plant be regarded as different

18. *See* 40 C.F.R. § 302.6(a).
19. *See* 50 Fed. Reg. 13,456, 13,460 (Apr. 4, 1985).
20. CERCLA § 101(9).

facilities, and, if not, should such releases be aggregated for purposes of determining if an RQ of a hazardous substance has been released? For reporting purposes, EPA interprets the term "facility" to apply to the "entire contiguous plant or installation . . . rather than each vent, pipe, or piece of equipment at the plant."[21] Moreover, EPA appears to favor aggregation of concurrent releases within a facility.[22] The latter point is far from a settled issue, but prudence in release reporting is the better course.

### 5. *"Person in Charge"*

Section 103(a) of CERCLA places the responsibility for reporting releases of hazardous substances on the person in charge of the facility from which the release occurs. The term "person in charge" is not defined by CERCLA. In theory, potential civil and criminal liability for nonreporting under CERCLA extends from the chairman of the board of a corporate parent to the individual operator on duty at the time a release occurs.

The broad range of individuals who may be regarded as "in charge" was made clear by a Second Circuit decision in *United States v. Carr*, upholding the criminal conviction of a civilian maintenance foreman at an army installation for failure to report the release of a reportable quantity of a hazardous substance.[23] The maintenance foreman had directed a work crew to dispose of cans of waste paint in small pits on a firing range. The court ruled that the person in charge includes lower-level supervisory employees such as the maintenance foreman:

> Congress intended the reporting requirements of CERCLA's Section 103 to reach a person—even if of relatively low rank—who, because he was in charge of a facility, was in a position to detect, prevent, and abate a release of hazardous substances.[24]

21. 50 Fed. Reg. 13,456, 13,459 (Apr. 4, 1985).
22. *See* 53 Fed. Reg. 12,868, 12,875 (Apr. 19, 1988).
23. 880 F.2d 1550 (2d Cir. 1989).
24. *Id.* at 1554.

The court rejected a more restrictive reading of "in charge," that it believed would frustrate congressional intent by exempting from the section 103 reporting requirement a large class of persons who are "uniquely qualified to assume the burden."[25]

Given the wide range of individuals who may be criminally liable for nonreporting in any particular situation, it is advisable that both corporations and individual plants develop procedures to ensure that releases of hazardous substances are identified and that informed decisions are made on the need to report those releases to the NRC.

### B. Reporting of Continuous Releases under CERCLA Section 103(f)(2)

Section 103(f)(2) of CERCLA modifies the general notification requirement of section 103 for certain releases. Releases that are "continuous" and "stable in quantity and rate" may be reported annually, rather than immediately, provided that notification has been given (1) under sections 103(a) and 103(b) "for a period sufficient to establish the continuity, quantity, and regularity" of the release; or (2) under section 103(c), which requires notification to the federal government of the existence of certain facilities that are or have been used for treatment, storage, or disposal of hazardous waste. Immediate notification of the continuous release must still be provided pursuant to section 103(a) whenever there is a "statistically significant increase" in the quantity of the hazardous substance released.

EPA has adopted regulations that implement section 103(f)(2).[26] Under EPA regulations, a release will be considered continuous not only when it occurs "without interruption or abatement" but also if the release is "routine, anticipated and intermittent and incidental to normal operations or treatment processes."[27] EPA defines "stable in quantity and rate"

25. *Id.*; *see also* United States v. Buckley, 934 F.2d 84, 86 (6th Cir. 1991) ("The reporting requirements of CERCLA apply to any person—even if of relatively low rank—who, because he was in charge of a facility, was in a position to detect, prevent, and abate a release of hazardous substances. The defendant's control over a facility need not be sole or exclusive. He need not be the only one in charge.")

26. 55 Fed. Reg. 30,166 (July 24, 1990).

27. 40 C.F.R. § 302.8(b).

to mean "predictable and regular in amount and rate of emission."[28] EPA has also issued a guidance document to help facilities understand the substance of the rule.[29]

To satisfy the continuous release reporting requirement, one must make (1) an initial telephone call and written notification, (2) a one-time follow-up report on the anniversary of the initial written notification, and (3) reports of statistically significant increases. Following is a summary of each of these requirements.

***Initial Notification.*** The initial notification required under the final rule consists of: "(1) a minimum of one telephone report to the NRC," the state emergency response commission (SERC) "of any State likely to be affected by the release," and the local emergency planning committee (LEPC) "for any local area likely to be affected by the release;" and (2) within 30 days of the initial telephone notification, submission of an initial written notification to the appropriate EPA Regional Office for the geographical region in which the facility or vessel is located, and the appropriate SERC and LEPC."[30]

During the initial telephone notification to the NRC, SERC, and LEPC, the person in charge must state that the release is continuous and that the facility is reporting the release under section 103(f)(2). In addition, the person in charge must identify (1) the name and location of the reporting facility, and (2) the identity of the hazardous substance(s) being released.[31]

The initial written report must include the following information:

1. The name of the facility or vessel; the location, including longitude and latitude; the case number assigned by the NRC or EPA; the Dun & Bradstreet number of a facility, if available; the port

28. 40 C.F.R. § 302.8(a)–(b).

29. *See* Reporting Requirements for Continuous Releases of Hazardous Substances, A Guide for Facilities on Compliance, EPA 540-R-97-047. The guidance is posted on the EPA website at http://www.epa.gov/superfund/policy/release/faciliti.htm.

30. 55 Fed. Reg. at 30,166, 30,171 (July 24, 1990).

31. 40 C.F.R. 302.8(d)(3).

of registration of the vessel (if applicable); and the name and telephone number of the person in charge of the facility or vessel.

2. The population density within a one-mile radius of the facility or vessel, described in terms of the following ranges: 0–50 persons, 51–100 persons, 101–500 persons, 501–1,000 persons, more than 1,000 persons.
3. The identity and location of sensitive populations and ecosystems within a one-mile radius of the facility or vessel (e.g., elementary schools, hospitals, retirement communities, or wetlands).
4. The name/identity of the hazardous substance(s), including the Chemical Abstract Service Registry Number for each substance. If the release is a mixture, the components of the mixture and their approximate concentrations and quantities, by weight.
5. The upper and lower bounds of the normal range of the release (in pounds or kilograms) over the previous year.[32]
6. The source(s) of the release (e.g., valves, pump seals, storage tank vents, stacks). If the source is a stack, the stack height (in feet or meters) must be provided.
7. The frequency of the release and the fraction of the release from each release source and the specific period over which it occurs.
8. A brief statement describing the basis for stating that the release is continuous and stable in quantity and rate.
9. An estimate of the total amount of the hazardous substance released in the previous year (in pounds or kilograms).
10. The environmental media affected by the release.[33]
11. A signed statement that the hazardous substance release(s) described is (are) continuous and stable in quantity and rate under

32. Following are examples of routine, anticipated, intermittent releases which EPA included in the preamble of the final rule: batch processes; shutdowns for scheduled maintenance; catalyst changeouts; loading and unloading; decompression of pressure vessels; drawing off of liquid at regular intervals during a production process; venting that occurs each time a storage tank is filled; removal of fly ash or dust from pollution control devices; and cyclical operations at production facilities. 55 Fed. Reg. at 30,170.

33. As noted below, a facility must include a statement of the basis for concluding that the release is continuous and stable in quantity and rate in the initial written and annual follow-up reports.

the definitions in 40 C.F.R. § 302.8(a) and that all reported information is accurate and current to the best knowledge of the person in charge.[34]

Recognizing that much of the information required in the initial written report may also be reported under section 313 of EPCRA, EPA is allowing facilities to utilize section 313 reports (Form R) to satisfy the initial written and follow-up reporting requirements, provided the section 313 report is supplemented with additional information required under the continuous release rule.[35] Specifically, the section 313 report must be supplemented with items (b) (population density), (c) (sensitive populations), (e) (normal range of release), (g) (frequency of release), and (h) (basis for reporting under section 103(f)(2)) from the above list.[36]

***One-Time Follow-Up Notification.*** Within 30 days of the first anniversary of the initial written notification, the reporting facility must evaluate the reported release and submit a one-time written follow-up report to the appropriate EPA regional office. The purpose of this report is to verify or update the information submitted in the initial written report. Accordingly, the follow-up report contains the same type of information required in the initial report, but reflects the evaluation of the release for the period following the initial report.[37]

Following submission of the follow-up report, the reporting facility must annually evaluate the continuous release and must document each annual evaluation. Although EPA need not be notified of each annual evaluation, records of the evaluation must be maintained for a period of one year for possible EPA inspection.[38] Changes in reported information discovered in the annual evaluation must be reported as discussed below.

***Statistically Significant Increase.*** Where there is a statistically significant release in the amount of hazardous substance being released, the release must be reported immediately pursuant to section 103(a).

34. *See* 40 C.F.R. § 302.8(e).
35. 55 Fed. Reg. at 30,180.
36. *See* 40 C.F.R. § 302.8(j).
37. 55 Fed. Reg. at 30,174.
38. *Id.* at 30,168.

EPA defines a statistically significant increase as any release "above the upper bound of the reported normal range of the release."[39] The specification of the "normal range" must be made in the initial written notification to the EPA region, SERC, and LEPC.[40] There is no requirement to install monitoring equipment to establish the normal range. Rather, it may be based on professional judgment, the operating history of the facility, experience with operating equipment and processes, and any existing data.[41] If a release reported as a statistically significant increase is likely to occur again in the future, the reporting facility may change the normal range of the release to include what was reported as a statistically significant increase.[42]

***Changes in Reported Information.*** In addition to statistically significant increases, other changes in the information submitted in written initial and follow-up notifications must be reported. Changes in the composition or source of a continuous release must be treated as a new release and, if appropriate, qualified as "continuous" by reporting to the NRC. Once qualified, such new releases are subject to the initial, follow-up, and statistically significant increase reporting described above.[43]

All other changes in reported information concerning a continuous release must be reported to the appropriate EPA region by letter within 30 days of the determination that previously submitted information is no longer valid. Notification must include the reason for the change, and the basis for stating that the release is continuous and stable under the changed conditions.[44]

## C. CERCLA Reporting Requirement under Section 103(c)

Section 103(c) of CERCLA requires the owner or operator of a facility where hazardous substances are treated, stored, or disposed of, and persons who transport hazardous substances to such facilities, to notify

39. 40 C.F.R. § 302.8(b).
40. 55 Fed. Reg. at 30,178.
41. 55 Fed. Reg. at 30,179.
42. *Id.*
43. 40 C.F.R. § 302.8(g)(1).
44. 40 C.F.R. § 302.8(g)(3).

EPA within 180 days after the enactment of CERCLA of (1) the existence of those facilities, (2) the amount and type of hazardous substances found there, and (3) any known, suspected, or likely releases of such substances from the facility. Notification is not required for facilities that are permitted or accorded interim status under RCRA. A knowing failure to report under this section is a criminal violation that is punishable by a fine of up to $10,000 and/or imprisonment for up to one year. The legislative history, although somewhat vague, indicates that section 103(c) was intended to create an *initial* inventory of sites for potential remediation under CERCLA, and did not intend to require reporting indefinitely.[45] EPA has likewise indicated that section 103(c) was intended to create an initial inventory of hazardous waste sites.[46]

The key question under section 103(c) is whether the reporting obligation extends beyond June 9, 1981, the date on which the original 180-day reporting period expired. While it is not clear that EPA agrees, the better interpretation is that the section 103(c) reporting obligation is a one-time obligation. The legislative history makes clear that the provision was intended to create "a *baseline* of information."[47]

### D. Exemptions from Release Reporting Requirements

Certain types of releases are exempt from the section 103 reporting requirements. However, the scope of several of these exemptions is ambiguous and requires careful analysis.

#### 1. *Federally Permitted Releases*

Section 103(a) of CERCLA exempts "federally permitted release(s)" from reporting requirements. Section 101(10) defines "federally permitted release(s)" to include releases subject to permits and other controls under nine federal environmental laws. This definition, which consumes scores

45. *See* S. REP. NO. 96-848, at 60 (1980) ("Thus the reporting provisions of this Act and existing law will serve as a baseline of information for compilation of priority lists.").

46. *See* EPA's Interim Interpretive Notice, and Policy Statement, 46 Fed. Reg. 22,144 (Apr. 15, 1981) ("The purpose of the notification, therefore, is to assist EPA in developing an inventory of hazardous waste sites and to help facilitate the development of priorities for attention and possible response action concerning those sites under the Act.").

47. S. REP. NO. 96-848, at 60 (1980) (emphasis added).

of lines in the statute, is rife with issues of interpretation but, boiled down to its essentials, the following releases are federally permitted:

1. Releases from point sources in compliance with National Pollutant Discharge Elimination System (NPDES) permits (which are issued pursuant to section 402 of the Clean Water Act);
2. Releases subject to permits issued under section 404 of the Clean Water Act ("dredge and fill" permits);
3. Releases from facilities with final RCRA permits;
4. Releases in compliance with a permit issued pursuant to section 102 of the Marine Protection, Research and Sanctuaries Act of 1972 (i.e., ocean dumping permits);
5. Injection of fluids in compliance with an underground injection permit issued pursuant to the Safe Drinking Water Act;
6. Air emissions subject to permit or control regulation under section 110, 111, or 112 of the Clean Air Act;
7. Injection of fluids or other materials for resource recovery;
8. Discharge into publicly owned treatment works (POTWs) in compliance with generic and industry-specific pretreatment standards; and
9. Releases of nuclear materials ("source, special nuclear, or by-product material") in compliance with a permit issued under the Atomic Energy Act of 1954.[48]

In 1988, EPA issued a proposed definition of federally permitted releases,[49] but the agency never promulgated a final rule. Instead, more than a decade later, EPA published its "Guidance on the CERCLA Section 101(10)(H) Federally Permitted Release Definition for Certain Air Emissions."[50] The guidance is in the form of questions and answers, and

48. *See* CERCLA § 101(10) (A)–(K).

49. 53 Fed. Reg. 27,268 (July 19, 1988).

50. 67 Fed. Reg. 18,899 (Apr. 17, 2002), *amended*, 67 Fed. Reg. 19,750 (Apr. 23, 2002). This guidance replaced interim guidance that EPA had issued in 1999, which various groups had challenged in *National Association of Manufacturers v. U.S. Environmental Protection Agency*, No. 00-1111 (D.C. Cir. 2000).

thus arguably would be limited to the premises of the questions. That said, a few principles do emerge from the guidance, specifically:

- A release of a hazardous substance "in excess of the CAA [Clean Air Act] control regulation" is not federally permitted.[51]
- Where a release of volatile organic compounds (VOCs) and particulate matter (PM) complies "with your federally enforceable CAA permit limit and or control regulation . . ." and those limits or controls include conditions that, when viewed together control the release of a constituent hazardous substance, such a release would likely qualify as a federally permitted release."[52]
- A release of NO and $NO_2$ that is in compliance with $NO_x$ permit limits and control regulations issued under CAA Title I is federally permitted.[53]
- CAA permit limits and control regulations "usually do not control or limit unanticipated releases such as accidents or malfunctions and for that reason such releases usually do not qualify" as federally permitted releases.[54]
- A release of a hazardous substance from normal operations of a "minor source" that complies with a federally enforceable threshold under a CAA section 112 as well as any federally enforceable technology requirement, operational requirement, work practice, or other control practice would generally meet the definition of a federally permitted release.[55]

Readers are advised to consult the guidance directly before concluding that a release is federally permitted.

51. 67 Fed. Reg. at 18,902.
52. *Id.*
53. *Id.*
54. *Id.*
55. *Id.* at 18,903.

### 2. *Application of Pesticides*

Under section 103(e) of CERCLA, the application of a "pesticide product registered under the Federal Insecticide, Fungicide, and Rodenticide Act" and the "handling and storage of such a pesticide product by an agricultural producer" are exempt from the CERCLA reporting requirements.[56] In the preamble to the April 4, 1985 regulations, EPA interprets this statutory language to require that the application of a pesticide be "generally in accordance with [its] purpose" to qualify for the exemption from reporting requirements.[57] The preamble goes on to state:

> While strict compliance with labeling direction is not a prerequisite for the exemption, the extent of compliance will be a critical factor in determining whether the application was generally in accordance with the pesticide's purpose. EPA does not consider a spill of a pesticide to be . . . an "application" of the pesticide . . . in accordance with the pesticide's purpose. Consequently, such spills must be reported.[58]

### 3. *The "Petroleum Exclusion"*

The statutory definition of "hazardous substance" expressly excludes

> petroleum, including crude oil or any fraction thereof which is not otherwise specifically listed or designated as a hazardous substance under subparagraphs (A) through (F) of this paragraph, and the term does not include natural gas, natural gas liquids, liquefied natural gas, or synthetic gas usable for fuel (or mixtures of natural gas and such synthetic gas).[59]

Because these materials are excluded from the statutory definition of hazardous substances, the release of such materials need not be reported under section 103(a) of CERCLA. The preamble to the April 4, 1985

56. CERCLA § 103(e).
57. 50 Fed. Reg. at 13,464.
58. *Id.*
59. CERCLA § 101(14).

regulations recognizes this "petroleum exclusion." However, the preamble states that EPA does not consider materials otherwise within the petroleum exclusion, but to which CERCLA hazardous substances *have been added*, to be within the petroleum exclusion.[60] A comprehensive discussion of the petroleum exclusion is provided in chapter 4.

### 4. *Workplace Exclusion*

Section 101(22)(A) excludes from CERCLA's definition of a release

> any release which results in exposure to persons solely within a workplace, with respect to a claim which such persons may assert against the employer of such persons.[61]

However, in the April 4, 1985, regulations EPA distinguishes the workplace exclusion as applicable only to claims compensable through workers' compensation. Accordingly, EPA concludes that the workplace exclusion is "not relevant in determining whether notification is appropriate."[62] Thus, EPA will not accept the workplace exclusion as a reason for not reporting an otherwise reportable release.

## III. EMERGENCY RELEASE NOTIFICATION UNDER SARA TITLE III

The emergency notification required by section 304 of SARA Title III covers more compounds and requires much more detailed information

60. 50 Fed. Reg. at 13,460.
61. CERCLA § 101(22) also excludes the following from the definition of a release:
    (B) [E]missions from the engine exhaust of a motor vehicle, rolling stock, aircraft, vessel, or pipeline pumping station engine,
    (C) [R]elease of source, byproduct, or special nuclear material from a nuclear incident, as those terms are defined in the Atomic Energy Act of 1954 . . . if such release is subject to requirements with respect to financial protection established by the Nuclear Regulatory Commission under section 170 of such Act . . . or, for the purposes of section [104] of this title or any other response action, any release of source byproduct, or special nuclear material from any processing site designated under section [102](a)(1) or [302](a) of [the Uranium Mill Tailings Radiation Control Act of 1978], and
    (D) [T]he normal application of fertilizer.
62. 50 Fed. Reg. at 13,462.

than the notification required to be given to the federal government under section 103 of CERCLA. These distinctions are discussed below.

## A. Releases Subject to Notification

Section 304 requires owners and operators of facilities where "a hazardous chemical is produced, used, or stored" to notify their LEPC of releases of certain categories of hazardous substances. Specifically, the owner or operator of a facility must report the release of a reportable quantity of *either* an extremely hazardous substance or a CERCLA-listed substance.[63]

The release of a reportable quantity of an extremely hazardous substance or CERCLA hazardous substance is subject to the section 304 reporting requirement only if it "occurs in a manner which would require notification under Section 103(a) of CERCLA."[64] Accordingly, the five elements of the CERCLA release reporting requirement—(1) release into the environment (2) of a hazardous substance (3) in a reportable quantity (4) from a facility (5) must be reported immediately by the person in charge—also apply to the SARA Title III release reporting requirement. For the same reason, the exemptions from reporting under section 103(a) of CERCLA also apply to release reporting under section 304 of SARA Title III. Thus, a continuous release of a CERCLA hazardous substance or extremely hazardous substance would be exempt from section 304 reporting requirements. Also, the "petroleum exclusion" and "application of pesticide" exemption apply to the SARA release reporting obligation. Additionally, releases that are "federally permitted . . . as defined in section 101(10) of CERCLA" are expressly exempt from the SARA Title III release reporting requirement.[65]

In addition to the CERCLA exemptions that apply to section 304 emergency reporting, an additional important exemption is found in section 304 itself. Releases resulting in exposure to persons solely within the

63. SARA § 304(b)(1). Reportable quantities for extremely hazardous substances are set forth in 40 C.F.R. pt. 355, app. A. Reportable quantities applicable to CERCLA hazardous substances are set forth in table 302.4 of 40 C.F.R. pt. 302.

64. SARA § 304(a)(2)(C).

65. SARA § 304(a)(2)(A).

facility site are exempt from reporting under section 304.[66] Thus, unlike the reporting requirement of section 103 of CERCLA, there is a "fence line" exemption to the release reporting requirement of section 304 of SARA Title III. As a consequence, a release of a hazardous substance that escapes past the fence line is reportable under both section 103 of CERCLA and section 304 of SARA Title III; however, if the same release of the same hazardous substance does not cross the fence line, reporting is required only under CERCLA section 103. It must be understood that the fence line rule does not apply to releases to air or groundwater, since such releases could result in exposure to persons beyond the facility fence line.

### B. Form of Notice

Immediately following a release, owners and operators of facilities must give notice (by telephone, radio, or in person) to the community emergency coordinator for any area likely to be affected and to the SERC. For releases arising out of transportation (or storage incident to such transportation) of extremely hazardous chemicals, the facility owner can give notice by dialing 911, or in the absence of a 911 emergency telephone number, by calling the operator.[67]

Under section 304(b) of SARA Title III, notice must include, to the extent known: (1) the name or identity of the chemical released; (2) whether the chemical is an extremely hazardous substance; (3) an estimate of the quantity released; (4) the time and duration of the release; (5) the environmental medium or media into which the release occurred; (6) any known or anticipated acute or chronic health risk associated with the release; (7) any precautions to be taken as a result of those risks; and (8) the name and telephone number of the individual (or individuals) to be contacted for further information.

The owner or operator of the facility also must provide a written follow-up emergency notice updating the information provided in the notification immediately following the release and describing actions taken to respond to and contain the release.[68]

---

66. SARA § 304(a)(4).
67. SARA § 304(b).
68. These notification requirements are codified at 40 C.F.R. § 355.40.

# CHAPTER 12

# Insurance

## I. INTRODUCTION

As we have seen throughout this book, CERCLA's liability scheme stacks the cards in EPA's favor in cost recovery actions and section 106(a) enforcement actions. Given the near certainty that *some* expenses will be incurred (at a minimum, the cost of defending the action, and, to the extent liability is found, the cost of remediation), potentially responsible parties (PRPs) must, as a matter of prudence, investigate whether they can recover some or all of their costs from their insurance carriers. Few, if any, insurance policies expressly provide coverage for "environmental impairment" or the equivalent. In most cases, a PRP must seek coverage under standard "comprehensive general liability" (CGL) policies that broadly provide coverage for personal injuries and property damage resulting from business activities.

An enormous body of case law has evolved concerning whether CERCLA response costs are recoverable under standard CGL policies. Unfortunately, the case law is as divided as it is voluminous, with courts reaching opposite conclusions on virtually every significant issue. The diversity of results stems in part from the fact that insurance coverage questions generally are governed by state law, rather than federal law. As a result, 50 different jurisdictions each can develop their own law on environmental coverage issues, so a PRP must be sure to consider coverage issues in light of applicable state law and precedent.

This chapter identifies and describes the critical issues that arise most frequently in Superfund insurance litigation, and will provide a general assessment of the state of the case law on each issue. Specifically, this chapter addresses typical CERCLA insurance issues in the order that a PRP seeking insurance coverage is likely to confront them: (1) when to provide initial notice of a claim to the insurer; (2) when and if the insurer's "duty to defend" the insured arises; (3) whether the costs of either performing remedial work or reimbursing the government or third parties for such work are recoverable under CGL policies; (4) when environmental damage occurs for purposes of determining which insurance policies provide coverage; (5) the scope of the "pollution exclusion"; (6) the scope of the "owned property" exclusion; and (7) whether an insured who obtains reimbursement of response costs from an insurer is barred from seeking contribution from PRPs.

## II. NOTICE TO THE INSURER

Most CGL insurance policies require, as a condition to coverage, that the insured promptly notify the insurer of any "occurrences," as defined in the policy, and of any claims or suits under the policy. Timely notice allows the insurer to investigate the occurrence or the basis for the suit. The earlier the notice is given, the better the insurer is able to protect its interests as against the insured or the insured's interests as against the adversary. Though a policy may specify a time period, for example, 90 days, usually the notice must be given "as soon as possible" or within a similarly worded reasonable time.

Timeliness of notice in the Superfund context has been addressed by several courts. Those courts based their determination of timeliness on one of two rules of law. In some states, including Delaware and Pennsylvania, the courts have allowed insurers to deny coverage only if the lack of, or delay in, notice actually prejudiced the insurer.[1] An insurer must prove that the delay in notice actually impaired its ability and opportunity to protect its and/or the insured's interests. In the absence of prejudice, delay in notification, no matter how long, is insufficient to negate the insurer's duties in the policy.

1. *See, e.g.*, Triangle Publ'ns, Inc. v. Liberty Mut. Ins. Co., 703 F. Supp. 367 (E.D. Pa. 1989); *see also* New Castle Cnty. v. Hartford Accident & Indem. Co., 685 F. Supp. 1321 (D. Del. 1988); Cessna Aircraft Co. v. Hartford Acc. & Indem. Co., 900 F. Supp. 1489, 1516–18 (D. Kan. 1995).

Under the other rule of law, which is used in New York, timely notice is a condition precedent to all of the duties of the insurer.[2] If the court determines that notice was untimely or that delay was unreasonable, the insurer can deny coverage because the insured did not comply with a condition precedent to the insurer's duties.[3]

Although the timeliness of the notice is not usually a crucial issue, an insured should endeavor to notify the insurer as soon as the insured becomes aware of an occurrence or a potential claim. It is better to err on the side of caution by giving early notice. Early notice will avoid one potential defense to coverage, will give the insurer the maximum time to investigate and decide on coverage, and will show that the insured is serious about expecting a defense and indemnification.[4]

## III. DUTY TO DEFEND

Under the standard CGL policy, the insurer has a duty to defend the insured against suits or similar claims and to indemnify the insured for any damages that result from such suits. With respect to the duty to defend, a frequently litigated issue in CERCLA insurance coverage cases is *when* the "duty to defend" arises. In the typical context (such as a personal injury lawsuit), the duty to defend arises when a suit or similar action is filed. Unfortunately, CERCLA cases do not follow the typical pattern of civil litigation; it can be late in the process before a suit is filed (if one is ever filed), and most or all critical decisions have been made before that time.

In a CERCLA case, the duty to defend could arise at a number of different points prior to the actual filing of suit against the insured:

2. *See* New York v. AMRO Realty Corp., 697 F. Supp. 99 (N.D.N.Y. 1988), *aff'd in part, rev'd in part*, 936 F.2d 1420 (2d 1991).

3. *Id.* at 105 (court determined that four years' delay in notice to insurer was unreasonable and, accordingly, denied coverage).

4. At least one court has examined the insured's general practice of notifying its insurer to determine if the notice in a particular instance was timely. Diamond Shamrock Chems. Co. v. Aetna Cas. & Sur. Co., No. C-3939-84, slip op. (N.J. Super. Ct. Apr. 12, 1989). In this case involving dioxin contamination, the court denied coverage based on the conduct of the insured. In all other contexts, the insured had promptly notified its insurers, but its notice was much later regarding its claim for coverage for the harm resulting from dioxin contamination. *Id.* at 34.

(1) when the insured first receives an information request; (2) when the insured receives a "general notice" or "special notice" letter; or (3) when the insured receives a section 106(a) order.

### A. Section 104(e) Information Requests

Many courts have held that the receipt of a CERCLA section 104(e) information request does not trigger a duty to defend. For example, in *Arco Industries Corp. v. The Travelers Insurance Co.*,[5] Arco had received an information request and sued for a declaratory judgment that its insurers had a duty to defend it in the EPA proceedings and that this duty included paying for the costs of responding to the information request. The court reasoned that no suit existed because EPA had not taken any action, such as file a complaint or issue a section 106(a) cleanup order, that could be regarded as adversarial: "[N]othing has occurred to date that resembles either a lawsuit or a formal administrative adjudication in any meaningful fashion."[6]

### B. General and Special Notice Letters

To initiate settlement negotiations regarding the cleanup of Superfund sites, EPA may issue a "general notice" letter to the PRPs it has identified. The general notice letter typically states that the recipient is potentially responsible for remediation of the site and solicits a commitment by the recipient to perform such remediation. In addition, or as an alternative, to a general notice letter, EPA may issue a special notice letter. Special notice letters, authorized under section 122(e) of CERCLA, inform the PRPs that they have 60 days to make a "good faith offer" to undertake specified response actions at a Superfund site, and that if no offers are submitted, EPA will undertake the required action itself and sue the

5. 730 F. Supp. 59 (W.D. Mich. 1989).

6. *Id.* at 68. Accord Aetna Cas. & Sur. Co. v. Gulf Res. & Chem. Corp., 709 F. Supp. 958, 960 (D. Idaho 1989), *rev'd on other grounds*, 948 F.2d 1507 (8th Cir. 1991); Detrex Chem. Indus., Inc. v. Emp'rs Ins. of Wausau, 681 F. Supp. 438, 446 (N.D. Ohio 1987); Technicon Elecs. Corp. v. Am. Home Assurance Co., 141 A.D.2d 124, (N.Y. 1988), *aff'd*, 74 N.Y.2d 66 (1989). *But see* Ash Grove Cement Co. v. Liberty Mut. Ins. Co., No. 09-239-KI, 2010 WL 3894119 (D. Or. Sept. 30, 2010) (insured's receipt of a CERCLA § 104(e) information request triggers an insurer's duty to defend under Oregon law).

PRPs for reimbursement.[7] Most courts have held that a policyholder's receipt of a PRP notice from EPA is the functional equivalent of a suit.[8] A sizeable minority, however, disagrees.[9]

## C. Section 106 Orders

Several courts have held that receipt of a section 106 administrative order triggers the insurer's duty to defend. In *Detrex Chemical Industries, Inc. v. Employers Insurance of Wausau*, the court stated, in dicta,

7. See chapter 5.

8. *See, e.g.*, Anderson Bros. v. St. Paul Fire & Marine Ins. Co., 729 F.3d 923 (9th Cir. 2013); Land O'Lakes, Inc. v. Emp'rs Mut. Ins. Co. of Wis., 846 F. Supp. 2d 1007 (D. Minn. 2012), *appeal dismissed*, 528 F. App'x 685 (8th Cir. 2013); Travelers Cas. & Sur. Co. v. Ala. Gas Corp., 117 So. 3d 695 (Ala. 2012); U.S. Fire Ins. Co. v. Estate of Campbell, No. 11-00006LEK-KSC, 2011 WL 693456 (D. Haw. Dec. 30, 2011); Dutton-Lainson Co. v. Cont'l Ins. Co., 778 N.W.2d 433 (Neb. 2010); Emhart Indus. v. Home Ins. Co., C.A. No. 02-53S, 2006 WL 2583759 (D.R.I. Sept. 7, 2006); R.T. Vanderbilt Co. v. Cont'l Cas. Co., 870 A.2d 1048 (Conn. 2005); Aetna Cas. & Sur. Co. v. Commonwealth, 179 S.W.3d 830 (Ky. 2005); Johnson Controls, Inc. v. Emp'rs Ins. of Wausau, 665 N.W.2d 257 (Wis. 2003); Compass Ins. Co. v. City of Littleton, 984 P.2d 606 (Colo. 1999); Travelers Indem. Co. v. Summit Corp. of Am., 715 N.E.2d 926 (Ind. Ct. App. 1999); Boardman Petrol. v. Federated Mut. Ins., 926 F. Supp. 1566 (S.D. Ga. 1995) *rev'd on other grounds*, 150 F.3d 1327 (11th Cir. 1998); Quaker State Minit-Lube v. Fireman's Fund Ins. Co., 868 F. Supp. 1278 (D. Utah 1994); Hardwick Recycling & Salvage, Inc. v. Acadia Ins. Co., 869 A.2d 82 (Vt. 2004); Hutchinson Oil Co. v. Federated Serv. Ins. Co., 851 F. Supp. 1546 (D. Wyo. 1994); Mich. Millers Mut. Ins. Co. v. Bronson Plating Co., 519 N.W.2d 864 (Mich. 1994), *overruled on other grounds*, Wilkie v. Auto-Owners Ins. Co., 664 N.W.2d 776 (Mich. 2003); Coakley v. Me. Bonding & Cas. Co., 618 A.2d 777 (N.H. 1992); Aetna Cas. & Sur. Co. v. Pintlar Corp., 948 F.2d 1507 (9th Cir. 1991); A.Y. McDonald Indus., Inc. v. Ins. Co. of N. Am., 475 N.W.2d 607 (Iowa 1991); Hazen Paper Co. v. U.S. Fid. & Guar. Co., 555 N.E.2d 576 (Mass. 1990); Boeing Co. v. Aetna Cas. & Sur. Co., No. C86-352WD, 1990 U.S. Dist. LEXIS 20231 (W.D. Wash. Apr. 17, 1990); C.D. Spangler Constr. Co. v. Indus. Crankshaft & Eng'g Co., 388 S.E.2d 557 (N.C. 1990); Harpool Bros., Inc. v. Trinity Universal Ins. Co., No. 89-C-702 (Kan. Dist. Ct. Nov. 17, 1989); Avondale Indus., Inc. v. Travelers Indem. Co., 887 F.2d 1200 (2d Cir. 1989).

9. *See, e.g.*, Foster-Gardner, Inc. v. Nat'l Union Fire Ins. Co., 959 P.2d 265 (Cal. 1998); Racal-Datacom Inc. v. Ins. Co. of N. Am., No. 95-1749, 1998 WL 1173527 (S.D. Fla. Feb. 11, 1998); Lapham-Hickey Steel Corp. v. Prot. Mut. Ins. Co., 655 N.E.2d 842 (Ill. 1995); Joslyn Mfg. Co. v. Liberty Mut. Ins. Co., 836 F. Supp. 1273 (W.D. La. 1993), *aff'd*, 30 F.3d 630 (5th Cir. 1994), *cert. denied*, 513 U.S. 1127 (1995); Harleysville Mut. Ins. Co. v. Sussex Cnty., 831 F. Supp. 1111 (D. Del. 1993), *aff'd*, 46 F.3d 1116 (3d Cir. 1994); Bausch & Lomb, Inc. v. Utica Mut. Ins. Co., 625 A.2d 1021 (Md. 1993); Aetna Cas. & Sur. Co. v. Gen. Dynamics Corp., 968 F.2d 707 (8th Cir. 1992); Prof'l Rental Inc. v. Shelby Ins. Co., 599 N.E.2d 423 (Ohio Ct. App. 1991); F.W. Scheper v. U.S. Fid. & Guar., No. 90-CP-07-879 (S.C. Ct. C.P. Oct. 29, 1991); Patrons Oxford Mut. Ins. Co. v. Marois, 573 A.2d 16 (Me. 1990).

that both a section 106(a) order and a section 107 cost recovery suit by EPA constituted "suits" that trigger the duty to defend.[10]

## IV. COVERAGE

Ultimately, the question in most CERCLA insurance litigation is whether coverage exists under a particular CGL policy. In general, an insured seeking indemnification for response costs under CGL policies can expect the insurer to raise four defenses: (1) response costs are not "damages because of property damage" (and thus, not covered); (2) the "occurrence" leading to the response costs did not take place within the period covered by a particular policy; (3) recovery of response costs is expressly precluded by the "pollution exclusion" clause; and (4) recovery is barred by a separate exclusion applicable to the "property of the insured." An insured must overcome *all* of these defenses to recover under a CGL policy.

### A. "As Damages"

Under a standard CGL policy, the insurer is required to pay all amounts the insured is "legally obligated to pay *as damages* because of property damage or bodily injury." In the Superfund context, the issue is whether "response costs," which are incurred by insureds either to clean up hazardous waste sites in response to government pressure under CERCLA authority or to reimburse federal or state governments or private parties for their response costs, constitute "damages." The case law on this issue is somewhat divided, and usually turns on the court's interpretation of the word "damages." Most courts have ruled that damages encompass CERCLA cleanup costs based on the rationale that if the term damages is ambiguous it should be interpreted from a layperson's perspective and thus reflect a layperson's expectations. A small minority of courts have adopted a narrower definition, including as damages legal remedies, but not "equitable" or "restitutionary" relief. The sharply divided reasoning

10. 681 F. Supp. at 446; *see also Technicon Elecs. Corp.*, 141 A.D.2d 124 (citing *Detrex*, 681 F. Supp. at 446).

is illustrated by two 1989 California state appellate court decisions, which reached opposite conclusions.

***"Damages" Cover Response Costs.*** The decision in *Aerojet-General Corp. v. Superior Court*[11] illustrates the reasoning of the majority of courts. The court interpreted "damages" from the standpoint of a reasonable layperson's expectations and found coverage for some of the insured's response costs.[12] In reaching this conclusion, the court applied three rules of interpretation for insurance policies: (1) terms not specifically defined in the policy are to be given common sense, laypersons' definitions; (2) ambiguities are to be construed against the drafter (i.e., the insurer); and (3) ambiguities are to be resolved in favor of finding policy coverage.[13]

The *Aerojet* court reasoned that a layperson would expect Superfund site remediation to be covered because "to the reasonable insured, [the] obligation is no less a legal sanction than that of a monetary judgment."[14] Responding to the insurers' arguments that response costs are equitable or restitutionary in nature, the court stated that an insured "would be surprised indeed to learn that coverage depended on whether the proceeding employed to obtain recompense was defined as 'legal' or 'equitable,'"[15] and further, that "[r]esponse costs, whether incurred directly by the polluter or paid to reimburse the government for its cleanup efforts, do not fit easily into [the definition of restitution—return of something wrongfully received]."[16] The court also stated that "the cost of restoration of property to its undamaged condition is one measure of compensatory property damages."[17]

***"Damages" Do Not Include Response Costs.*** The decision in *Continental Insurance Cos. v. Northeastern Pharmaceutical & Chemical Co. (NEPACCO)*[18] illustrates the minority view. Contrary to the majority

11. 211 Cal. App. 3d 216 (1989).
12. *Id.*
13. *Id.* at 224–25.
14. *Id.* at 228
15. *Id.* (citation omitted).
16. *Id.* at 231.
17. *Id.*
18. 842 F.2d 977 (8th Cir. 1988), *cert. denied*, 488 U.S. 821 (1988).

view that the term "damages" is ambiguous, the court in this case stated that in the insurance context, the term is not ambiguous:

> Although not defined in the CGL policies, "[t]he word 'damages' is not ambiguous in the insurance context. Black letter insurance law holds that claims for equitable relief are not claims for 'damages' under liability insurance contracts."[19]

The court viewed federal and state actions for recovery of response costs under CERCLA as a form of equitable relief: "These lawsuits are essentially equitable actions for monetary relief in the form of restitution or reimbursement of costs."[20] Based on that conclusion, the court denied coverage to the insured.

In summary, the case law on the "as damages" issue is split and still evolving. The vast majority of states to address the issue have held that response costs under CERCLA qualify as damages covered by CGL policies: Alabama,[21] Alaska,[22] California,[23] Colorado,[24] Delaware,[25]

19. *Id.* at 986 (citation omitted).

20. *Id.* at 987.

21. Ala. Plating Co. v. United States Fid. & Guar. Co., 690 So. 2d 331, 336 (Ala. 1996).

22. Mapco Alaska Petrol., Inc. v. Cent. Nat'l Ins. Co., 795 F. Supp. 941, 952 (D. Alaska 1991).

23. Powerline Oil Co. v. Superior Ct., 27 Cal. 4th 377 (Cal. 2005); Certain Underwriters at Lloyd's of London v. Superior Ct., 24 Cal. 4th 945 (Cal. 2001); AIU Ins. Co. v. Superior Ct., 51 Cal. 3d 807 (Cal. 1990).

24. Compass Ins. Co. v. City of Littleton, 984 P.2d 606, 622–23 (Colo. 1999); Metro Wastewater Reclamation Dist. v. Cont'l Cas. Co., 834 F. Supp. 1254, 1259 (D. Colo. 1993).

25. New Castle Cnty. v. Hartford Accident & Indem. Co., 933 F.2d 1162, 1191 (3d Cir. 1991), *rev'd on other grounds*, 970 F.2d 1267 (3d Cir. 1992); Nat'l Union Fire Ins. Co. v. Rhone-Poulenc Basic Chems. Co., No. 87C-SE-11, 1992 WL 22690, at *12 (Del. Super. Ct. Jan. 16, 1992, *aff'd*, 616 A.2d 1192 (Del. 1992).

Florida,[26] Georgia,[27] Idaho,[28] Illinois,[29] Indiana,[30] Iowa,[31] Kansas,[32] Kentucky,[33] Maryland,[34] Massachusetts,[35] Michigan,[36] Minnesota,[37] Missouri,[38] Nebraska,[39] New Hampshire,[40] New Jersey,[41] New York,[42] North Carolina,[43] Ohio,[44] Oklahoma,[45] Oregon,[46] Pennsylvania,[47] South Carolina,[48] Texas,[49] Utah,[50] Vermont,[51] Virginia,[52] Washington,[53]

26. United States v. Pepper's Steel & Alloys, Inc., 823 F. Supp. 1574, 1583–84 (S.D. Fla. 1993), *aff'd in part and rev'd in part*, 87 F.3d 1329 (11th Cir. 1996).

27. Atl. Wood Indus. v. Lumbermen's Underwriting Alliance, 396 S.E.2d 541, 543 (Ga. App. 1990), *cert. denied*, 498 U.S. 1085 (1991).

28. Aetna Cas. & Sur. Co. v. Pintlar Corp., 948 F.2d 1507, 1513 (9th Cir. 1991).

29. Outboard Mar. Corp. v. Liberty Mut. Ins. Co., 607 N.E.2d 1204, 1216 (Ill. 1992).

30. Travelers Indem. Co. v. Summit Corp. of Am., 715 N.E.2d 926, 9234 (Ind. App. 1999); Hartford Accident & Indem. Co. v. Dana Corp., 690 N.E.2d 285 (Ind. App. 1997).

31. A.Y. McDonald Indus., Inc. v. Ins. Co. of N. Am., 475 N.W.2d 607, 624 (Iowa 1991).

32. Cessna Aircraft Co. v. Hartford Accident & Indem. Co., 900 F. Supp. 1489, 1497 (D. Kan. 1995).

33. Aetna Cas. & Sur. Co. v. Commonwealth of Ky., 179 S.W.3d 830 (Ky. 2005).

34. Bausch & Lomb Inc. v. Utica Mut. Ins. Co., 625 A.2d 1021, 1032 (Md. 1993).

35. Hazen Paper Co. v. United States Fid. & Guar. Co., 555 N.E.2d 576, 583 (Mass. 1990).

36. United States Aviex Co. v. Travelers Ins. Co., 336 N.W.2d 838, 843 (Mich. Ct. App. 1983).

37. Minn. Mining & Mfg. Co. v. Travelers Indem. Co., 457 N.W.2d 175, 184 (Minn. 1990).

38. Farmland Indus., Inc. v. Republic Ins. Co., 941 S.W.2d 505, 512 (Mo. 1997).

39. Lindsay Mfg. Co. v. Hartford Accident & Indem. Co., 118 F.3d 1263, 1270 (8th Cir. 1997).

40. Coakley v. Me. Bonding & Cas. Co., 618 A.2d 777, 787 (N.H. 1992).

41. Morton Int'l, Inc. v. Gen. Accident Ins. Co. of Am., 629 A.2d 831, 846–47 (N.J. 1993).

42. *Avondale Indus.*, 887 F.2d at 1207.

43. C.D. Spangler Constr. Co. v. Indus. Crankshaft & Eng'g Co., 388 S.E.2d 557, 569 (N.C. 1990).

44. Sanborn Plastics Corp. v. St. Paul Fire & Marine Ins. Co., 616 N.E.2d 988, 998 (Ohio Ct. App. 1993).

45. Nat'l Indem. Co. v. U.S. Pollution Control, Inc., 717 F. Supp. 765, 767 (W.D. Okla. 1989).

46. St. Paul Fire & Marine Ins. Co. v. McCormick & Baxter Creosoting Co., 870 P.2d 260, 267 (Or. Ct. App. 1994).

47. Fed. Ins. Co. v. Susquehanna Broad. Co., 727 F. Supp. 169, 174 (M.D. Pa. 1989).

48. Helena Chem. Co. v. Allianz Underwriters Ins. Co., 594 S.E.2d 455 (S.C. 2004).

49. Snydergeneral Corp. v. Century Indem. Co., 113 F.3d 536, 539 (5th Cir. 1997).

50. Quaker State Minit-Lube, Inc. v. Fireman's Fund Ins. Co., 868 F. Supp. 1278, 1293–94 (D. Utah 1994), *aff'd*, 52 F.3d 1522 (10th Cir. 1995).

51. Gerrish Corp. v. Universal Underwriters Ins. Co., 947 F.2d 1023, 1030 (2d Cir. 1991).

52. Morrow Corp. v. Harleysville Mut. Ins. Co., 101 F. Supp. 2d 422 (E.D. Va. 2000).

53. Boeing Co. v. Aetna Cas. & Sur. Co., 784 P.2d 507, 516 (Wash. 1990).

Wisconsin,[54] and Wyoming.[55] The courts in each of these cases gave the term "damages" its plain and ordinary meaning.

The minority view—that CGL policies do not to cover response costs—has been embraced by Arkansas,[56] Connecticut,[57] and Maine.[58]

## B. "Occurrence"

Standard CGL policies require that the property or bodily damage be "caused by an occurrence" for coverage to apply.[59] "Occurrence" is generally defined as "an accident, including continuous or repeated exposure to conditions, injury or property damage neither expected nor intended from the standpoint of the insured."[60] A major issue is whether the occurrence occurred during the term of the policy, triggering coverage.

In the Superfund context, this "trigger" determination is complicated by the fact that the environmental contaminants that injure property often do so gradually and without detection. In addressing trigger issues, courts have devised at least four theories to analyze the various scenarios in which the occurrence issue arises: (1) exposure; (2) injury in fact; (3) manifestation; and (4) continuous trigger.

Under the exposure theory, the occurrence occurs when the hazardous wastes are initially released into the environment. For example, in *NEPACCO*,[61] drums of liquid wastes were dumped into a shallow trench and buried. The court ruled that "[M]issouri would adopt the

54. Johnson Controls, Inc. v. Emp'rs Ins. Co. of Wausau, 665 N.W.2d 257, 264 (Wis. 2003).

55. Compass Ins. Co. v. Cravens, Dargan & Co., 748 P.2d 724, 728 (Wyo. 1988).

56. Parker Solvents Co. v. Royal Ins. Cos. of Am., 950 F.2d 571, 571 (8th Cir. 1991); Grisham v. Commercial Union Ins. Co., 951 F.2d 872, 875 (8th Cir. 1991).

57. Linemaster Switch Corp. v. Aetna Life & Cas. Corp., No. CV-91-03964325, 1995 WL 462270, at *23 (Conn. Super. Ct. July 25, 1995).

58. Patrons Oxford Mut. Ins. Co. v. Marois, 573 A.2d 16, 19 (Me. 1990).

59. A typical CGL policy provides insurance coverage for "all sums which the insured shall become legally obligated to pay as damages because of . . . bodily injury or . . . property damage to which this insurance applies, caused by an occurrence, and the company shall have the right and duty to defend any suit against the insured seeking damages on account of such bodily injury or property damage." *See* Mraz v. Canadian Universal Ins. Co., 804 F.2d 1325, 1327 (4th Cir. 1986).

60. *NEPACCO*, 842 F.2d at 980 (quoting one of the insurance policies at issue).

61. 842 F.2d at 981.

'exposure' theory," and that under the exposure theory, the release of chemicals into the environment (in this case, the burying of the drums) constituted exposure that triggered coverage.[62]

Under the injury-in-fact theory, coverage is triggered at the time the "injury" took place. The injury need not have been discovered to have existed.[63] The court in *Triangle Publications* denied the insurer's motion for summary judgment because the insured could potentially prove that injury took place during the policy period, even though the injury was discovered after the policy expired.[64]

Under the manifestation theory, "the occurrence is deemed to take place when the injuries first manifest themselves."[65] The *Mraz* court adopted the manifestation theory because "[d]etermining exactly when damage begins can be difficult, if not impossible."[66] Applying the manifestation theory to disposal in a hypothetical landfill, the occurrence is neither the disposal into the landfill nor the leaching of contaminants from the landfill, but instead is the *detection* of the contamination, for example, in nearby wells.

The fourth theory, the "continuous trigger," addresses the situation in which "new injury is, in fact continuously occurring."[67] The *New Castle* court was confronted with the issue of whether injury to adjacent property caused by the gradual leaching of pollutants from the county-run Tybouts Corner Landfill was a covered occurrence. Given the gradual and continuous nature of the contamination at issue, the court decided that "injury" could be described as an "injurious process" and any policy in effect during the process would be triggered:

62. *Id.* at 984; *see also* Fireman's Fund Ins. Cos. v. Ex-Cell-O Corp., 685 F. Supp. 621, 626 (E.D. Mich. 1987).

63. *See Triangle Publ'ns*, 703 F. Supp. at 370–71; *see also AMRO Realty Corp.*, 697 F. Supp. at 102 n.5.

64. 703 F. Supp. at 370–71.

65. *Mraz*, 804 F.2d at 1328.

66. *Id.*

67. *See* New Castle Cnty. v. Cont'l Cas. Co., 625 F. Supp. 800, 812 (D. Del. 1989), *aff'd in part and rev'd in part*, 933 F.2d 1162 (3d Cir. 1991).

> The continuing damage caused by the gradual spread of leachate from Tybouts supports the analogy to the development of an insidious disease. The application of the continuous trigger to the present case is, therefore, appropriate. Thus, every policy, from the start of the injurious process, is triggered.[68]

From the standpoint of the insured, the continuous trigger theory is by far the best because it effectively finds coverage under all policies in effect from the date of the initial disposal or release, through the dates when leakage began and was discovered, right up until the time the leakage or damage was remediated. As a practical matter, this can mean that several years or even decades of policies can be "stacked" to collectively provide substantial coverage.

### C. Pollution Exclusion

Most CGL policies issued since 1972 contain a clause excluding from coverage damage caused by pollution. This "pollution exclusion" clause does *not* apply if the release that caused the damages was "sudden and accidental." A standard clause is as follows:

> [This insurance does not apply] to bodily injury or property damage arising out of the discharge, dispersal, release or escape of smoke, vapors, soot, fumes, acid, alkalis, toxic chemicals, liquids or gases, waste materials or other irritants, contaminants or pollutants into or upon land, the atmosphere or any water course or body of water, but this exclusion does *not apply* if such discharge, dispersal, release or escape is *sudden and accidental.*[69]

Obviously, the critical issue in determining the scope and application of this exclusion is the interpretation of its exception for "sudden and accidental" releases. There is relatively little dispute about "accidental":

68. *Id.* at 813.

69. Claussen v. Aetna Cas. & Sur. Co., 865 F.2d 1217, 1218–19 (11th Cir. 1989) (emphasis added) (alteration in original) (quoting the insurance policy at issue).

it means "unexpected and unintended." Most of the controversy concerns the term "sudden."

Because CGL policies do not define the critical term "sudden," courts interpret the term using general contract principles. Insureds invariably argue that the term is ambiguous and, thus, must be construed against the drafter (the insurer) and in such a way to promote coverage. The insurers typically argue that the term is clear and unambiguous: "sudden" has a temporal meaning, as in "brief." Because most Superfund cases and claims involve gradual pollution (such as unseen leakage from tanks, impoundments, etc.), courts adopting the insurers' interpretation of the term "sudden" usually find no policy coverage.[70] Such courts reason that the pollution exclusion clause is not ambiguous and that the term "sudden" must be given its ordinary meaning; for these courts, the ordinary meaning of "sudden" includes a temporal element.[71] The courts also point out that to leave the temporal element out of "sudden" would render the word surplusage because the other meanings of "sudden"—specifically "unexpected" and "unintended"—are included in the definition of "accidental."[72]

Courts adopting the insured's interpretation conclude that "sudden" *is* an ambiguous term, capable of more than one reasonable interpretation, and that because the insurers drafted the policy and had an opportunity to define the term, it should be interpreted in the way most favorable to the insured.[73] The courts that adopt this approach generally do not

70. *See* U.S. Fid. & Guar. Co. v. Murray Ohio Mfg. Co., 693 F. Supp. 617, 620–21 (M.D. Tenn. 1988), *aff'd*, 875 F.2d 868 (6th Cir. 1989); *see also* Borden, Inc. v. Affiliated FM Ins. Co., 682 F. Supp. 927, 929–30 (S.D. Ohio 1987), *aff'd*, 865 F.2d 1267 (1989); Centennial Ins. Co. v. Lumbermens Mut. Cas. Co., 677 F. Supp. 342, 347–49 (E.D. Pa. 1987); *Technicon Elecs. Corp.*, 141 A.D.2d 124. *See also* Fireman's Fund Ins. Cos. v. Ex-Cell-O Corp., 702 F. Supp. 1317, 1322–27 (E.D. Mich. 1988) (suit for declaratory judgment as to meaning of terms).

71. *See Murray Ohio Mfg. Co.*, 693 F. Supp. at 621; *Borden, Inc.*, 682 F. Supp. at 930; *Centennial Ins. Co.*, 677 F. Supp. at 349.

72. *Fireman's Fund*, 702 F. Supp. at 1326; *Centennial*, 677 F. Supp. at 348–49.

73. *See* Ross Dev. Corp. v. Fireman's Fund Ins. Co., 910 F. Supp. 2d 828 (D.S.C. 2012), *aff'd*, 526 F. App'x 299 (4th Cir. 2013); *see also* Claussen v. Aetna Cas. & Sur. Co., 888 F.2d 747 (11th Cir. 1989); New Castle Cnty. v. Hartford Accident & Indem. Co., 673 F. Supp. 1359, 1362–64 (D. Del. 1987); Broadwell Realty Servs., Inc. v. Fid. & Cas. Co., 528 A.2d 76, 83 (N.J. Super. App. 1987); CPS Chem. Co. v. Cont'l Ins. Co., 487 A.2d 1256, 1270–71 (N.J. Super. App. 1984) ("It may well be that the drafters of this [pollution exclusion] clause believed that 'sudden and accidental' connoted a sense of a dramatic catastrophe, limited in

attribute a temporal aspect to "sudden," which would bar recovery for gradual contamination. Accordingly, these courts find that pollution that was unexpected and unintended is covered by the policy. These courts are not troubled by the "surplusage" argument, and hold that "sudden and accidental" means the same as "accidental," that is, neither expected nor intended from the standpoint of the insured.

As noted above, courts uniformly define "accidental" as "unexpected and unintended." Thus, a court can deny coverage if the insured in some way knew or should have known that its activities were causing environmental contamination. The court in *Borden, Inc. v. Affiliated FM Insurance Co.* pointed out that the insured's regular business practice had caused the contamination; the company "regularly deposited radioactive wastes on its property as part of its production of phosphoric acid."[74] Accordingly, the *Borden* court held that the pollution was "precisely the type of activity which the pollution exclusion was drafted to preclude."[75]

Similar to the case law interpreting damages, the case law on the pollution exclusion clause is so deeply split among the various jurisdictions that coverage often depends on where the contaminated site is located, where the insured's headquarters is located, where the insurance policy was written, and, ultimately, where coverage litigation is filed. The ambiguity, however, was largely eliminated in policies written after 1987, when the insurance industry revised the standard CGL language to exclude coverage for *all* pollution damage claims ("absolute" pollution exclusion).[76]

duration and immediate in its consequences, but it cannot fairly be said that this was unambiguously expressed.").

74. 682 F. Supp. at 930.

75. *Id.* Accord Great Lakes Container Corp. v. Nat'l Union Fire Ins. Co., 727 F.2d 30, 33–34 (1st Cir. 1984); Fisher & Porter Co. v. Liberty Mut. Ins. Co., 656 F. Supp. 132, 140 (E.D. Pa. 1986); Am. States Ins. Co. v. Md. Cas. Co., 587 F. Supp. 1549, 1552–54 (E.D. Mich. 1984).

76. *See* Jonathan C. Averback, *Comparing the Old and the New Pollution Exclusion Clauses in General Liability Insurance Policies: New Language—Same Results?* 14 B.C. ENVTL. AFFAIRS L. REV. 4 (Aug. 1, 1987).

### D. Owned Property Exclusion

CGL policies typically exclude coverage for damage to property owned or controlled by the insured. The purpose of this exclusion is to ensure that an insured does not have an incentive to be careless when insurance exists.[77] Insureds whose Superfund liability is not based on property ownership—generators or transporters—need not be concerned with this clause, given that the contaminated property is owned or controlled by a third party. However, even in cases in which the insured owns or controls the contaminated site, some courts have found that the owned property exclusion does not bar coverage for cleanup costs on the ground that other property also will or may be affected by the contamination.

The courts differ as to which costs—those for cleaning up the insured's property, restoring the property of third parties, or preventing further contamination of either type of property—are covered. Some courts have found coverage for expenditures affecting third parties' property, either cleaning it up or preventing future contamination, while excluding from coverage costs related to the insured's land.[78] Other courts have found broader coverage. For example, the court in *Intel Corp. v. Hartford Accident & Indemnity Co.*[79] stated that the exclusion did not bar coverage for costs incurred to restore, or prevent damage to, the groundwater beneath the insured's property (which the court found to be the property of the state, not of the insured), and declared that all costs consistent with the EPA's consent decree, including those related to the insured's site, and to expenses covering third parties' property, were recoverable from the insurance company.[80] The court also ruled

77. *See* United States v. Conservation Chem. Co., 653 F. Supp. 152, 199–200 (W.D. Mo. 1986).

78. *See Conservation Chem. Co.*, 653 F. Supp. at 200; *see also* Allstate Ins. Co. v. Quinn Constr. Co., 713 F. Supp. 35, 41 (D. Mass. 1989) (insureds should not be given an incentive to wait for the contamination to spread to others' property so cleanup of their property will then be covered by insurance), *opinion vacated*, 784 F. Supp. 927 (D. Mass. 1990).

79. 692 F. Supp. 1171 (N.D. Cal. 1988), *aff'd in part and rev'd in part*, 952 F.2d 1551 (9th Cir. 1991).

80. 692 F. Supp. at 1194–95.

that preconsent decree costs that were consistent with the consent decree were covered.[81]

Accordingly, even if an insured has an "owned property exclusion" in the policy, recovery for cleanup costs may not be barred. The insured may be able to argue that other property has been or is in danger of being contaminated. This claim is particularly viable when the property is owned by, or considered the property of, the state.[82]

## V. DIRECT ACTION AGAINST INSURERS

Section 108(c) of CERCLA allows certain entities to proceed directly against persons who provide evidence of financial responsibility for a facility:

> In the case of a release or threatened release from a facility, any claim authorized by section 9607 or 9611 of this title may be asserted directly against any guarantor providing evidence of financial responsibility for such facility under subsection (b) of this section, if the person liable under section 9607 of this title is in bankruptcy reorganization, or arrangement pursuant to the Federal Bankruptcy Code, or if, with reasonable diligence, jurisdiction in the Federal courts cannot be obtained over a person liable under section 9607 of this title who is likely to be solvent at the time of judgment. In the case of any action pursuant to this paragraph, the guarantor shall be entitled to invoke all rights and defenses which would have been available to the person liable under section 9607 of this title if any action had been brought against such person by the claimant and all rights and defenses which would have been available to the guarantor if an action had been brought against the guarantor by such person.[83]

81. *See* Twp. of Gloucester v. Md. Cas. Co., 668 F. Supp. 394, 400 (D.N.J. 1987) (coverage found where "costs [which] are inextricably linked to damage claims of a third party [even though] the expenditures will be made in part to repair property owned by the insured.").

82. *See Intel Corp.*, 692 F.2d 1171; *see also* Riehl v. Travelers Ins. Co., 772 F.2d 19, 22 (3d Cir. 1985); *Fireman's Fund Ins. Cos.*, 685 F. Supp. at 625.

83. CERCLA § 108(c)(1).

This section authorizes direct action under CERCLA against guarantors. The question is whether insurers are subject to liability as guarantors. "Guarantor" is defined in CERCLA as "any person, other than the owner or operator, who provides evidence of financial responsibility for an owner or operator under this chapter." Until the passage of the Superfund Amendments and Reauthorization Act of 1986 (SARA),[84] Congress did not clarify further the application of section 108 as to the persons included within the definition of guarantor with respect to facilities. In SARA, however, Congress amended section 108(b)(2) of CERCLA to provide that "[f]inancial responsibility may be established by . . . insurance, guarantee, surety bond, . . . or qualification as a self-insurer."[85] Based upon this straightforward reading of section 108, the definition of guarantor would appear to include insurers. The court in *Port Allen Marine Services, Inc. v. Chotin*,[86] however, considered the liability of guarantors under section 108(c) and held that insurers are not guarantors.

This case involved a contribution action brought by Chotin and Port Allen Marine against Chotin's insurer under CERCLA and various Louisiana state law theories. In granting the insurer's motion to dismiss for failure to state a claim, the court relied on the fact that the list of persons liable under section 107 of CERCLA did not include insurers:

> To read [section 107(a) to include insurers] would require the Court to ignore and disregard the clear language of § 9607(a). If insurers are to be sued directly under § 9607(a), the Congress, and not this Court, must amend the § 9607(a) to provide this right.[87]

84. Pub. L. No. 99-499, 100 Stat. 1613 (1986) (codified as amended at 42 U.S.C. §§ 9607–9675 (1988 & Supp. 1991)).

85. Pub. L. No. 99-499, 100 Stat. 1672 (1986) (codified as amended at 42 U.S.C. § 9608 (1988)). Section 108(a) of CERCLA, which is the provision that enables EPA to promulgate financial assurance requirements for vessels, includes identical language. 42 U.S.C. § 9608(a)(1).

86. 765 F. Supp. 887 (M.D. La. 1991).

87. *Id.* at 889 (footnote omitted). *See also* City of New Orleans v. Kernan, 933 F. Supp. 565 (E.D. La. 1996); Allied Corp. v. Envtl. Purification Advancement Corp., 848 F. Supp. 67 (M.D. La. 1994).

As a practical matter, most insurance companies began writing absolute pollution exclusion policies after passage of CERCLA in 1980; thus instances where an insurer might be said to have financial responsibility for releases from a facility are likely to be few and far between.

## VI. SUBROGATION CLAIMS BY INSURERS

Section 112(c)(2) of CERCLA provides a right of subrogation to parties that have paid response costs incurred by another party:

> Any person . . . who pays compensation pursuant to this chapter to any claimant for damages or costs resulting from a release of a hazardous substance shall be subrogated to all rights, claims, and causes of action for such damages and costs of removal that the claimant has under this chapter or any other law.[88]

Pursuant to this provision, insurance companies that paid claims by insureds relating to liability for response costs under CERCLA have sought recovery from other PRPs. Based on case law, however, parties asserting subrogation claims under section 112(c) must meet rather specific requirements.

The Ninth Circuit decision in *Chubb Custom Insurance Co. v. Space Systems/Loral, Inc.*[89] is instructive. In this case, an insurance company sought to recover money that it paid to an insured from other PRPs in a CERCLA case under a subrogation theory. At the direction of the California Regional Water Quality Control Board, a property owner, Taube Koret, had performed remediation on property it owned. The contamination was caused by adjacent property owners, as well as former owners of Taube Koret's property. Taube Koret presented a $2.4 million claim to its insurer, Chubb. After paying the claim, Chubb filed a claim against other PRPs under both section 107 and section 112(c)(2) of CERCLA. The district dismissed both claims. The Ninth Circuit affirmed the district court ruling. According to the court, Chubb lacked standing to

88. CERCLA § 112(c)(2).
89. 710 F.3d 946 (9th Cir. 2013), *cert. denied*, 134 S. Ct. 906 (2014).

assert a claim under section 107 because Chubb did not incur response costs within the meaning of CERCLA. As for the claim under section 112(c)(2), the court held that a condition of a subrogation claim is that one must have paid damages to a claimant. Here, Taube Koret had never presented a demand to the other PRPs, so it was not a "claimant" within the meaning of section 112(c)(2) of CERCLA. Although this outcome might seem harsh, the Ninth Circuit noted that it is not unduly burdensome to require the insured to pursue a claim before an insurer's subrogation rights vest: "Insurance companies may write their policies in a way as to require reasonable cooperation from their insureds."[90] Moreover, the court reasoned that requiring a policyholder to make a demand before the insurer's subrogation rights ripen prevents the policy holder from obtaining an improper double recovery from its insurer and other PRPs.

Since the decision in *Chubb*, there have been several other cases presenting a similar issue. In *Chartis Specialty Co. v. United States*,[91] the Chartis alleged that it had reimbursed the policyholder for response costs incurred by the policyholder, and had also paid in excess of $30 million in response costs directly. The district court dismissed Chartis's claim under section 107 finding that payments made pursuant to a contractual obligation in an insurance policy do not qualify as response costs under CERCLA, even when those payments are made directly to contractors instead of the policyholder. As for the section 112(c)(2) claim, however, the court permitted Chartis to proceed with that claim, finding that Chartis had alleged in its complaint that the insured had made a sufficient prior demand against a PRP (namely, the United States) to qualify it as a claimant within the meaning of section 112(c)(2).

In *Carolina Casualty Ins. Co. v. Oahu Air Conditioning Service, Inc.*,[92] the plaintiff insurance company sued two defendants it alleged were PRPs for a hazardous waste spill. The dispute stemmed from the delivery of a trailer containing hazardous refrigerant waste oil to a disposal site

90. *Id.* at 970.
91. No. C-13-1527, 2013 WL 3803334 (N.D. Cal. July 19, 2013).
92. 994 F. Supp. 2d 1082 (E.D. Cal. 2014).

in Sacramento. Shortly after delivery, the trailer caught fire and released hazardous vapors and contaminated water runoff into the environment.

As a result of the incident, the City and County of Sacramento, among others, issued notices of violation against the company that transported the trailer and demanded payment for cleanup of the hazardous waste spill. Pursuant to the transportation company's insurance policy, the plaintiff insurance company settled and paid several of the claims.

The plaintiff insurance company then sought reimbursement from two PRPs that it alleged supplied, packed, and transported the refrigerant waste oil. Specifically, the insurance company sought: (1) subrogation pursuant to section 112(c) of CERCLA; (2) contribution pursuant to section 113(f) of CERCLA; and (3) contribution, indemnity, allocation, and apportionment under California state laws. The defendants sought to dismiss the CERCLA section 112(c) claims because the insurer had not alleged in its complaint that its insured had made a formal claim against the PRPs.

In evaluating the sufficiency of the insurer's complaint, the court first discussed the availability of subrogation under CERCLA section 112(c)(2), which provides:

> Any person . . . who pays compensation pursuant to this chapter to any claimant for damages or costs resulting from a release of a hazardous substance shall be subrogated to all rights, claims, and causes of action for such damages and costs of removal that the claimant has under this chapter or any other law.

The court noted that the Ninth Circuit decision in *Chubb* restricts the availability of CERCLA section 112(c) subrogation to situations where the insured has already made a claim against the PRP against whom its insurer seeks subrogation.

The court, however, distinguished the facts of *Chubb* by noting that the present plaintiff has alleged in its complaint that its insured "made demand and claim upon and against" the PRPs. The PRP defendants argued that this allegation was insufficient as a matter of law and that the insurer's complaint needed more detailed allegations confirming that a formal claim had been submitted against them. The court disagreed and noted that neither *Chubb* nor CERCLA section 112(c) specified the precise form that a written claim for reimbursement must take. Accordingly, the court denied the defendants' motion to dismiss plaintiff's CERCLA section 112(c) claim.

## VII. COLLATERAL SOURCE RULE IN CERCLA CONTEXT

The "collateral source" rule is a common law rule that "allows a plaintiff to seek full recovery [from a defendant] even though an independent source has compensated the plaintiff in full or in part for the loss."[93]

As the court in *Quinones v. Pennsylvania General Insurance Co.* stated, the collateral source rule "evolved around the commonsense notion that a tortfeasor ought not be excused because the victim was compensated by another source, often by insurance."[94] A question that has arisen in the CERCLA context is whether a party that incurs response costs and is reimbursed by an insurance company can nevertheless seek contribution from PRPs who caused or contributed to the contamination. In other words, is "double recovery" permissible under CERCLA? The courts have said "no."

The decision in *Friedland v. TIC—The Industrial Co.*[95] is instructive. The plaintiff in this case, Robert Friedland, was the former director and president of Summitville Consolidated Mining Company, which had operated a gold mine in Colorado before abandoning it and declaring bankruptcy. The gold mining process involved applying cyanide solution to gold-bearing ore to extract the gold from the ore matrix. The cyanide solution had contaminated groundwater under the property.

EPA cleaned up the site, and filed a CERCLA cost recovery action against Friedland to recover its cleanup costs. Friedland in turn filed a third-party contribution action against 12 defendants. Friedland settled the EPA cost recovery case for approximately $20 million. In the third-party action, Friedland disclosed in discovery that he had already

93. *See* Green v. Denver & Rio Grande W. R.R. Co., 59 F.3d 1029, 1032 (10th Cir. 1995); *see also* Restatement (Second) of Torts § 920A(2) & cmt. d (1979).

94. 804 F.2d 1167, 1171 (10th Cir. 1986).

95. 566 F.3d 1203 (10th Cir. 2009).

recovered $20 million from an indemnitor and an insurance company. The defendants moved for summary judgment on the ground that Friedland had no right to contribution as he had already recovered his response costs from other parties. Friedland contended that the collateral source rule prohibits crediting the defendants in the amount he received from insurers.[96] The district court granted the defendants' summary judgment motion.

The Tenth Circuit affirmed the district court, ruling that the collateral source rule is inapplicable in the context of CERCLA:

> We have explained that a [CERCLA] claim for contribution is a claim "by and between jointly and severally liable parties for an appropriate division of the payment one of them has been compelled to make. . . ." Thus, a CERCLA contribution action is not a personal injury action by an innocent plaintiff. Instead, it is a claim between two or more culpable tortfeasors, and the policy underlying the collateral source rule—to provide the innocent party with the benefit of any windfall—is simply not advanced in such cases.[97]

The decision in *Freidland* is consistent with a sizeable body of cases, which have reached the same conclusion (although some of these cases do not expressly reference the collateral source rule).[98]

96. *Id.* at 1205.

97. *Id.* at 1206–07 (citations omitted).

98. *See, e.g.*, N.Y. Elec. & Gas Corp. v. FirstEnergy Corp., 766 F.3d 212 (2d Cir. 2014) (collateral source rule barred by section 114(b) of CERCLA); Boeing Co. v. Cascade Corp., 207 F.3d 1177, 1189 (9th Cir. 2000) ("[O]ne equitable factor is preventing someone from recovering for the same harm twice."); Basic Mgmt., Inc. v. United States, 569 F. Supp. 2d 1106, 1125 (D. Nev. 2008) (refusing to apply the collateral source rule because "[e]quity and common sense . . . dictate that Plaintiffs cannot recover the remediation costs paid for by their insurance policies"); Vine St., LLC v. Keeling, 460 F. Supp. 2d 728, 765 (E.D. Tex. 2006) (refusing to apply the collateral source rule, reasoning that because the "Court has a broad duty to consider facts bearing on the proper equitable allocation of response costs, the monies [plaintiff] has already recovered are relevant").

# CHAPTER 13

# Bankruptcy

## I. INTRODUCTION

Lawyers practicing in the Superfund area must have a working knowledge of bankruptcy law. In cases involving a large site and numerous potentially responsible parties (PRPs), it is hardly unusual for several PRPs to be in bankruptcy. Indeed, it is possible for CERCLA liability to precipitate a PRP's decision to declare bankruptcy. It is important, therefore, to understand how the bankruptcy proceedings will affect the rights and liabilities of other parties to the cleanup action. This chapter addresses those provisions of the Bankruptcy Code most relevant to Superfund actions, discusses how the courts reconcile the competing goals of the bankruptcy and environmental statutes, and provides guidance regarding the assertion or preservation of cost recovery claims against bankrupt PRPs.

### A. Overview of Bankruptcy Proceedings

The purpose of a bankruptcy proceeding is to provide a debtor with a fresh start.[1] Broadly speaking, the bankruptcy proceeding identifies the debtor's prebankruptcy debts, distributes the debtor's remaining assets to settle such debts, and allows the debtor to emerge from the proceeding with a chance to regain profitability.

1. Grogan v. Garner, 498 U.S. 279 (1991).

A bankruptcy proceeding may be initiated either voluntarily by the debtor or involuntarily by creditors. Bankruptcy proceedings generally fall into one of three categories:[2]

***Individual Chapter 7.*** Under Chapter 7 of the Bankruptcy Code,[3] the debtor's nonexempt assets are liquidated and distributed among the creditors based on a priority scheme set forth in the Code, after which the debtor receives a discharge from all prebankruptcy debts, subject to certain exclusions. A trustee is appointed to take charge of and liquidate the debtor's property.

***Business Liquidation.*** Business liquidations generally proceed under Chapter 7, but some Chapter 11 plans may provide for liquidation. A Chapter 7 liquidation requires the appointment of a trustee. A nonindividual Chapter 7 debtor receives no discharge and all of its assets are either (1) distributed to creditors or (2) abandoned.

***Business Reorganization.*** Chapter 11[4] permits reorganization of an insolvent business, which allows it to continue operating and gives the business a fresh start. Chapter 11 plans may provide for liquidation of the debtor. To emerge from a Chapter 11 proceeding requires the proposal and confirmation of a plan of reorganization that, in part, establishes classes of creditors and specifies the distribution of assets to each class. Pending confirmation of the reorganization plan, the debtor usually remains in possession of its assets and manages itself. In liquidating Chapter 11 cases, the debtor receives no discharge and reclaims no property, except as property may be abandoned. In reorganizing Chapter 11 cases, the debtor receives a discharge of all prepetition and postpetition claims, and reclaims such property as is provided by the plan.

## B. Competing Goals and Interests

There is an inherent conflict between the goals of the bankruptcy laws and the goals of CERCLA. The Bankruptcy Code is designed to facilitate

2. Other types of proceedings are encountered less frequently in the environmental context, including individual Chapter 11 proceedings and Chapter 13 proceedings (wage earner plans).

3. 11 U.S.C. §§ 701–784.

4. 11 U.S.C. §§ 1101–1174.

and expedite the discharge of debts and the distribution of assets in Chapter 7 and liquidating Chapter 11 proceedings and to provide reorganizing Chapter 11 debtors with a fresh start.[5] The purpose of CERCLA, on the other hand, is to ensure environmental cleanup in the interest of public health and safety, with PRPs bearing the cost. A natural tension exists between these two statutory goals when a PRP in bankruptcy is seeking to discharge a cleanup obligation or to abandon a contaminated site as burdensome to the estate.

## II. EPA VERSUS THE DEBTOR

The tension between the Bankruptcy Code and CERCLA is squarely raised where EPA has a CERCLA claim against a PRP who is a debtor in bankruptcy, or where a PRP attempts to avoid liability to the United States under CERCLA by filing for bankruptcy under Chapter 7 or 11. The principal issues, as discussed below, are whether a debtor PRP may receive a discharge in bankruptcy as to CERCLA claims of the United States; whether the "automatic stay" provisions of the Bankruptcy Code apply to efforts to compel debtor PRPs to clean up Superfund sites; whether the trustee in bankruptcy may "abandon" contaminated property that is or may be subject to EPA's response authority under CERCLA; and the "priority" of CERCLA claims by the United States in bankruptcy proceedings.

### A. Discharge of Federal Cost Recovery Claim

With some exceptions,[6] a "discharge in bankruptcy" discharges the debtor from all debts that arose before bankruptcy.[7] For example, in corporate reorganizations under Chapter 11, "the confirmation of a plan" of reorganization "discharges the debtor from any debt that arose before the date of such confirmation."[8] Similarly, in individual bankruptcies under Chapter 7, an individual may be discharged "from all debts that

5. Smith-Douglass, Inc. v. Borden, Inc. (*In re* Smith-Douglas, Inc.), 842 F.2d 729 (4th Cir. 1988).

6. 11 U.S.C. §§ 523(a), 727(a), 1141.

7. 11 U.S.C. § 727(a).

8. *Id.* § 1141(d)(1)(A).

arose before the date of the order for relief."[9] For bankruptcy purposes, a debt is a liability on a "claim," which is broadly defined under the Bankruptcy Code to include virtually any "right to payment" or "equitable remedy for breach of performance" giving rise to a right to payment.[10] Liability for response costs at a Superfund site under CERCLA is no different from liability on any other kind of claim. Thus, a claim by the United States against a PRP for response costs, like any other claim, would be subject to a PRP's discharge in bankruptcy.

***When Does a "Claim" Arise under CERCLA?*** This is one of the key questions at the intersection of CERCLA and bankruptcy law. As one bankruptcy court observed, "Because bankruptcy discharges liability only for claims that arose before the bankruptcy petition was filed, the focus of the Court's inquiry must be on when the CERCLA obligations arose."[11] Stated another way, "at what point does a party have a CERCLA claim for purposes of bankruptcy so that the failure to raise the claim before bankruptcy bar dates forever bar the claim from being brought against the successors to a reorganized company?"[12] Courts have articulated three different approaches or tests for answering this question.

The first approach, articulated in *United States v. Union Scrap-Iron & Metal*, is as follows: a claim exists only if the prebankruptcy relationship between the debtor and third party contained all the elements necessary to give rise to a legal obligation under CERCLA.[13] In *Union Scrap*, the court held that "[b]ecause the EPA had incurred no response costs" at the time of confirmation, EPA "could have no claim in the bankruptcy proceedings."[14]

The second approach, articulated in *In re Chateaugay*, ties the existence of a dischargeable claim to the date when the release, or threat

9. *Id.* § 727(b).
10. 11 U.S.C. §§ 502(c)(2).
11. United States v. Chateaugay Corp. (*In re* Chateaugay), 112 B.R. 513, 520 (S.D.N.Y. 1990), *aff'd sub nom.* United States v. LTV Corp., 944 F.2d 997 (2d Cir. 1991).
12. *In re* Chi., Milwaukee, St. Paul & Pac. R.R., 974 F.2d 775, 779 (7th Cir. 1992).
13. 123 B.R. 831, 835 (D. Minn. 1990).
14. *Id.* at 836.

of release, of hazardous substances first occurred.[15] Some courts refer to this approach as the "conduct" test.[16] At least one circuit court has rejected the conduct test as incompatible with CERCLA's remedial objectives because it sets such an early date for purposes of discharge in bankruptcy.[17]

The third test, a middle ground between the approaches described in *Union Scrap* and *Chateaugay*, is often referred to as the "fair contemplation" test[18] because it dates the dischargeable CERCLA claim to the moment when the parties could be deemed to "fairly" contemplate the existence and incurrence of costs in response to environmental contamination.[19] This test views CERCLA claims as dischargeable in bankruptcy "when a potential CERCLA claimant can tie [the] bankruptcy debtor to a known release of a hazardous substance which the potential claimant knows will lead to CERCLA response costs."[20]

Where a court concludes that the government's claim for cleanup of a Superfund site arose prior to the filing of the petition for bankruptcy, the court must determine the amount of the claim. At sites for which EPA has yet to select a remedial plan, the court may end up engaging in what is tantamount to pre-enforcement review of the agency's proposed remedy, notwithstanding that section 113(h) of CERCLA bars pre-enforcement review. This unusual situation is illustrated by *In re National Gypsum, supra*. In this case, one of the government's claims related to a site for which it had not yet selected a remedy. To "fix" the United States' claim for purposes of sharing in the assets of the estate, the bankruptcy court convened an "estimation proceeding." The purpose of this proceeding

15. 944 F.2d 997, 1005 (2d Cir. 1991).

16. *See, e.g.*, Hassanally v. Republic Bank (*In re* Hassanally), 208 B.R. 46, 51 (9th Cir. 1997).

17. Cal. Dep't of Health Servs. v. Jensen (*In re* Jensen), 995 F.2d 925, 930–31 (9th Cir. 1993).

18. *See, e.g.*, Alan N. Resnick, *Bankruptcy as a Vehicle for Resolving Enterprise-Threatening Mass Tort Liability*, 148 U. PA. L. REV. 2045, 2072 (2000) (noting that "[s]everal courts have adopted the 'fair contemplation test' to determine when a future claim arises").

19. *In re* Nat'l Gypsum Co., 139 B.R. 397, 409 (N.D. Tex. 1992).

20. *In re* Chi., Milwaukee, St. Paul & Pac. R.R., 974 F.2d 775, 786 (7th Cir. 1992). *See In re* Nat'l Gypsum, 139 B.R. 397, 407–08 (N.D. Tex. 1992) (CERCLA claim arises when "fair contemplation of future costs based on pre-petition conduct can occur at any particular site by the parties"); *see also In re* Jensen, 995 F.2d 925, 930 (9th Cir. 1993).

was to produce a reasonable estimate of the likely value of the government's unmatured claim—in other words, to predict EPA's most likely remedy and its associated cost. In the proceeding, both the government and National Gypsum presented evidence of the remedy they believed the agency would ultimately select. Unlike a CERCLA cost recovery proceeding in which a government's remedy must be accepted by the court unless shown to be arbitrary and capricious and inconsistent with the National Contingency Plan (NCP), in this estimation proceeding there are no such presumptions. Rather, each side presented their view of a remedy that would be produced as a result of EPA's typical remedy selection process. The bankruptcy court ultimately gave greater weight to National Gypsum's evidence and estimated the cost of a remedy proposed by National Gypsum's expert.[21]

***Does a Cleanup Order Constitute a "Claim" under the Code?*** This is another complicated issue at the intersection of the Bankruptcy Code and CERCLA, namely, whether an injunction directing a PRP to perform response action at a site constitutes a claim that would be discharged in bankruptcy. As background, section 101(5) of the Bankruptcy Code broadly defines claim to include

> (A) right to payment, whether or not such right is reduced to judgment, liquidated, unliquidated, fixed, contingent, matured, unmatured, disputed, undisputed, legal, equitable, secured, or unsecured; or
> (B) right to an equitable remedy for breach of performance if such breach gives rise to a right of payment, whether or not such right to an equitable remedy is reduced to judgment, fixed, contingent, matured, unmatured, disputed, undisputed, secured, or unsecured.[22]

21. The *In re National Gypsum* estimation proceeding is described in detail in Kirsch, Van Horn & Williams, *A Whole New Ballgame: Judicial Review and Estimation of CERCLA Claims in Bankruptcy*, 22 ENVTL. L. REP. 10,785 (1992).
22. 11 U.S.C. § 101(5)(A), (B).

The Supreme Court addressed the question of whether a CERCLA cleanup order constitutes a claim in *Ohio v. Kovacs*.[23] The State of Ohio sued William Kovacs, the chief executive officer of Chem-Dyne Corp., which operated a landfill, for polluting public waters, maintaining a nuisance, and causing fish kills, in violation of state environmental laws. To settle the lawsuit, Kovacs signed a stipulation that enjoined Chem-Dyne from causing further pollution, prohibited disposal of additional wastes in the landfill, compelled the defendants to clean up the site, and required payment of $75,000 for injury to wildlife.[24] Kovacs and Chem-Dyne thereafter failed to comply with the terms of the stipulation, and the State placed all of Kovacs's property and assets in receivership. The receiver took possession and was in the process of cleaning up the site when Kovacs filed for bankruptcy. Ohio then filed a motion in state court to discover Kovacs's income and assets. The bankruptcy court stayed the action, and the State filed a complaint in the bankruptcy court seeking a declaration that Kovacs's obligations under the cleanup order were not a liability on a claim subject to discharge. The bankruptcy court ruled against Ohio, as did the district court. The Sixth Circuit affirmed, holding that Ohio essentially was seeking a monetary payment, which was dischargeable.[25]

The Supreme Court affirmed the Sixth Circuit's ruling that the obligation constituted a dischargeable claim under the Bankruptcy Code. Specifically, the Court found that the state had a "right to an equitable remedy for breach of performance [giving] rise to a right to payment," the definition of a "claim" under the Code.[26]

The Court rejected the State's claim that it had no right to enforce an alternative right to payment. With the appointment of a receiver, Kovacs was disabled from personally performing the cleanup. The State conceded that after the appointment, "the only performance sought from Kovacs was the payment of money."[27] Thus, the Court concluded

23. 469 U.S. 274 (1985).
24. *Id.* at 276.
25. *Id.* at 277.
26. *Id.* at 280.
27. *Id.* at 283.

that "the cleanup order had been converted into an obligation to pay money."[28]

Cases decided since *Kovacs* have for the most part limited the holding to its facts. For example, the court in *United States v. F.E. Gregory & Sons, Inc.* permitted the United States to enforce an injunction under section 106 of CERCLA against a debtor PRP; the court distinguished the *Kovacs* ruling as inapplicable to the facts of the case since, unlike *Kovacs*, the debtor in *F.E. Gregory* had access to the property.[29]

*In re Chateaugay Corp.*[30] is also instructive. In this case, the Second Circuit held that injunctive remedies may be dischargeable if the government has the option to perform the remediation and recover costs from the debtor. The court ruled, however, that "a cleanup order that accomplishes the dual objectives of removing accumulated wastes and stopping or ameliorating ongoing pollution emanating from such wastes is not a dischargeable claim."[31] When there is no option for the enforcing agency to accept payment in lieu of ongoing pollution, any "order that to any extent ends or ameliorates continued pollution is not an order for breach of an obligation that gives rise to a right of payment and is for that reason not a 'claim.'"[32] On the other hand, a cleanup order that imposes obligations distinct from the obligation to stop ongoing pollution is a claim if the agency had the option to do the cleanup work and sue the corporation for response costs.[33] The Second Circuit predicted that most injunctions would "fall on the non-'claim' side of the line," because most cleanup orders include obligations to remove or remediate contaminated soil or other sources from which pollution continues to emanate.[34]

28. *Id.*

29. 58 B.R. 590, 592 (W.D. Pa. 1986). *See also* Cournoyer v. Town of Lincoln, 790 F.2d 971, 975–76 (1st Cir. 1986).

30. 944 F.2d 997 (2d Cir. 1991).

31. *Id.* at 1008.

32. *Id.*

33. *Id.*

34. *Id.*; *see also* United States v. Apex Oil Co., 579 F.3d 734 (7th Cir. 2009) (reorganized debtor's obligation to pay for an environmental cleanup of property that it did not own was not a claim and was not dischargeable), *cert denied*, 131 S. Ct. 67 (2010); Torwico Elecs., Inc. v. N.J. Dep't of Envtl. Prot. (*In re* Torwico Elecs., Inc.), 8 F.3d 146 (3d Cir. 1993) (debtor

## B. The Automatic Stay

Section 362 of the Bankruptcy Code provides that the filing of a petition for bankruptcy automatically stays the commencement or continuation of a judicial, administrative, or other action against the debtor that was or could have been brought before the date of filing, or the recovery of a claim against the debtor that arose before the filing.[35] The stay also applies to the enforcement of a judgment obtained before the filing date.[36]

The automatic stay provision does not, however, apply to the "commencement or continuation of an action or proceeding by a government unit to enforce such government unit's police or regulatory power."[37] Nor does it apply to the enforcement of a judgment, "other than a money judgment," obtained in an action or proceeding by a government unit to enforce that unit's police or regulatory power.[38] Thus, governmental actions to enforce money judgments *are* subject to the automatic stay, even if they are in furtherance of a state's police powers.

The meaning of "money judgment" has been at the center of litigation between government regulators and debtors. Several courts have held, and EPA takes the position, that the automatic stay provision does not apply to actions by the United States to prevent environmental harm, including, actions under CERCLA for injunctive relief or to impose a fine or penalty, because such actions fall within the bankruptcy exception for a government's exercise of its police power and do not constitute money judgments. In *Penn Terra Ltd. v. Department of Environmental Resources*,[39] the district court and the bankruptcy court had held that an action by the state to enforce an order requiring corrective actions at the debtor's surface mine operation was an action to enforce a money judgment and, thus, subject to the automatic stay. The Third Circuit

---

must comply with prepetition cleanup order at a site it did not own or operate because waste posed continuing environmental hazard and statute at issue did not authorize State to perform cleanup).

35. 11 U.S.C. § 362(a)(1).
36. *Id.* § 362(a)(2).
37. *Id.* § 362(b)(4).
38. *Id.*
39. 733 F.2d 267 (3d Cir. 1984).

reversed, stating that the state's actions fell "squarely within Pennsylvania's police and regulatory powers."[40] In this regard, the court noted that the legislative history of section 362(b) of the Bankruptcy Reform Act "explicitly acknowledge[d] environmental protection as a part of the State's police power" and, as such, exempt from the automatic stay.[41]

The court distinguished between the enforcement of a money judgment and the enforcement of an injunction requiring the expenditure of funds. Rejecting the bankruptcy court's interpretation of "money judgment" as "anything which costs money to enforce," the court stated that

> the inquiry is more properly focused on the nature of the injuries which the challenged remedy is intended to redress—including whether plaintiff seeks compensation for past damages or prevention of future harm—in order to reach the ultimate conclusion as to whether these injuries are traditionally rectified by a money judgment and its enforcement.[42]

The court further observed that "in contemporary times, almost everything costs something. An injunction which does not compel some expenditure or loss of monies may often be an effective nullity."[43]

Courts have also found that cost recovery actions brought by the United States under sections 104 and 107 of CERCLA are exempt from the automatic stay provision, at least to the extent that the claims are for past response costs. In *United States v. Nicolet, Inc.*,[44] the debtor petitioned for Chapter 11 reorganization after the United States had initiated a cost recovery action under section 107 of CERCLA, and sought to stay the CERCLA proceeding. The court viewed as meritorious the debtor's argument that the government's action to recover past costs

40. *Id.* at 274.
41. *Id.*
42. *Id.* at 278.
43. *Id. See also* United States v. Nicolet, Inc., 857 F.2d 202 (3d Cir. 1990) (actions for the collection of CERCLA cleanup costs fall into this regulatory power exception to a bankruptcy stay); United States v. Alsol Corp., 2014 U.S. Dist. LEXIS 1179 (D.N.J. Jan. 2, 2014) (same); *cf.* Ohio v. Kovacs, 469 U.S. 274 (1985).
44. 857 F.2d 202 (3d Cir. 1988).

could be viewed as a simple claim for a monetary judgment "materially indistinguishable" from debts owed to other creditors subject to the stay. Nevertheless, the court determined that section 362(b) of the Bankruptcy Code "embod[ies] Congress' recognition that enforcement of the environmental protection laws merits a higher priority than the debtor's rights to a 'cease fire' or the creditors' rights to an orderly administration of the estate."[45] Accordingly, the court held that CERCLA cost recovery actions brought by the United States are exempt from the automatic stay provision. According to the court, the fact that CERCLA cost recovery actions serve as a deterrent by ensuring that responsible parties "will be held accountable for their environmental misdeeds" brought the action within the "police or regulatory powers" exception to the automatic stay.[46]

The police power exception to the automatic stay has also been held to apply to actions brought by cities.[47]

## C. Abandonment

The conflict between CERCLA and the Bankruptcy Code is directly raised by the power of the trustee in bankruptcy to abandon contaminated property. The trustee may be able to avoid liability in excess of the value of the property that is subject to EPA's response authority by abandoning the property. Section 554 of the Bankruptcy Code authorizes a trustee, after notice and a hearing, to "abandon any property of the estate that is burdensome to the estate or that is of inconsequential value and benefit to the estate."[48] Although at one time trustees could rely on this provision as authority for abandoning contaminated property, thereby avoiding liability under state and federal hazardous waste laws, the 1986 Supreme Court decision in *Midlantic National Bank v.*

45. *Id.* at 207. The government in fact sought both past and future response costs, but the court decided the case "on the premise that only expenses for past activities are at issue." *Id.*

46. 857 F.2d at 209.

47. *See, e.g.*, City of New York v. Exxon, 932 F.2d 1020 (2d Cir. 1991).

48. 11 U.S.C. § 554(a).

*New Jersey Department of Environmental Protection* made it more difficult for a trustee to do so.[49]

In *Midlantic*, the Supreme Court created an exception to the trustee's broad authority to abandon property, holding that "a trustee may not abandon property in contravention of a state statute or regulation that is reasonably designed to protect the public health or safety from identified hazards."[50] The Court, however, qualified the holding with the following footnote:

> This exception to the abandonment power vested in the trustee by § 554 is a narrow one. It does not encompass a speculative or indeterminate future violation of such laws that may stem from abandonment. *The abandonment power is not to be fettered by laws or regulations not reasonably calculated to protect the public health or safety from imminent and identifiable harm.*[51]

The circuits are split on whether the *Midlantic* holding should be construed broadly or narrowly in the CERCLA context. Those courts advocating a narrow construction focus on the footnote language in *Midlantic* and suggest that the exception applies only when there is an immediate threat to public health and safety, and that it does not apply when the hazards are speculative or may await appropriate action by an environmental agency.[52] In addition, there is some support for the

49. 474 U.S. 494 (1986).
50. *Id.* at 507.
51. *Id.* at 507 n.9 (emphasis added).
52. *See, e.g.*, N.M. Env't Dep't v. Foulston (*In re* L.F. Jennings Oil Co.), 4 F.3d 887, 890 (10th Cir. 1993) (abandonment allowed where there was no immediate threat at the time of the abandonment); *In re* Smith-Douglass, Inc., 856 F.2d 12 (4th Cir. 1988) (permitting unconditional abandonment of contaminated property where estate lacked unencumbered assets to finance cleanup and plant did not present any imminent harm or danger to public); Minn. Pollution Control Agency v. Gouveia (*In re* Globe Bldg. Materials, Inc.), 345 B.R. 619, 636 (Bankr. N.D. Ind. 2006) (construing *Midlantic* to be inapplicable where a Chapter 7 trustee is merely liquidating environmentally contaminated property on which trustee did not operate facility giving rise to environmental problems and to be applicable only where actions are necessary to abate conditions that pose an imminent and identifiable harm to public health or safety); *In re* Unidigital, Inc., 262 B.R. 283, 286–87 (Bankr. D. Del. 2001) ("Since the *Midlantic* decision, the majority of courts have read the exception to abandonment . . . only where there is

notion that where abandonment would result in reversion of possession to a third party with resources to address environmental concerns, abandonment will not be barred under *Midlantic*.[53] Other courts, however, have interpreted the *Midlantic* holding as generally prohibiting the abandonment of any contaminated property.[54]

It should be noted that where a debtor continues to own contaminated property after confirmation of a reorganization plan, the debtor may be held liable under section 107(a)(1) of CERCLA as the current owner.[55]

### D. Federal "Superiority" Lien Authority

Prior to the 1986 amendments to CERCLA, the priority afforded prepetition CERCLA claims of the United States in bankruptcy proceedings was the same as that of any other unsecured creditors. As an unsecured

an imminent and identifiable harm to the public health or safety"); *In re* Doyle Lumber, Inc., 137 B.R. 197 (Bankr. W.D. Va. 1992).

53. *See* White v. Coon (*In re* Purco, Inc.), 76 B.R. 523 (Bankr. W.D. Pa. 1987) (allowing abandonment of inventory where there was insufficient evidence of a threat to public health and safety and noting that landlord was able to provide any remedy necessary to protect public health); *In re* Franklin Signal Corp., 65 B.R. 268 (Bankr. D. Minn. 1986) (trustee permitted to abandon 14 drums of hazardous waste); *In re* Okla. Ref. Co., 63 B.R. 562 (Bankr. W.D. Okla. 1986) (court permitted trustee to abandon site contaminated with heavy metals, since there was no immediate danger at the site); *In re* Smith-Douglass, 856 F.2d 12, 16 (4th Cir. 1988) (Fourth Circuit affirmed the lower court's decision to allow trustee to abandon a plant where several drums of hazardous waste were leaking into a creek); *In re* Microfab, Inc., 105 B.R. 161, 166 (D. Mass. 1989) (if the trustee lacks "sufficient resources to achieve appreciable results" in a remediation effort, it is not required to expend limited resources on that effort). *Cf. In re* FCX Inc., 96 B.R. 49, 55 (Bankr. E.D.N.C. 1989) (court allowed trustee to abandon site, but only so long as a $250,000 set-aside was provided to EPA to finance cleanup).

54. *See, e.g.*, Texas v. Lowe (*In re* H.L.S. Energy Co.), 151 F.3d 434, 438 (5th Cir. 1998) ("A bankruptcy trustee may not abandon property in contravention of a state law reasonably designed to protect public health or safety") (citation omitted); *In re* Am. Coastal Energy Inc., 399 B.R. 805, 810 (Bankr. S.D. Tex. 2009) ("[A] trustee cannot abandon contaminated estate property if state health or safety law prohibits such abandonment") (citations omitted); *In re* Insilco Techs., Inc., 309 B.R. 111, 114 (Bankr. D. Del. 2004) ("property of the estate may not be abandoned if the abandonment will act to contravene laws designed to protect public health and safety and will pose an imminent threat to the public's welfare"); *In re* Peerless Plating Co., 70 Bankr. 943 (Bankr. W.D. Mich. 1987) (trustee may not abandon property in contravention of environmental law designed to protect public health from identifiable and imminent harm); *In re* Stevens, 68 B.R. 774, 781 (D. Me. 1987) (trustee may not abandon polychlorinated biphenyls (PCBs) stored in a trailer box absent the "acquiescence of the public authorities" and only if adequate conditions were formulated).

55. *See, e.g.*, *In re* CMC Heartland Partners, 966 F.2d 1143 (7th Cir. 1992); *In re* Indus. Salvage, Inc., 196 B.R. 784 (Bankr. S.D. Ill. 1996).

creditor, the United States' claims were subordinated to those of secured creditors and, thus, the likelihood of receiving full compensation was small.[56]

The Superfund Amendments and Reauthorization Act of 1986 (SARA) amended section 107 to impose a lien in favor of the United States on all real property belonging to a party liable for CERCLA costs and subject to or affected by a removal or remedial action.[57] The lien operates only on the real property at which the removal or remedial action took place. The lien arises on the date costs are first incurred with respect to a response action or the date notice of potential liability is given to the responsible party, whichever is later. This "superlien" remains in effect so long as costs and natural resource damages for which the party is liable remain unpaid or until the statute of limitations has passed.[58]

The rights of EPA under the superlien are subordinate to those of other creditors whose interests were perfected under applicable state law before recording of the lien.[59] After EPA records the lien, the federal government establishes its priority ahead of purchasers, holders of security interests, and judgment lien creditors that have not perfected their interests. EPA may execute the lien in an action in rem in the district court for the district where the property is located or in a cost recovery action.[60]

One effect of the federal lien is to give EPA's environmental claims the status of judgment liens under the Bankruptcy Code's priority scheme, thereby enabling EPA to bypass arguments for priority based on administrative expense. This recharacterization of EPA's interest in environmental cleanup obligations may facilitate EPA's recovery of response costs from private parties.

---

56. *See* 11 U.S.C. § 507. EPA attempted to enhance the priority of its CERCLA claims by classifying them as "administrative expenses," which are defined under section 503(b)(1)(A) of the Bankruptcy Code as "the actual, necessary costs and expenses of preserving the estate." *See, e.g.*, *In re* T.P. Long Chem., Inc., 45 B.R. 278, 282 (Bankr. N.D. Ohio 1985).

57. CERCLA § 107(1).

58. CERCLA § 107(1)(2).

59. CERCLA § 107(1)(3).

60. CERCLA § 107(1)(4).

# III. PRPS VERSUS THE DEBTOR

## A. Disallowance of PRP Claims as "Contingent"

Up until the U.S. Supreme Court's 2007 decision in *United States v. Atlantic Research Corp.*,[61] the conventional wisdom was that claims by PRPs against other PRPs arose under CERCLA § 113, the contribution provision. (See discussion of *Atlantic Research* in chapter 9.) This had significant implications in bankruptcy proceedings. Bankruptcy Code section 502(e)(1)(B) mandates disallowance of claims for reimbursement or contribution of a party that is co-liable with the debtor to a third-party creditor.[62] Specifically, under section 502(e)(1)(B), the court must disallow "any claim for reimbursement or contribution . . . to the extent that . . . such claim . . . is contingent as of the time of allowance or disallowance of such claim for reimbursement or contribution."[63]

Thus, the crucial issue to a party seeking to assert a CERCLA contribution claim against a bankrupt PRP is whether the claim is "contingent." For example, in *Syntex Corp. v. Charter Co.* (*In re Charter Co.*),[64] the plaintiffs brought a contribution action for damages that might be assessed against them in a pending cost recovery case. The court disallowed the claim under section 502(e)(1)(B) of the Bankruptcy Code, holding that the plaintiff's rights were contingent because they "depend entirely on the success of the plaintiffs against the appellants in the still pending Missouri Litigation."[65] This is just one of many cases in which courts disallowed PRP contribution claims against bankrupt PRPs, pursuant to section 502(e)(1)(B) of the Bankruptcy Code.[66]

61. 551 U.S. 128 (2007).

62. 11 U.S.C. § 502(e)(1)(B); *see also In re* Pinnacle Brands, Inc., 259 B.R. 46, 55 (Bankr. D. Del. 2001).

63. 11 U.S.C. § 502(e)(1).

64. 862 F.2d 1500 (11th Cir. 1989).

65. *Id.* at 1503. The claim might not have been regarded as contingent if the parties had spent money or performed work on the cleanup. *Id.* at 1504.

66. *See also* Norpak Corp. v. Eagle-Picher Indus. (*In re* Eagle-Picher), 131 F.3d 1185, 1190 (6th Cir. 1997); *In re* APCO Liquidating Trust, 370 B.R. 625 (Bankr. D. Del. 2007); *In re* Tri-Union Dev. Corp., 314 B.R. 611 (Bankr. S.D. Tex. 2004); *In re* G-I Holdings, Inc., 308 B.R. 196, 212 (Bankr. D.N.J. 2004) ("Here, the funds have been expended and thus the claim to that extent is not contingent").

In 2007, however, the Supreme Court decided *United States v. Atlantic Research Corp.*, holding that a PRP was not limited to contribution under CERCLA § 113, and can also seek cost recovery under section 107: "A private party may recover under § 107(a) without any establishment of liability to a third party."[67] In theory, this opens the doors for private parties to circumvent section 502(e)(1)(B) of the Bankruptcy Code and preserve their claims against PRPs seeking protection from liability under the Bankruptcy Code.

Post-*Atlantic Research* case law, however, has shown that PRPs still face an uphill battle in prevailing on section 107 claims against bankrupt PRPs. First, a number of post-*Atlantic Research* cases have rejected attempts by PRPs to recast classic contribution claims as cost recovery in order to assert the claim under section 107.[68] Second, in a number of post-*Atlantic Research* bankruptcy proceedings, courts have disallowed CERCLA claims asserted by co-liable PRPs.[69] These cases further doubt on the ability of PRPs to circumvent Bankruptcy Code section 502(e)(1)(B) when pursuing claims against a debtor in bankruptcy.

### B. Disallowance Because Claim Has Not "Arisen"

In order for the bankruptcy court to allow a claim, it must have "arisen" prior to the filing of the petition in bankruptcy.[70] This poses another obstacle to PRP contribution claims against bankrupt PRPs. The question of when a claim for contribution under CERCLA "arises" for purposes of bankruptcy has been the subject of litigation, just like disputes between debtor PRPs and EPA. Does it arise when a generator PRP

67. 551 U.S. at 139.

68. *See* Niagara Mohawk Power Corp. v. Chevron U.S.A., Inc., 596 F.3d 112, 128 (2d Cir. 2010) ("Congress recognized the need to add a contribution remedy for PRPs similarly situated to NiMo. To allow NiMo to proceed under § 107(a) would in effect nullify the SARA amendment and abrogate the requirements Congress placed on contribution claims under § 113"); Agere Sys., Inc. v. Advanced Envtl. Tech. Corp., 602 F.3d 204 (3d Cir. 2010).

69. *See In re* Lyondell Chem. Co., 442 B.R. 236, 257 (Bankr. S.D.N.Y. 2011) (disallowing PRP's section 107 "cost recovery" claim under section 502(e)(1)(B) of the Bankruptcy Code, reasoning that co-liability to the government can still exist whether the claimant's claim against the debtor is a direct claim rather than one for contribution); *In re* Chemtura Corp., 443 B.R. 601, 626 (Bankr. S.D.N.Y. 2011) (same).

70. 11 U.S.C. § 362(a)(1).

disposes waste at the site or when the plaintiff cleans up the site or somewhere between these two events? The Second Circuit, in *Carter Day Industries, Inc. v. U.S. Environmental Protection Agency (In re Combustion Equipment Associates, Inc.)*,[71] framed the possibilities as follows: a contribution claim may arise after "EPA seeks enforcement";[72] or may arise "in stages as the EPA incurs costs";[73] or may "mature into a claim in stages, but accrues as a whole at some point between deposit of toxic waste and final cleanup."[74]

Because the issue of when a contribution claim arises under CERCLA remains uncertain, a creditor PRP's ability to obtain allowance is problematic. Even if the court disallows a claim, however, the PRP may still be entitled to contribution at a later date to the extent that their claims are deemed to be claims that do not preexist confirmation and, therefore, cannot be discharged. For example, in *AT&T Co. v. Chateaugay Corp.*, the court stated:

> If a CERCLA claim does not pre-exist confirmation, it would appear that it cannot be discharged upon confirmation of [the debtor's] Chapter 11 plan. If the CERCLA claim should arise after confirmation of [the debtor's] Chapter 11 plan, then like many liabilities that are incurred by Chapter 11 debtors for the first time after confirmation of their plans, a post-confirmation CERCLA liability must be performed by [the debtor]. A debtor cannot hide behind confirmation of its plan to escape from liabilities that do not exist prior to confirmation. See 11 U.S.C. § 1141(d)(1).[75]

As a practical matter, however, creditor PRPs are unlikely to recover any funds from a debtor PRP after it has gone through a bankruptcy proceeding unless the debtor emerges from reorganization as a financially strong, ongoing business. In a Chapter 7 or liquidating Chapter 11

71. 838 F.2d 35 (2d Cir. 1988).
72. *Id.* at 39.
73. *Id.* at 40.
74. *Id.*
75. 88 B.R. 581, 586 (S.D.N.Y. 1988).

case, PRPs are better off with a preexisting claim entitling them to some distribution rather than a claim that has not yet arisen.

### C. PRP Actions to Preserve Claim against Debtors

Given the prospect of a PRP claim being discharged in bankruptcy, PRPs must determine which parties in a Superfund case are in bankruptcy or are threatening to go into bankruptcy, and take certain actions to preserve their ability to recover some portion of their CERCLA claims.

#### 1. *Filing Proofs of Claim*

A PRP does not have a secured interest in the estate of a bankrupt PRP. Accordingly, in order to obtain part of a bankrupt estate, a PRP must file a proof of claim.[76] The claim may be filed by any party, including the claimant, grantor, debtor, trustee, or co-debtor.[77]

In Chapter 7 cases, a proof of claim must be filed within 90 days after the first meeting of creditors. The first meeting of creditors will be held between 20 and 40 days after the order for relief, which is the filing of the petition in a voluntary case.[78] In Chapter 11 cases, the court will fix a deadline for filing proofs of claim, called a bar date. The court can extend the bar date for cause shown.[79]

#### 2. *Postconfirmation Action*

There will be instances in which a PRP is not aware of a pending bankruptcy case involving a debtor PRP until after a claims bar date or plan confirmation in a Chapter 11 case. To the extent that the claims against the debtor can be shown to have arisen postconfirmation, it may be advisable to bring an action for declaratory relief to determine discharge as soon as possible after the bankruptcy case is discovered.

Alternatively, if a PRP did not receive notice of a prior claims bar date (or of the pending bankruptcy case), the PRP may be entitled to a determination that their claims were not discharged even if they arose

76. See discussion *infra*.
77. The form for filing a proof of claim is form B10 of the federal bankruptcy forms.
78. FED. R. BANKR. P. 3002(c).
79. *Id.* 3003(c)(3).

preconfirmation or prepetition under Rule 2002 and under constitutional due process requirements. (If the debtor is an individual, PRPs that were not scheduled as creditors and did not have actual knowledge of the bankruptcy case may be entitled to have their claims, even if preconfirmation, excepted from discharge.) PRPs that did not take timely action during the case may, if appropriate, seek leave to file late proofs of claim. Under certain circumstances it may be preferable to take appropriate legal action against the reorganized debtor without regard to the prior grant of a discharge, especially if the prior bankruptcy case is closed.

# CASES

C

D



E

**F**

L

## M

New Castle Cnty. v. Hartford Accident & Indem. Co., 673 F. Supp. 1359 (D. Del. 1987), 355 n.73

New Castle Cnty. v. Hartford Accident & Indem. Co., 685 F. Supp. 1321 (D. Del. 1988), 344 n.1

New Castle Cnty. v. Hartford Accident & Indem. Co., 933 F.2d 1162 (3d Cir. 1991), *rev'd on other grounds*, 970 F.2d 1267 (3d Cir. 1992), 350 n.25

New Jersey Society for Environmental & Economic Development v. Campbell, N.J. Sup. Ct., Law Div. (Docket No. MER-L-34-04), 320

New Orleans Public Service, Inc. v. United Gas Pipe Line Co., 732 F.2d 452 (5th Cir. 1984), 215

New York State Elec. & Gas Corp. v. FirstEnergy Corp., 808 F. Supp. 2d 417 (N.D.N.Y. 2011), *aff'd in part, vacated in part*, 766 F.3d 212 (2d Cir. 2014), 130 n.174

New York v. Almy Bros., Inc., 90-CV-818, 1996 U.S. Dist. LEXIS 341 (N.D.N.Y. Jan. 8, 1996), 262 n.29

New York v. AMRO Realty Corp., 697 F. Supp. 99 (N.D.N.Y. 1988), *aff'd in part, rev'd in part*, 936 F.2d 1420 (2d Cir. 1991), 345 n.2, 353 n.63

New York v. Gen. Elec. Co., 592 F. Supp. 291 (N.D.N.Y. 1984), 8 n.25, 100 n.54, 304 n.30

New York v. Lashins Arcade Co., 856 F. Supp. 153 (S.D.N.Y. 1994), *aff'd*, 91 F.3d 353 (2d Cir. 1996), 128 n.166

New York v. Nat'l Serv. Indus., Inc., 352 F.3d 682 (2d Cir. 2003), 119 n.130, 120 n.134

New York v. Shore Realty Corp., 759 F.2d 1032 (2d Cir. 1985), 5 n.12, 86 n.1, 115 n.117, 129 n.169, 289 n.132

Niagara Mohawk Power Corp. v. Chevron U.S.A., Inc., 596 F.3d 112 (2d Cir. 2010), 277 n.82, 380 n.68

Niagara Mohawk Power Corp. v. Consol. Rail Corp., 565 F. Supp. 2d (N.D.N.Y. 2008), *aff'd in part and rev'd in part on other grounds by* Niagara Mohawk Power Corp. v. Chevron U.S.A., Inc., 596 F.3d 112 (2d Cir. 2010), 275 n.75

Niagara Mohawk Power Corp. v. Jones Chem., Inc., 315 F.3d 171 (2d Cir. 2003), 98 n.46

O

## P

## Q



## R

S

# STATUTES

**Regulations**

**Rules**



**Executive Orders**

# INDEX

## C